Ultrametric Banach Algebras

Ultrametric Banach Algebras

Alain Escassut
Université Blaise Pascal, France

World Scientific
New Jersey • London • Singapore • Hong Kong

Published by

World Scientific Publishing Co. Pte. Ltd.

5 Toh Tuck Link, Singapore 596224

USA office: Suite 202, 1060 Main Street, River Edge, NJ 07661

UK office: 57 Shelton Street, Covent Garden, London WC2H 9HE

British Library Cataloguing-in-Publication Data
A catalogue record for this book is available from the British Library.

ULTRAMETRIC BANACH ALGEBRAS

ISBN 981-238-194-5

This book is printed on acid-free paper.

Printed in Singapore by Mainland Press

Contents

Introduction

Thanks to Gelfand's theory, semi-simple commutative complex Banach algebras may be seen as algebras of functions on a compact set whose points characterize all maximal ideals, which all are of codimension 1. Moreover, it is known that all multiplicative semi-norms, continuous with respect to the $\mathbb{C}$-algebra norm, are of the form $|\chi|$, with χ a $\mathbb{C}$-algebra homomorphism. Given an element $x \in A$, it is possible to define a holomorphic mapping from its spectrum to the algebra: this is called the holomorphic functional calculus.

Now, consider an algebraically closed complete ultrametric field K and a commutative ultrametric Banach K-algebra A with unity. Such an algebra may have maximal ideals of infinite codimension. Anyway, it is not possible to make it isomorphic to an algebra of functions (with values in K) defined on a compact set. Trying to construct a (basic) spectral theory on A is then much more difficult than in Archimedean analysis. Following basic works by Monna and T.A. Springer, B. Guennebaud saw the importance of continuous multiplicative semi-norms on such an ultrametric Banach K-algebras [34]. More recently, V. Berkovich also considered the theory of multiplicative semi-norms and enjoyed it to construct an improved generalization of Tate's theory [1,49]. Here we particularly make a kind of Gelfand transform which associates to any element f of the algebra A a function f^* defined on the set of multiplicative continuous semi-norms of A, with values in the set $Mult(K[x])$ of multiplicative semi-norms of $K[x]$, also called the one dimensional Berkovich affine space. Thus, among continuous multiplicative semi-norms of A, we find those whose kernel is a maximal ideal of codimension 1, and then the restriction of f^* to them appears similar to the classical definition in Archimedean analysis.

Our way to examine semi-norms consists of characterizing them by circular filters [32, 30]. Moreover, the set of circular filters has a tree structure which is easily described: every pair admits a supremum, and if two elements b, c are bigger than another a, then b and c are comparable. The set is provided with two topologies: the simple convergence, induced by the topology on continuous multiplicative semi-norms on $K[x]$ (for which it is locally compact and every compact is sequentially compact), and a metric topology whose definition is closely linked to the tree structure. This metric defines a topology which is strictly stronger than the simple convergence: the set of circular filers is not locally compact for the metric topology, but it is complete. Following a study by K. Boussaf, an analytic element or a meromorphic function roughly transforms a circular filter into another one. Since the field K appears as a subset of the set of circular filters, such mappings are continuous for both topologies on the set of circular filters. A meromorphic function is increasing with respect to the order on the set of circular filters if and only if it is an entire function. And when two functions are close enough, then give the same image of a circular filter. On the other hand, when the field K is topologically separable i.e. has a countable dense subset, it is possible to show that the simple convergence is metrizable and then we construct another metric defining the simple convergence [39].

In a Banach K-algebra A there does exist a spectral semi-norm, defined as $\|x\|_{si} = \lim_{n\to\infty} \|x^n\|^{\frac{1}{n}}$, and it is equal to the supremum of all continuous multiplicative semi-norms. It is easy to construct examples where this spectral norm is strictly superior to the supremum of multiplicative semi-norms whose kernel is a maximal ideal of codimension 1, which gives an idea of the problems we risk encountering. Meanwhile, every maximal ideal is the kernel of (at least) one continuous multiplicative semi-norm (this is not the case of certain closed prime ideals [7]). On the other hand, a Shilov boundary exists for continuous multiplicative semi-norms, relatively to the spectral semi-norm and we construct it by following a method due to B. Guennebaud. In the particular case of a Krasner algebra $H(D)$, the Shilov boundary consists of circular filters secant with both D and $K \setminus D$ [8].

Since 1975, it has been possible to define a kind of holomorphic functional calculus in A [23]: given $x \in A$, thanks to the ultrametric property of the norm, we can show that if a, b are in a same hole of the spectrum of x, and such that $|b-a| < \frac{1}{\|(x-a)^{-1}\|}$, then $\|(x-a)^{-1}\| = \|(x-b)^{-1}\|$.

Thanks to this property, combined with the Krasner-Mittag-Leffler Theorem, it is possible to define, on the algebra of rational functions with no poles in the spectrum of x, an ultrametric norm that makes continuous the natural homomorphism from this algebra into A. As a consequence, many problems on ultrametric Banach algebras may be studied through the set $Mult(K[x])$.

So, in a few words, we can say that the book consists of associating methods based on affinoid algebras (also called "Tate algebras") and methods based on holomorphic functional calculus involving very thin property on analytic functions in one variable. Particularly, we can consider the behaviour of multiplicative semi-norms on the algebra of rational functions with no pole in the spectrum of x by using properties of analytic functions. In such considerations, T-filters play a central role. This K-subalgebra B of $K(x)$ is then provided with an ultrametric norm defined with help of the relation above, and its completion $\widehat{B}$ looks like a Krasner algebra. This way, we can define a natural continuous homomorphism from $\widehat{B}$ into A, and therefore find in A properties known in $\widehat{B}$.

The problem whether a maximal ideal of infinite codimension may be the kernel of several continuous multiplicative semi-norms had an answer by 1976 [21]: thanks to properties of T-filters, when the field K is "weakly valued" i.e. both its residue class field and its value group are countable, then it is possible to construct an example where an infinity of continuous multiplicative semi-norms admit for kernel a same maximal ideal: for instance $\mathbb{C}_p$ is weakly valued. However, if at least one of these two sets is not countable, then each maximal ideal is the kernel of only one continuous multiplicative semi-norm [23].

Another problem arose: is the spectral norm equal to the supremum of all multiplicative semi-norms whose kernel is a maximal ideal? When the field K is strongly valued, the answer is yes [23]. But when it is weakly valued, if there is no particular hypothesis on the norm of A, the answer is "no": there even exist certain ultrametric Banach K-algebras admitting elements in the Jacobson radical, which have a spectral norm different from 0 [23]. However, when the algebra is complete with respect to its spectral norm, the answer is yes [29]. Certain other spectral properties appear to be linked to integrity.

The set of continuous multiplicative semi-norms whose kernel is a maximal ideal is known to be dense in the whole set of all continuous multiplicative semi-norms in many simple examples, such as Krasner algebras and affinoid algebras. As a consequence of properties of the spectral norm described above, it is at least *pseudo-dense* when either the norm is the spectral one, or the field is strongly valued [29].

Affinoid algebras form a large panel of commutative ultrametric Banach K-algebras [49]. We recall the proofs of their algebraic and topological properties: they are noetherian Jacobson rings whose maximal ideals have codimension 1, and, in a reduced affinoid algebra, the spectral semi-norm is equivalent to the Banach algebra norm (a property which is not easily proven in characteristic $p \neq 0$). Moreover, topologically pure extensions are factorial [45]. By showing that the set of continuous multiplicative semi-norms whose kernel is a maximal ideal is dense in the whole set of all continuous multiplicative semi-norms [34], we have an easier proof of the Jacobson ring property.

An infraconnected subset of K is called affinoid [2] (or ultracirconférencié [18]) if it has finitely many holes and if its envelope and all its holes have a diameter in the value set of K. More generally, a subset of K is called affinoid (or ultracirconférencié) if it is a finite union of infraconnected affinoid subsets of K. Given an element x of an affinoid algebra, Y. Morita showed that the spectrum of x is an affinoid subset of K [37]. Here we give a new and more simple proof of this property, based on elementary considerations. This is useful to characterize algebras that are both Krasner algebras and Tate algebras: they are called *Krasner-Tate algebras* [18]. They are characterized among Krasner algebras (as the algebras $H(D)$ when D is affinoid), and among Tate algebras, as a quotient of the topologically pure extension $T_2 = K\{T, Y\}$ by a principal ideal of a particular form. They are also characterized among ultrametric Banach algebras, by a set of properties, where each is logically independent from all others. A particular interest of Krasner-Tate algebras is that they make a very typical and basic class of affinoid algebras that one can easily represent as an algebra of functions defined in a set of K. Thanks to the mapping on the set of circular filters described above, we can characterize the expansions of the Gauss norm defined on $K\{T\}$ to continuous absolute values on a Krasner-Tate algebra $H(D) = K\{T\}[x]$ (this involves the Shilov boundary of $H(D)$).

With help of Krasner-Tate algebras, we can also find fine conditions to obtain an idempotent associated to a subset of the spectrum of an element in an ultrametric Banach algebra. Particularly, if the spectral semi-norm is equal to the upper bound of the spectrum, we can show that if A has no non trivial idempotents, then the spectrum of every element is infraconnected. And particularly, this holds for a Banach algebra complete for its spectral norm. Several examples and counter-examples, provided with non usual ultrametric algebra norms, show the relative importance of such properties involved in theorems. They also give opportunities to test certain other claims or conjectures.

So far, books related to affinoid theory mainly ignored general ultrametric Banach algebra, and books dedicated to ultrametric functional analysis ignored affinoid algebras. This is why it seemed useful to gather both properties and methods. Many theorems proven in [30] are indispensable in the present book and are recalled with references, without proving them again.

I am very grateful to Jesus Araujo for reading my drafts, and noticing misprints, and to Bertin Diarra and Gabriel Picavet for advising me in basic algebra.

Chapter 1

Basic Properties in Commutative Algebra

In this first chapter, we have to collect several classic results in commutative algebra that will be indispensable when studying affinoid algebras.

Definitions and notation Throughout the chapter, B will denote a commutative ring with unity, E will denote a field, F will denote an algebraically closed field, and A will be a commutative E-algebra with unity.

Given an ideal I of B, we call *radical* of I the ideal of the $x \in B$ such that $x^n \in I$ for some $n \in \mathbb{N}^*$. We call *nilradical* of B the radical of $\{0\}$, i.e. the intersection of all prime ideals of B and *Jacobson radical* the intersection of all maximal ideal of B. Moreover, B is said to be *reduced* if its nilradical is equal to $\{0\}$, and *semi-simple* if its Jacobson radical is equal to $\{0\}$. Moreover, B is called a *Jacobson ring* if every prime ideal of B is equal to the intersection of all maximal ideals that contain it.

$Max(B)$ is the set of all maximal ideals of B. Henceforth, $Max_a(A)$ will denote the subset of $Max(A)$ which consists of all maximal ideals M such that $\frac{A}{M}$ is an algebraic extension of E. We will denote by $Max_\infty(A)$ the set of maximal ideals of infinite codimension. In particular, if E is algebraically closed, $Max_a(A)$ is the set of maximal ideals of codimension 1 and $\{Max_a(A), Max_\infty(A)\}$ makes a partition of $Max(A)$.

We will denote by $\mathcal{X}(A, E)$ the set of E-algebra homomorphisms from A onto E, and by $\mathcal{X}(A)$ the set of all E-algebra homomorphisms from A onto various fields $\frac{A}{\mathcal{M}}$ when $\mathcal{M} \in Max(A)$. Given $x \in A$, we denote by $sp_A(x)$ the set of $\lambda \in E$ such that $x - \lambda$ is not invertible in A, and by $sa_A(x)$ the set of $\lambda \in sp_A(x)$ which are images of x by E-algebra homomorphisms from A

onto algebraic extensions of E. So, in particular, when E is algebraically closed, we have $sa_A(x) = \{\chi(x) \mid \chi \in \mathcal{X}(A, E)\}$.

When there is no risk of confusion on the E-algebra A, we will only write $sp(x)$ instead of $sp_A(x)$, and $sa(x)$ instead of $sa_A(x)$.

Propositions 1.1, 1.2 and 1.3 are classical in commutative algebra.

Proposition 1.1: *Let $I_1, \ldots, I_n$ be ideals of B such that $I_k + I_l = B\ \forall k \neq l$. Then $\frac{B}{I_1 \ldots I_n}$ is isomorphic to $\frac{B}{I_1} \times \ldots \times \frac{B}{I_n}$.*

Lemma 1.2: *Let B be a noetherian integral domain. Then B is factorial if and only if for every irreducible element f of B, the ideal fB is prime.*

Lemma 1.3: *Let B be factorial and let $f \in B$. The ideal fB is equal to its radical if and only if any factorization of f into irreducible factors admits no factors with a power $q > 1$.*

Proposition 1.4: *Let F be a field which is an integral ring extension of B. Then B is a field.*

Proposition 1.5: *Suppose that for every $x \in B \setminus \{0\}$, the quotient ring $\frac{B}{xB}$ is noetherian. Then B is noetherian.*

Proof: Let I be an ideal of B. Let $t \in I$, $t \neq 0$ and let $\widetilde{B} = \frac{B}{tB}$. Then tB is also an ideal of I, and $\frac{I}{tB}$ is an ideal of $\frac{B}{tB}$. For every $x \in B$, we denote by $\tilde{x}$ its class in $\frac{B}{tB}$. Since $\widetilde{B}$ is noetherian, we find $x_1, \ldots, x_q \in I$ such that

$$\frac{I}{tB} = \widetilde{B}\tilde{x}_1 + \ldots + \widetilde{B}\tilde{x}_q.$$

Then $tB + \sum_{i=1}^q x_i B$ is an ideal of B included in I. Let $x \in I$. Since $\tilde{x}$ belongs to $\frac{I}{tB}$, we can find $a_1, \ldots, a_q \in B$ such that $\tilde{x} = \sum_{i=1}^q \tilde{a}_i \tilde{x}_i$. Therefore $x - \sum_{i=1}^q \tilde{a}_i \tilde{x}_i \in tB$, and then is equal to some ta, with $a \in B$. Thus, $x = tb + \sum_{i=1}^q a_i x_i$, and then $I \subset tB + \sum_{i=1}^q Ax_i$. Since by definition I contains $tB + \sum_{i=1}^q Ax_i$, this finishes showing that I is finitely generated, and thereby A is noetherian. $\square$

Proposition 1.6 is a consequence of Kronnecker's Theorem:

Proposition 1.6: *Let B be an integral domain, let B' be an integrally closed subring B containing the unity u of B, and let $x \in B$, be integral over B'. Then, the minimal polynomial of x over the field of fractions of B' belongs to $B'[X]$.*

Notation Let B' be an integrally closed subring of B containing the unity of B and let $x \in B$ be integral over B'. Then, the minimal polynomial of x over the field of fractions of B will be denoted by $irr(x, B')$. Hence by Proposition 1.6, $irr(x, B') \in B'[X]$.

Proposition 1.7: *Let B be an integral domain and integrally closed and let $R \supset B$ be a commutative ring with unity which is integral over B. Let $x \in B$ and let $irr(x, B) = x^n + \sum_{i=0}^{n-1} a_i x^i$. Then x is invertible in R if and only if a_0 is invertible in B.*

Proof: We have $x^n + \sum_{i=0}^{n-1} a_i x^i = 0$. If a_0 is invertible in B, then $x(\sum_{i=0}^{n-1} a_i(a_0)^{-1}x^{n-i-1}) = -1$, hence x is invertible in R. Now, suppose that x is invertible in R, and let $u = x^{-1}$. Then u satisfies $1 + \sum_{i=1}^{n} a_{n-i}u^i = 0$. Let $P(X) = irr(u, B)$. Since B is integrally closed, x is of degree n over the field of fractions of B, and so is u, hence P divides the polynomial $G(X) = 1 + \sum_{i=1}^{n} a_{n-i}X^i$, and is of same degree. Consequently $G = a_0 P$, and therefore $P(0) = \frac{1}{a_0}$, which shows that a_0 is invertible in B. □

Definition: We will call *a Luroth F-algebra* a F-algebra of the form $F[h, x]$, with $h \in F(x)$ and x transcendental over F. Given $h(x) = \frac{P(x)}{Q(x)} \in F(x)$, we call *degree of h* the number $\deg(P) - \deg(Q)$, and we will denote it by $\deg(h)$.

Lemma 1.8 is immediate:

Lemma 1.8: *Let $B = F[h, x]$ be a Luroth F-algebra. For every pole α of h, $x - \alpha$ is invertible in B.*

Corollary 1.9: *Let $h_j = \frac{P_j(x)}{Q_j(x)} \in F(x)$, $(1 \leq j \leq n)$, with P_j and Q_j relatively prime. Let $B = F[h_1, \ldots, h_n, x]$. Then $\frac{1}{Q_j(x)}$ belongs to B for all $j = 1, \ldots, n$.*

Propositions 1.10 and 1.12 are classical in commutative algebra [6]:

Theorem 1.10: *The set of prime ideals of B is inductive with respect to relation $\supset$.*

Corollary 1.11: *Every prime ideal of B contains a minimal prime ideal.*

Theorem 1.12: *If B is Noetherian it has finitely many minimal prime ideals.*

Lemma 1.13 is immediate:

Lemma 1.13: *Let I be an ideal of A and let $A' = \frac{A}{I}$. Let θ be the canonical surjection of A onto A'. The maximal ideals of A' are the $\theta(\mathcal{M})$ whenever $\mathcal{M}$ runs through maximal ideals of A containing I. Moreover, the maximal ideals of codimension 1 of A' are the $\theta(\mathcal{M})$ whenever $\mathcal{M}$ runs through maximal ideals of codimension 1 of A containing I.*

Lemma 1.14: *Let S be a subset of $\mathcal{X}(A)$. If there exists an idempotent u of A satisfying $\chi(u) = 1\ \forall \chi \in S$, and $\chi(u) = 0\ \forall \chi \in \mathcal{X}(A) \setminus S$, then it is unique.*

Proof: Suppose there exist two idempotents u and u' satisfying $\chi(u) = \chi(u') = 1\ \forall \chi \in S$, and $\chi(u) = \chi(u') = 0\ \forall \chi \in \mathcal{X}(A) \setminus S$. We notice

$$(u - u')(1 - u - u') = 0. \tag{1.1}$$

Let $\chi \in S$. Then $\chi(u') = 1$, hence $\chi(1 - u') = 0$, and therefore $\chi(1 - u - u') = \chi(-u) = -1$. Now, let $\chi \in \mathcal{X}(A) \setminus S$. Since $\chi(u') = 0$, we have $\chi(1 - u - u') = \chi(1) = 1$. Consequently, $\chi(1 - u - u') \neq 0$ whenever $\chi \in \mathcal{X}(A)$, and therefore $1 - u - u'$ is invertible in A. Hence by (1) we have $u - u' = 0$. □

Lemma 1.15: *Let A be a F-algebra and have finitely many minimal prime ideals $\mathcal{P}_1, \dots, \mathcal{P}_q$, and for each $j = 1, \dots, q$, let θ_j be the canonical surjection from A onto $\frac{A}{\mathcal{P}_j}$. Then for every $x \in A$, $sp(x) = \bigcup_{j=1}^{q} sp(\theta_j(x))$.*

Proof: For each $j = 1, \dots, q$, we put $A_j = \frac{A}{\mathcal{P}_j}$. Given $\chi \in \mathcal{X}(A_j)$ then $\chi \circ \theta_j$ belongs to $\mathcal{X}(A)$. In particular this true when $\chi(\theta_j(x)) \in E$. Therefore we have $\bigcup_{j=1}^{q} sp(\theta_j(x)) \subset sp(x)$. Conversely, let $\phi \in \mathcal{X}(A)$. By Corollary 1.11 $Ker(\phi)$ contains at least one minimal prime ideal $\mathcal{P}_h$. Consequently, ϕ factorizes in the form $\chi \circ \theta_h$, with $\chi \in \mathcal{X}(A_j)$. Particularly this is true when $\chi(x) \in E$. Consequently, $sp(x) \subset \bigcup_{j=1}^{q} sp(\theta_j(x))$, which ends the proof. □

Normalization Lemma is classical:

Theorem 1.16 (Normalization lemma): Let A be a finite type E-algebra. There exists a finite algebraically free set $\{y_1, \dots, y_s\} \subset A$ such that A is finite over $E[y_1, \dots, y_s]$.

When studying Krasner-Tate algebras, we will consider universal generators.

Definition: Let A be a F-algebra. An element $x \in A$ will be said to be *spectrally injective* if, for every $\lambda \in sp(x)$, $x - \lambda$ belongs to a unique maximal ideal. Moreover, x will be called *a universal generator* if, for every $\lambda \in sp(x)$, $(x-\lambda)A$ is a maximal ideal.

Lemma 1.17: *Let B be an integral domain integrally closed F-algebra of finite type the field of fractions of which is a degree one pure transcendental extension of F. Then, there exists $x \in B$ and $f \in F(x)$ of degree > 0 such that $B = F[f, x]$.*

Proof: Let G be the field of fractions of B. By hypothesis, there exists $z \in G$ such that $G = F[z]$. By Theorem 1.16 there exist $t, y_1, \ldots, y_n \in B$ such that $B = F[t, y_1, \ldots, y_n]$ and that B be integral over $F[t]$. Let $t = g(z)$ and let us show that there exists $x \in B$, integral over $F[t]$ such that $t \in F(x)$. We can write $g(z) = \frac{P(z)}{Q(z)}$, with $P,\ Q \in F[X]$ relatively prime, and P monic. Let $k = \deg(P)$, $l = \deg(Q)$. If $k > l$, we can just take $x = z$ because z is clearly integral over $F[t]$. So, we have to assume $k \leq l$. Let a be a zero of Q, and let $P_0(X) = P(X+a)$, $Q_0(X) = Q(X+a)$. Like P and Q, P_0 and Q_0 are relatively prime. Consequently, $P_0(0) \neq 0$ because by definition $Q_0(0) = 0$. Now, let $x = \frac{1}{z-a}$. We check that

$$t = \frac{P_0\left(\frac{1}{x}\right)}{Q_0\left(\frac{1}{x}\right)} = \frac{(a_0x^k + \ldots + a_k)x^{l-k}}{(b_1x^{l-1} + \ldots + b_l)}.$$

And since $k \leq l$, t appears as a rational function in x, of degree > 0, hence x is integral over $F[t]$. On the other hand, since B is integrally closed, we have $x \in B$, hence $F[t, x] \subset B$. So it just remains us to show that $B \subset F[t, x]$. Since $F(x)$ is clearly equal to G, each y_j lies in $F(x)$, and therefore is of the form $h_j(x)$, with $h_j(X) = \frac{S_j(X)}{W_j(X)} \in F(X)$, and S_j, W_j relatively prime. Now, we only have to show that each $W_j(x)$ is invertible in $F[t, x]$. For each $j = 1, \ldots, n$, W_j has obviously an inverse Z_j in $F(x)$. Consider any $j = 1, \ldots, n$. Then, since x lies in B, by Corollary 1.9 we know that so does Z_j and consequently Z_j is integral over $F[t]$. Let

$$A_j(X) = \sum_{i=0}^{u_i} \lambda_{i,j} X^i \in F[t][X]$$

be a monic polynomial such that $A_j(Z_j) = 0$. Then we see that

$$(W_j)^{u_j} F_j(Z_j) = 1 + W_j(\lambda_{j,u_j-1} + \ldots + \lambda_{j,0} S^{u_j-1}) = 0$$

and this shows that W_j is invertible in $F[t, x]$. □

Lemma 1.18 is immediate by considering the field of fractions of A and B.

Lemma 1.18: *Let A be a F-algebra finite over a F-subalgebra B. For all $\chi \in \mathcal{X}(B)$ there exists $\widehat{\chi} \in \mathcal{X}(A)$ satisfying $\widehat{\chi}(x) = \chi(x)\ \forall x \in B$.*

Theorem 1.19: *Let A be a F-algebra without non zero divisors of zero admitting a spectrally injective element x. Let B be an integrally closed F-subalgebra of A containing x and the unity of A, such that A is finite over B. Assume that all maximal ideals of A and B have codimension 1. Then for every $y \in B$, we have $sp_B(y) = sp_A(y)$, and x is a spectrally injective element of B. Moreover for every $\mathcal{M} \in Max(A)$, $\mathcal{M} \cap B$ belongs to $Max(B)$ and the mapping ϕ from $Max(A)$ into $Max(B)$ defined as $\phi(\mathcal{M}) = \mathcal{M} \cap B$ is an injection from $Max(A)$ into $Max(B)$. Further, assume that B is semi-simple and let p be the characteristic of F: either $p = 0$, then $A = B$, or $p \neq 0$, and then for every $y \in A$, $irr(y,B)$ is of the form $X^{p^t} - f$, $f \in B$.*

Proof: Since the unity of A lies in B it is easily seen that for each maximal ideal $\mathcal{M}$ of A, the F-algebra homomorphism from A to F whose kernel is $\mathcal{M}$ has a restriction to B which is surjective on F, so $\mathcal{M} \cap B$ is a maximal ideal of B. We will deduce that the mapping ϕ is injective. Indeed, let $\mathcal{M}_1$ and $\mathcal{M}_2$ be two different maximal ideals of A, and let $\lambda_1,\ \lambda_2 \in K$ be such that $x - \lambda_1 \in \mathcal{M}_1,\ x - \lambda_2 \in \mathcal{M}_2$. Then $x - \lambda_1$ lies in $\mathcal{M}_1 \cap B$, but does not lie in $\mathcal{M}_2$, hence $\mathcal{M}_1 \cap B \neq \mathcal{M}_2 \cap B$, thereby ϕ is injective. The mapping is also surjective: let $\mathcal{J}$ be a maximal ideal of B and let $\chi \in \mathcal{X}(B)$ admit $\mathcal{J}$ for kernel, then by Lemma 1.18 there exists $\widetilde{\chi} \in \mathcal{X}(A)$ whose restriction to B is equal to χ, and therefore $\mathcal{J} = Ker(\widetilde{\chi}) \cap B = \phi(Ker(\widetilde{\chi}))$. Consequently $sp_A(y) = sp_B(y)\ \forall y \in B$. Then it is clear that x is spectrally injective in B: given $\lambda \in sp_B(x) = sp_A(x)$, then there is a unique maximal ideal $\mathcal{M}$ of A which contains $x - \lambda$ and then $\phi(\mathcal{M})$ is the unique maximal ideal of B which contains $x - \lambda$. We now suppose that B is semi-simple. Let $y \in A \setminus B$, let $B' = B[y]$ and let $P(X) = \sum_{j=0}^{n} f_j X^j = irr(y, B)$. Then x is a spectrally injective element of B'. Let $\chi \in \mathcal{X}(B)$. Since ϕ is bijective there exists a unique $\widetilde{\chi} \in \mathcal{X}(A)$ whose restriction to B is equal to χ. Consequently, the polynomial $G(X) = \sum_{j=0}^{n} \chi(f_j) X^j$ admits a unique zero α of order n. Therefore we have

$$\chi(f_j) = (-\alpha)^j \binom{n}{j}, \quad \forall j = 0, \ldots, n-1,$$

and in particular $\chi(f_{n-1}) = -n\alpha$. Suppose p does not divide n. Since this is true for all $\chi \in \mathcal{X}(B)$ and since the Jacobson radical of B is null, this shows

$$f_j = \binom{n}{j}\left(\frac{-f_{n-1}}{n}\right)^j, \quad \forall j = 1, \ldots, n-1.$$

Consequently, $G(X) = (X - \frac{f_{n-1}}{n})^n$, therefore $n = 1$. Thus, if $p = 0$ then $B = A$. Similarly, if p divides n, we can see that P is of the form $X^{p^t} - f, f \in B$. □

Chapter 2

Tree Structure

Trees will be very helpful when studying multiplicative semi-norms on polynomials and rational functions in one variable. Most of results given here were published in [9]. In this chapter we define a basic notion of tree such that, given a strictly increasing function from the tree to $\mathbb{R}_+$, this function provides the tree with a distance.

Definition: Let E be a set provided with an order relation $\leq$. Then E will be called *a tree* if:

(i) For every $a,\ b \in E$ there exists $\sup(a,b) \in E$.
(ii) For every $a \in E$ and $b,\ c \in E$ satisfying $a \leq b,\ a \leq c$, then b and c are comparable with respect to the order $\leq$.

Theorem 2.1: *Let $(E,\ \leq)$ be a tree and let f be a strictly increasing function from E to $\mathbb{R}$. For all $a,\ b \in E$ we put $\sigma_1(a,b) = (f(\sup(a,b)) - \min(f(a), f(b))$ and $\sigma_2(a,b) = 2f(\sup(a,b)) - f(a) - f(b)$. Then σ_1 and σ_2 are two equivalent distances such that $\sigma_1 \leq \sigma_2 \leq 2\sigma_1$. Moreover, if $b, c \in E$ satisfy $a \leq b,\ a \leq c$ and if $\sigma_j(a,b) = \sigma_j(a,c)$ for some j, then $b = c$.*

Proof: It is easily seen that $\sigma_j(a,b) \geq 0\ \forall a, b \in E,\ (j = 1,2)$ and that $\sigma_j(a,b) = 0$ if and only if $a = b\ (j = 1,2)$ because f is strictly increasing. It is clear that $\sigma_1 \leq \sigma_2$ so the same properties hold for σ_2. Now, let $a, b, c \in E$. By definition of both σ_1 and σ_2 it is obviously seen that they satisfy $\sigma_j(a,b) = \sigma_j(b,a)$. Thus, it only remains us to check the triangular inequality.

Since E is a tree, $\sup(a,b)$ and $\sup(b,c)$ are comparable. For instance, suppose first $\sup(b,c) \geq \sup(a,b)$. Then it is easily seen that $\sup(b,c) = \sup(a,b,c) \geq \sup(a,c)$, and therefore we obtain

$$\begin{aligned}\sigma_1(a,b)+\sigma_1(b,c) &\geq f(\sup(a,b)) - f(a) + f(\sup(b,c)) - f(b)\\ &\geq f(\sup(b,c)) - f(a)\\ &\geq f(\sup(a,c)) - f(a). \qquad (2.1)\end{aligned}$$

$$\begin{aligned}\sigma_1(a,b)+\sigma_1(b,c) &\geq f(\sup(a,b)) - f(a) + f(\sup(b,c)) - f(c)\\ &\geq f(\sup(b,c)) - f(c)\\ &\geq f(\sup(a,c)) - f(c). \qquad (2.2)\end{aligned}$$

In the same way, suppose now $\sup(b,c) \leq \sup(a,b)$. Then we obtain

$$\begin{aligned}\sigma_1(a,b)+\sigma_1(b,c) &\geq f(\sup(a,b)) - f(a) + f(\sup(b,c)) - f(b)\\ &\geq f(\sup(a,b)) - f(a)\\ &\geq f(\sup(a,c)) - f(a). \qquad (2.3)\end{aligned}$$

$$\begin{aligned}\sigma_1(a,b)+\sigma_1(b,c) &\geq f(\sup(a,b)) - f(a) + f(\sup(b,c)) - f(c)\\ &\geq f(\sup(a,b)) - f(c)\\ &\geq f(\sup(a,c)) - f(c). \qquad (2.4)\end{aligned}$$

So, by (2.1),(2.2) whenever $\sup(b,c) \geq \sup(a,b)$ and by (2.3), (2.4) whenever $\sup(b,c) \geq \sup(a,b)$, we obtain $\sigma_1(a,b)+\sigma_1(b,c) \geq \sigma_1(a,c)$. This finishes proving that σ_1 is a distance.

We now consider σ_2. We first suppose again $\sup(b,c) \geq \sup(a,b)$. Then

$$\begin{aligned}\sigma_2(a,b)+\sigma_2(b,c) &= 2f(\sup(a,b)) - f(a) - f(b)\\ &\quad + 2f(\sup(b,c)) - f(b) - f(c)\\ &= 2f(\sup(a,b,c)) - f(a) - f(c) + 2f(\sup(a,b)) - 2f(b)\\ &\geq 2f(\sup(a,b,c)) - f(a) - f(c)\\ &\geq 2f(\sup(a,c)) - f(a) - f(c)\\ &= \sigma_2(a,c).\end{aligned}$$

And in the same way, when

$$\sup(b,c) \geq \sup(a,b)$$

then

$$\begin{aligned}\sigma_2(a,b)+\sigma_2(b,c) &= 2f(\sup(a,b,c)) - f(a) - f(c) \\ &\quad + 2f(\sup(b,c)) - 2f(b) \\ &\geq 2f(\sup(a,b,c)) - f(a) - f(c) \\ &\geq 2f(\sup(a,c)) - f(a) - f(c) \\ &= \sigma_2(a,c).\end{aligned}$$

This finishes proving that σ_2 is also a distance. Next, it is easily seen that the two distances are equivalent because they satisfy $\sigma_1(a,b) \leq \sigma_2(a,b) \leq 2\sigma_1(a,b)$.

Now, suppose $a \leq b$ and $a \leq c$. Since E is a tree, b and c are comparable. Without loss of generality we can assume $b \leq c$. Then we have $\sigma_j(a,b) = j(f(b)-f(a))$, $\sigma_j(a,c) = j(f(c)-f(a))$, (j=1,2). Consequently, if $\sigma_j(a,b) = \sigma_j(a,c)$ for some j then $f(b) = f(c)$. But since f is strictly increasing, that means $b = c$. □

Definition: Let E be a tree and let f be a strictly increasing mapping from E into $\mathbb{R}$. We will call *metric topology associated to f on E* the topology defined by the distances introduced in Theorem 2.1. The distance denoted by σ_1 in Theorem 2.1 will be called *supremum distance associated to f* and this denoted by σ_2 in Theorem 2.1 will be called *whole distance associated to f*.

A topological space E is said to be *pathwise connected* or *arcwise connected* if for every $a, b \in E$ there exists a continuous mapping ϕ from an interval $[\alpha, \beta]$ into E such that $\phi(\alpha) = a$, $\phi(\beta) = b$.

Theorem 2.2 is classical in basic topology:

Theorem 2.2: *A pathwise connected topological space is connected.*

Theorem 2.3: *Let $(E, \leq)$ be a tree and let f be a strictly increasing function from E to $\mathbb{R}$. Let $a,\ b \in E$ be such that $a \leq b$ and $\{f(x) \mid a \leq x \leq b\} = [f(a), f(b)]$. Then there exists a mapping from $[f(a), f(b)]$ into E bicontinuous with respect to the metric topology of E associated to f.*

Proof: For each $r \in [f(a), f(b)]$, we denote by $\psi(r)$ the unique $x \in E$ such that $a \leq x$ and $f(x) - f(a) = r$. Let σ_1 be the supremum distance associated to f. Then we have $\sigma_1(\psi(r), \psi(s)) = r - s\ \forall r,\ s \in [f(a), f(b)]$, which clearly shows that ψ is bicontinuous. □

Theorem 2.4: *Let $(E, \leq)$ be a tree and let f be a strictly increasing function from E to $\mathbb{R}$. If for all $a, b \in E$ such that $a \leq b$, the equality $\{f(x) \mid a \leq x \leq b\} = [f(a), f(b)]$ holds, then E is arcwise connected with respect to its metric topology associated to f.*

Proof: Indeed, let a, $b \in E$ and let $s = \sup(a, b)$. By Theorem 2.3 there exists a bicontinuous mapping ϕ_1 from $[f(a), f(s)]$ into E such that $\phi_1(f(a)) = a$, $\phi_1(f(s)) = s$ and a bicontinuous mapping ϕ_2 from $[f(b), f(s)]$ into E such that $\phi_1(f(b)) = b$, $\phi_1(f(s)) = s$. Then we can obviously define a bicontinuous mapping ψ from an interval $[\alpha, \beta]$ into E such that $\psi(\alpha) = a$, $\psi(\beta) = b$. □

Chapter 3

Ultrametric Absolute Values

In this chapter, we have to recall many basic definitions and results about valuations in ultrametric fields. Most of them were given in [30] with full proofs. Here, when we state one with some additional considerations, it is followed by a new proof.

Notation: Throughout the book L will denote a field provided with a non trivial ultrametric absolute value $| \,.\, |$ and K will be an algebraically closed complete ultrametric field with a non trivial ultrametric absolute value $| \,.\, |$. It is convenient and useful to define the valuation v associated to the absolute value $| \,.\, |$. Let $\omega \in]1, +\infty[$ and let log be the real logarithm function of base ω. We put $v(x) = -\log |x|$, and v is named *the valuation associated to the absolute value* $| \,.\, |$.

The archimedean absolute value defined on $\mathbb{R}$ will be denoted by $| \,.\, |_\infty$.

Lemmas 3.1 and 3.2 are classical, and proven in the same way regardless the absolute value.

Lemma 3.1: *Let E be a field provided with two ultrametric absolute values whose associated valuations are v and w, respectively. They are equivalent if and only if there exists $r > 0$ such that $w(x) = rv(x)$ whenever $x \in E$.*

Lemma 3.2: *Let V be an L-vector space of finite dimension provided with two norms. These two norms are equivalent to one of the following two cases:*

(*1*) *The dimension of V is one;*
(*2*) *L is complete.*

Theorem 3.3 is an easy corollary:

Theorem 3.3: *Let L be complete and let F be an algebraic extension of L, provided with two absolute values extending this of L. These absolute values are equal.*

We will sometimes use Banach's Theorem for non archimedean vector spaces:

Theorem 3.4: *Let L be complete and let F be a L-vector space which is complete for two norms $\| \,.\, \|$ and $\| \,.\, \|'$ such that $\|x\| \leq C\|x\|'$ $\forall x \in F$, where C is a positive constant. Then the two norms are equivalent.*

Notation: The set of $x \in L$ such that $|x| \leq 1$ will be denoted by U_L, and the set of $x \in L$ such that $|x| < 1$ will be denoted by M_L.

Lemma 3.5: *U_L is a local subring of L whose maximal ideal is M_L.*

Lemma 3.6: *Let B be a commutative ring with unity provided with an absolute value $| \,.\, |$ such that $|x| \leq 1$ $\forall x \in B$, and such that every x satisfying $|x| = 1$ is invertible in B. The set of $x \in B$ such that $|x| < 1$ is the unique maximal ideal of B. Let F be the field of fractions of B. Then the absolute value of B has continuation to F and $B = U_F$.*

Definitions and notation: We call *a valuation ring* a commutative ring with unity provided with an absolute value $| \,.\, |$ such that $|x| \leq 1$ $\forall x \in B$, and such that every x satisfying $|x| = 1$ is invertible in B, and then the maximal ideal of B is called *the valuation ideal of B*. In a the field L, U_L is called *the valuation ring of L*. The maximal ideal M_L of U_L is called the *valuation ideal* and the field $\mathcal{L} = \frac{U_L}{M_L}$ is called *the residue class field of L*. For any $a \in L$, the residue class of a will be denoted by $\overline{a}$. If D is a set in L we put $|D| = \{|x| \mid x \in D\}$, and $v(D) = \{v(x) \mid x \in D\}$.

The characteristic of $\mathcal{L}$ is named *the residue characteristic of L* and will be denoted by p.

We put $L^* = L \setminus \{0\}$. Then $|L^*|$ is a subgroup of the multiplicative group $(\mathbb{R}_+^*, .)$. The image of L^* by the valuation v associated to L is then a subgroup of the additive group $(\mathbb{R}, +)$ called *the valuation group of L*. The valuation of L is said to be *discrete* if its valuation group is a discrete subgroup of $\mathbb{R}$. Else, the valuation group is dense in $\mathbb{R}$, and then the valuation is said to be *dense*.

The field L will be said to be *strongly valued* if at least one of the two sets $|L|$, $\mathcal{L}$, is not countable. If L is not strongly valued it will be said to be *weakly valued.*

Lemma 3.7 is well known:

Lemma 3.7: *If L is algebraically closed, its valuation group is dense in $\mathbb{R}$.*

Remark: The most classical example of an ultrametric complete algebraically closed field is the field $\mathbb{C}_p$ (one can find a description in [30, Chapter 8]). Then $\mathbb{C}_p$ is weakly valued. Now, consider a field L with a dense valuation, and let A be the L-algebra B of bounded analytic functions in $d(0, 1^-)$, provided with the norm of uniform convergence in $d(0, 1^-)$. It is known and easily seen that $\{\|f\| \mid f \in B\} = [0, +\infty[$. Therefore, its field of fractions is strongly valued. When $p \neq 0$ it is useful to take $\omega = p$.

Theorems 3.8 and 3.9 are well known (proofs are given in [30, Theorems 6.3 and 6.10]).

Theorem 3.8: *Let L be complete and let F be an algebraic extension of L. There exists a unique ultrametric absolute value on F extending this of L. Moreover, given $x \in F$, $P(X) = irr(x, F)$ and $q = \deg(P)$, then $|x| = \sqrt[q]{|P(0)|}$.*

Theorem 3.9: *Let F be an algebraically closed field provided with a non trivial ultrametric absolute value. The completion of F is algebraically closed.*

Example: $\mathbb{C}_p$ is the completion of an algebraic closure of $\mathbb{Q}_p$, the completion of $\mathbb{Q}$ with respect to the p-adic absolute value. Then $\mathbb{C}_p$ is algebraically closed.

Chapter 4

L-Productal Vector Spaces

In characteristic $p > 0$, studying the spectral norm of an affinoid algebra will require properties of L-productal vector spaces.

Definitions and notation: Let E be a normed L-vector space. We will denote by E^* its algebraic dual and by E' the L-vector space of its continuous linear forms.

Lemma 4.1: *Let E be a normed L-vector space of dimension n. Then E is homeomorphic to L^n if and only if for every $x \in E$, there exists $\phi \in E'$ satisfying $\phi(x) \neq 0$. Moreover, if E is homeomorphic to L^n, then $E' = E^*$.*

Proof: If E is homeomorphic to L^n for every $x \in E$, there obviously exists $\phi \in E'$ satisfying $\phi(x) \neq 0$ because, given $u \in E$, $u \neq 0$, we can take a basis $(e_j)_{1\leq j\leq n}$ of E such that $u = e_1$ and such that $\| \sum_{j=1}^{n} \lambda_j e_j \| \leq \max_{1\leq j\leq n} |\lambda_j|$, and then, in the dual basis $(\phi_j)_{1\leq j\leq n}$ of the basis $(e_j)_{1\leq j\leq n}$, we check that ϕ_1 satisfies $|\phi_1(\sum_{j=1}^{n} \lambda_j e_j)| = |\lambda_1| \leq \| \sum_{j=1}^{n} \lambda_j e_j \|$.

Conversely, we assume that for every $x \in E$, there exists $\phi \in E'$ satisfying $\phi(x) \neq 0$. We will show the existence of a basis of E^* consisting of elements of E'. Let n be the dimension of E. By hypothesis, we can find $\phi_1 \in E'$, $\phi_1 \neq 0$. Suppose we have already constructed $\phi_1, \ldots, \phi_m \in E'$, linearly independent, with $m < n$. Then the intersection of the m hyperplanes $Ker(\phi_j)$ is not reduced to the null subspace. Hence, there exists $u \in \bigcap_{j=1}^{m} Ker(\phi_j)$, with $u \neq 0$. By hypothesis, there exists $\phi_{m+1} \in E'$ such that $\phi_{m+1}(u) \neq 0$. Then, since $\phi_j(u) = 0\ \forall j = 1, \ldots, m$, and $\phi_{m+1} \neq 0$, we easily check that $\phi_1, \ldots, \phi_{m+1}$ are linearly independent. Thus, by induction, we can obtain $\phi_1, \ldots, \phi_n \in E'$ which are linearly independent. Consequently, this is a basis of E^* and therefore $E^* = E'$.

Now, consider the mapping ψ from E into L^n defined as $\psi(x) = (\phi_1(x), \ldots, \phi_n(x))$. We can easily check that ψ is an isomorphism from E onto L^n, and is continuous by definition because so are the ϕ_j. We have to check that ψ^{-1} is also continuous. By considering the bidual of E we can obtain a basis $\{e_1, \ldots, e_n\}$ of E such that $\{\phi_1, \ldots, \phi_n\}$ is the dual basis of $\{e_1, \ldots, e_n\}$. Let $M = \max_{1 \le j \le n} \|e_j\|$, $(\lambda_1, \ldots, \lambda_n) \in L^n$ and $x = \sum_{j=1}^{n} \lambda_j e_j$. Then $\psi(x) = (\lambda_1, \ldots, \lambda_n)$, hence $\psi^{-1}(\lambda_1, \ldots, \lambda_n) = x$, and therefore $\|\psi^{-1}(\lambda_1, \ldots, \lambda_n)\| \le M \max_{1 \le j \le n} |\lambda_j|$. Thus ψ^{-1} is continuous, and therefore is a homeomorphism. □

Corollary 4.2: *A normed L-vector space E is L-productal if and only if for every $x \in E$, there exists $\phi \in E'$ satisfying $\phi(x) \neq 0$.*

Proof: If E is L-productal, obviously for every $x \in E$, there exists $\phi \in E'$ satisfying $\phi(x) \neq 0$. Conversely, if for every $x \in E$, there exists $\phi \in E'$ satisfying $\phi(x) \neq 0$, this property is true in each finite dimensional L-subspace F of dimension n of E, and proves that F is homeomorphic to L^n. □

Lemma 4.3: *Let E be a L-productal normed L-vector space and let $q \in \mathbb{N}^*$. Then E^q, provided with the product topology, is L-productal.*

Proof: Let W be a subspace of E^q of finite dimension n. For each $h = 1, \ldots, q$, we denote by E_h the subspace of E^q consisting of the $(x_1, \ldots x_q)$ such that $x_j = 0 \ \forall j \neq h$ and we put $W_j = W \cap E_j$. Then W is the direct sum of the W_j, and for each $j = 1, \ldots, q$, W_j has finite dimension n_j. Let $n = \sum_{j=1}^{q} n_j$. By hypothesis, W_j is homeomorphic to L^{n_j}. Let $\| \,.\, \|$ be the norm of E, let $\| \,.\, \|$ be the product norm on E^q defined as $\|(x_1, \ldots x_q)\| = \max_{1 \le j \le q} \|x_j\|$. Now, in each subspace W_j, we denote by $| \,.\, |_j$ the norm induced by L^{n_j} on W_j. Thus, there exist positive constants A and B such that $A|a_j|_j \le \|a_j\| \le B|a_j|_j \ \forall a_j \in W_j, \forall j = 1, \ldots, q$. But since E^q is provided with the product topology, given $a \in W$, a is of the form $\sum_{j=1}^{q} a_j$, with $a_j \in W_j$, $(j = 1, \ldots, q)$, and therefore we have $A \max_{1 \le j \le q} \|a_j\| \le \|a\| \le B \max_{1 \le j \le q} \|x_j\|$. But it is seen that $\max_{1 \le j \le q} \|x_j\|$ is a norm equivalent to the product norm on L^n, and this lets us conclude that W, provided with the norm $\| \,.\, \|$, is homeomorphic to L^n. □

Lemma 4.4: *Let F be a subfield of L such that L is a F-productal F-vector space with respect to its absolute value. Then, every L-productal L-vector space is F-productal.*

Proof: Let E be a L-productal L-vector space. Then E is a F-vector space. Let W be a F-subspace of finite dimension over F and let $\{e_1, \dots, e_q\}$ be a basis of W. Let V be the L-subvector space generated by $\{e_1, \dots, e_q\}$ in E. By hypothesis, W is homeomorphic to L^q. But since L is F-productal, by Lemma 4.3 so is L^q. Hence V is F-productal, and therefore, so is every F-subspace of V, hence so is W. □

Before ending this chapter, we can also recall this immediate and classical lemma:

Lemma 4.5: *Let $(E_n)_{n\in\mathbb{N}}$ be a sequence of Banach L-vector spaces and let E be the L-vector space of all sequences $(a_n)_{n\in\mathbb{N}}$, $a_n \in E_n$, $(n \in \mathbb{N})$, such that $\lim_{n\to\infty} \|a_n\| = 0$. Then E is a Banach L-vector space with respect to the norm $\|(a_n)_{n\in\mathbb{N}}\| = \sup_{n\in\mathbb{N}} \|a_n\|$.*

Chapter 5

Multiplicative Semi-Norms and Shilov Boundary

The idea of studying systematically the multiplicative semi-norms of an ultrametric normed algebra is due to Bernard Guennebaud [8]. This theory found many applications in the study of analytic elements and actually gave the good explanation of several basic properties [6]. More generally, this is also true in ultrametric Banach algebras, where multiplicative semi-norms play a role somewhat similar to that of maximal ideals in complex Banach algebras [5,7,8]. Next, Vladimir Berkovich also used the set of multiplicative semi-norms in order to reconstruct Tate's Theory [1,10]. In [4], Kamal Boussaf showed the existence of a Shilov boundary for the norm of an algebra of analytic elements and described it with help of circular filters that characterize multiplicative semi-norms. In [1], V. Berkovich showed the existence of a Shilov boundary for the spectral semi-norm of an affinoid algebra. This chapter is aimed at proving, in the general case, the existence of a Shilov boundary for a semi-multiplicative semi-norm of a L-algebra, i.e. the existence of a closed subset F of the set of continuous multiplicative semi-norms, minimal for inclusion, such that for every $x \in A$, there exists $\varphi \in F$ satisfying $\varphi(x) = \|x\|$. Such a process was outlined by B. Guennebaud in his unpublished Thèse d'Etat [34], and our way widely takes from it. However, certain intermediate results were missing, or suffered a lack of rigour, while hypotheses and definitions were often confusing. Besides, there was a confusion between two different problems: on one hand, finding a *Shilov boundary* for a semi-multiplicative function defined on a semi-group, inside a set of multiplicative functions, and on the other hand, finding a *Shilov boundary* for a semi-multiplicative semi-norm defined on an algebra, inside a set of multiplicative semi-norms. Here, we have reconstructed the framework of the proof and only considered the problem of a Shilov boundary for a semi-multiplicative semi-norm.

Definitions and notation: Throughout this chapter, E will denote a field provided with a non trivial absolute value $| \, . \, |$ which makes it complete, hence is either $\mathbb{R}$ or $\mathbb{C}$, or a field L provided with a non trivial ultrametric absolute value for which it is complete.

Given sets X, Y we will denote by $\mathcal{F}(X,Y)$ the set of mappings from X to Y. Let D a be subset of E and let F be a E-vector space of bounded functions from D to E. In the sequel, we will denote by $\| \, . \, \|_D$ the norm of uniform convergence on D. The $\mathbb{R}$-vector space $\mathcal{F}(D,\mathbb{R})$ is provided with the topology of simple convergence: the filter of neighborhoods of $\varphi \in \mathcal{F}(D,\mathbb{R})$ admits as a generating system the family of sets $\mathcal{V}(\varphi, a_1, \ldots, a_n, \epsilon) = \{\psi \in \mathcal{F}(D,\mathbb{R}) \mid |\psi(a_i) - \varphi(a_i)|_\infty \leq \epsilon\}$, with $a_1, \ldots, a_n \in D$ and $\epsilon > 0$. Given a topological space D we denote by $\mathcal{C}(D,\mathbb{R})$ the algebra of continuous functions from D into $\mathbb{R}$.

We call *semi-group* a set provided with an associative internal law and of a neutral element for this law, and we will denote by B, B' semigroups. We will denote by A (resp. A') a commutative E-algebra with unity.

Now, let θ be a function from D to $\mathbb{R}_+$ and let $\mathcal{B}(D,\theta)$ be the subset of $\mathcal{F}(D,\mathbb{R})$ consisting of the functions f from D to $\mathbb{R}_+$ such that $f(x) \leq \theta(x)$ $\forall x \in D$.

Let $\mu \in \mathcal{F}(B,\mathbb{R}_+)$. The function μ will be said to be *submultiplicative* (resp. *multiplicative*) if $\mu(xy) \leq \mu(x)\mu(y)$, $\forall x, y \in B$ (resp. $\mu(xy) = \mu(x)\mu(y)$, $\forall x, y \in B$). And μ will be said to be *semi-multiplicative* if $\mu(x^n) = \mu(x)^n$, $\forall x \in B, \forall n \in \mathbb{N}^*$.

Let S be a subsemi-group of B. A submultiplicative function θ on B will be said to be *S-multiplicative* if $\theta(xy) = \theta(x)\theta(y)$, $\forall x, y \in S$.

Let us recall that *a semi-norm* (resp. *a norm*) of E-algebra defined on A is a semi-norm (resp. a norm) of E-linear space φ that satisfies $\varphi(x.y) \leq \varphi(x)\,\varphi(y)$, $\forall x,\ y \in A$. Given a semi-norm of E-algebra φ, we will denote by $Ker(\varphi)$ the set of the $x \in A$ such that $\varphi(x) = 0$. Henceforth, we will denote by $\| \, . \, \|$ (resp. $\| \, . \, \|'$) a E-algebra semi-norm on A (resp. A').

We will denote by $SM(A)$ the set of non identically zero semi-multiplicative semi-norms of A, by $Mult(A)$ the set of non identically zero multiplicative semi-norms of A, by $Mult_m(A)$ the set of the $\varphi \in Mult(A)$ such that $Ker(\varphi) \in Max(A)$, by $Mult_a(A)$ the set of the $\varphi \in Mult(A)$

such that $Ker(\varphi) \in Max_a(A)$ and by $Mult'(A)$ the set of multiplicative norms of A [32].

Let A be provided with a topology T of E-vector space. We will denote by $SM(A,T)$ (resp. $Mult(A,T)$, $Mult'(A,T)$, $Mult_m(A,T)$, $Mult_a(A,T)$) the set of the $\varphi \in SM(A)$ (resp. $\varphi \in Mult(A)$, $\varphi \in Mult'(A)$, $\varphi \in Mult_m(A)$, $\varphi \in Mult_a(A)$) that are continuous with respect to the topology T. Particularly, if the topology T on A is defined by a norm or a semi-norm $\| \, . \, \|$, we will denote by $SM(A, \| \, . \, \|)$ (resp. $Mult(A, \| \, . \, \|)$, $Mult'(A, \| \, . \, \|)$, $Mult_m(A, \| \, . \, \|)$, $Mult_a(A, \| \, . \, \|)$, $Mult_m(A, \| \, . \, \|)$) the set $SM(A,T)$ (resp. $Mult(A,T)$, $Mult'(A,T)$, $Mult_m(A,T)$, $Mult_a(A,T)$).

By Tykhonov's Theorem, we have Lemma 5.1:

Lemma 5.1: *Let g be a function from D to $\mathbb{R}_+$. Then $\mathcal{B}(D,g)$ is compact with respect to the topology of simple convergence.*

Statements of Lemma 5.2 are classical, and come from properties in commutative algebra and in topological vector spaces theory.

Lemma 5.2: *Let φ be a semi-norm of E-algebra of A. Then $Ker(\varphi)$ is an ideal of A. If φ is multiplicative, $Ker(\varphi)$ is a prime ideal of A.*

Let I be an ideal of A, let A' be the quotient E-algebra $\frac{A}{I}$ and let Θ be the canonical homomorphism from A onto A'. Let φ' be the mapping from A' to $\mathbb{R}_+$ defined as $\varphi'(\Theta(x)) = \inf\{\varphi(t) | \ \Theta(t) = \Theta(x)\}$. Then φ' is a E-algebra semi-norm on A'. If φ is ultrametric, so is φ'. Moreover, if φ is a norm, and if I is closed with respect to φ, then φ' also is a norm.

Lemma 5.3 is Lemma 1.7 in [30]:

Lemma 5.3: *If A is a L-algebra, every element of SM(A) is ultrametric.*

Lemmas 5.4 is obvious.

Lemma 5.4: *Let S be a subset of $SM(A)$. If $\sup\{\varphi(x) \mid \varphi \in S\}$ is pointwise bounded, then the mapping ϕ defined on A by $\phi(x) = \sup\{\varphi(x) \mid \varphi \in S\}$, belongs to $SM(A)$.*

Lemma 5.5: *Both Mult(A) and SM(A) are closed in $\mathcal{F}(A, \mathbb{R}_+)$ with respect to the topology of simple convergence.*

Proof: Let ϕ belong to the closure of $Mult(A)$ (resp. of $SM(A)$) in $Mult(A)$ and let ϵ be > 0. Let $x, y \in A$ and let $\lambda \in L$, let $n \in \mathbb{N}$ and let

$m = \max(|x|, |y|, |\lambda x|)$. There exists $\psi \in Mult(A)$ such that

$$\max(|\phi(x) - \psi(x)|_\infty, |\phi(x^n) - \psi(x^n)|_\infty, \\ |\phi(y) - \psi(y)|_\infty, |\phi(\lambda x) - \psi(\lambda x)|_\infty, |\lambda|) < \epsilon.$$

Then

$$\begin{aligned} |\phi(\lambda x) - |\lambda|\phi(x)|_\infty &\leq |\phi(\lambda x) - \psi(\lambda x)|_\infty + |\psi(\lambda x) - |\lambda|\phi(x)|_\infty \\ &\leq \epsilon(1 + |\lambda|). \end{aligned}$$

Since ϵ is arbitrary, we have $\phi(\lambda x) = |\lambda|\phi(x)$. In the same way,

$$\phi(x+y) \leq \psi(x+y) + \epsilon \leq \max(\psi(x), \psi(y)) + \epsilon \leq \max(\phi(x), \phi(y)) + 2\epsilon.$$

Consequently, $\phi(x+y) \leq \max(\phi(x), \phi(y))$. Thus, ϕ is an ultrametric semi-norm of L-vector space. Now, suppose first that ϕ belongs to the closure of $Mult(A)$. Then

$$\begin{aligned} |\phi(xy) - \phi(x)\phi(y)|_\infty &\leq |\phi(xy) - \psi(xy)|_\infty + |\psi(x)\psi(y) - \phi(x)\psi(y)|_\infty \\ &\quad + |\phi(x)\psi(y) - \phi(x)\phi(y)|_\infty \\ &\leq \epsilon(1 + 2m + \epsilon), \end{aligned}$$

therefore $|\phi(xy)|_\infty = |\phi(x)\phi(y)|_\infty$. This finishes showing that ϕ belongs to $Mult(A)$. Consequently, $Mult(A)$ is closed in $\mathcal{F}(A, \mathbb{R}_+)$. Now, assume that ϕ belongs to the closure of $SM(A)$. Similarly,

$$|\phi(x^n) - \phi(x)^n|_\infty \leq |\phi(x^n) - \psi(x^n)|_\infty + |\psi(x)^n - \phi(x)^n|_\infty \leq 2\epsilon$$

therefore $|\phi(x^n)|_\infty = |\phi(x)^n|_\infty$. Consequently, $SM(A)$ is closed in $\mathcal{F}(A, \mathbb{R}_+)$. □

Lemma 5.6: *Let A be provided with a norm $\| \,.\, \|$ of E-algebra and let $\varphi \in SM(A)$. If φ belongs to $SM(A, \| \,.\, \|)$ then $\varphi(x) \leq \| x \|$ whenever $x \in A$.*

Proof: Suppose that for some $x \in A$ we have $\varphi(x) > \| x \|$. Since the absolute value E is not trivial, it contains a subgroup of the form $\{a^n \mid n \in \mathbb{Z}\}$, with $a \in R_+^*$, $a > 1$. We can find $q \in \mathbb{N}$ such that $q \log(\|x\|) < \log a \leq \log(\varphi(x))$, thereby there exists $\lambda \in E$ such that $\|x^q\| < \lambda \leq \varphi(x^q)$. Putting $t = \frac{x^q}{\lambda}$, we have $\varphi(t) \geq 1 > \|t\|$, and therefore φ is not continuous. □

Corollary 5.7: *Let A be provided with a semi-norm $\| \,.\, \|$ of E-algebra. Then $Mult(A, \| \,.\, \|)$ is closed in $\mathcal{F}(A, \mathbb{R}_+)$.*

Proof: Indeed, by Lemma 5.5 $Mult(A)$ is closed in $\mathcal{F}(A,\mathbb{R})$. But consider ϕ in the closure of $Mult(A, \| \, . \, \|)$. Let $x \in A$ and let ϵ be > 0. There exists $\psi \in Mult(A, \| \, . \, \|)$ such that $\phi(x) \leq \psi(x) + \epsilon \leq \|x\| + \epsilon$, hence finally $\phi(x) \leq \|x\|$. □

Definitions and notation: A set F provided with an order relation $\leq$ is said to be *well ordered* with respect to the inverse order $\geq$ if every subset of F admits a maximum element, and then the order $\geq$ is called *a good order.* Let $\geq$ be a good order on the set F and let $x \in F$. If there exists $y < x$, we call *follower of* x the element $\sup\{y \in F \mid y < x\}$. If there exists $z \in F$ such that x is the follower of z, then z is called *precedent of* x.

Let A be provided with a E-algebra semi-multiplicative semi-norm $\| \, . \, \|$. A subset F of $Mult(A, \| \, . \, \|)$ will be called a *boundary for* $(A, \| \, . \, \|)$ if for every $x \in A$, there exists $\psi \in F$ such that $\psi(x) = \|x\|$. A closed boundary (with respect to the topology of simple convergence on A) is called *Shilov boundary for* $(A, \| \, . \, \|)$ if it is the smallest of all closed boundaries for $(A, \| \, . \, \|)$ with respect to inclusion.

Let S be a subsemi-group of B. A submultiplicative function θ on A will be said to be *S-multiplicative* if $\theta(xy) = \theta(x)\theta(y)$ $\forall x, y \in S$. Let μ be a S-multiplicative function such that $\mu(s) \neq 0$ $\forall s \in S$. We will denote by μ^S the function defined in A by $\mu^S(x) = \inf\{\frac{\mu(sx)}{\mu(s)} \mid s \in S\}$.

Let N be a subset of $\mathcal{F}(B, \mathbb{R}_+)$. Then N will be said to be *contructible* if it satisfies the following three properties:

(i) N is well ordered with respect to the order $\geq$.

(ii) For all $\mu \in N$ whose follower is μ' (with respect to the good order $\geq$), there exists a subsemi-group S of $(A, .)$ such that μ is S-multiplicative and such that $\mu' = \mu^S$.

(iii) For every $\mu \in N$ having no precedent, μ is equal to $\inf\{\nu \in N \mid \mu < \nu\}$.

Let θ, ν be semi-multiplicative functions on B. ν will be said to be *constructible from* μ if there exists a constructible set N admitting μ as a its first element and ν as its last element, with respect to the good order on N.

We will denote by $Min(B, \theta)$ the set of multiplicative functions $\phi \in \mathcal{B}(B, \theta)$ which are constructible from θ and by $Z(B, \theta)$ the set of the $\phi \in \mathcal{B}(B, \theta)$ which are constructible from θ. Given be a subsemigroup S of B we will denote by $Z(B, S, \theta)$ the set of $\phi \in Z(B, \theta)$ which are S-multiplicative.

Lemma 5.8 is obvious, and comes from the definition of constructible sets.

Lemma 5.8: *Let X be a constructible subset of $\mathcal{F}(B, \mathbb{R}_+)$ and let $\zeta = \inf\{\phi \in X\}$. Then $X \cup \{\zeta\}$ is a constructible subset of $\mathcal{F}(B, \mathbb{R}_+)$. Let $\psi \in X$ and let Y be a subset of X such that $\inf(Y) \geq \psi$. Then $\inf(Y)$ belongs to X.*

Lemma 5.9: *Let S be a subsemi-group of $(A, .)$ and let θ be a S-multiplicative E-algebra semi-norm on A. Then θ^S also is a S-multiplicative E-algebra semi-norm on A.*

Proof: We have to show that θ is an E-algebra semi-norm. Let $x, y \in A$, and let ϵ be > 0. We can find $s, t \in S$ such that

$$\frac{\theta(sx)}{\theta(s)} < \theta^S(x) + \epsilon, \quad \frac{\theta(ty)}{\theta(t)} < \theta^S(y) + \epsilon.$$

On the other hand, since θ is a E-algebra semi-norm and is S-multiplicative, we check that

$$\theta^S(x+y) \leq \frac{\theta(st(x+y))}{\theta(st)} \leq \frac{\theta(stx) + \theta(sty)}{\theta(st)} \leq \frac{\theta(t)\theta(sx) + \theta(s)\theta(by)}{\theta(s)\theta(t)}$$
$$= \frac{\theta(sx)}{\theta(s)} + \frac{\theta(ty)}{\theta(t)}.$$

Consequently, we have $\theta^S(x+y) \leq \theta^S(x) + \theta^S(y) + 2\epsilon$. Since ϵ is arbitrary, we have proven that $\theta^S(x+y) \leq \theta^S(x) + \theta^S(y)$. Next,

$$\theta^S(xy) \leq \frac{\theta(stxy)}{\theta(st)} \leq \left(\frac{\theta(sx)}{\theta(s)}\right)\left(\frac{\theta(ty)}{\theta(t)}\right) \leq (\theta^S(x) + \epsilon)(\theta^S(y) + \epsilon),$$

hence $\theta^S(xy) \leq \theta^S(x)\theta^S(y)$. Thus, we have proven that θ^S is an E-algebra semi-norm. Then, it is obviously seen that it is S-multiplicative in the same way as θ. □

Proposition 5.10: *Let $\mathcal{N}$ be a totally ordered family of E-algebra semi-norms of A and let ϕ be the function defined on A as $\phi(x) = \inf\{\varphi(x) \mid \varphi \in \mathcal{N}\}$. Then ϕ is a E-algebra semi-norm.*

Proposition 5.11: *Let $\| \, . \, \|$ be a semi-multiplicative E-algebra semi-norm of A. Then every element of $Z(A, \| \, . \, \|)$ is a E-algebra semi-norm.*

Proof: Suppose Proposition 5.11 is not true. Let $\phi \in Z(A, \| \, . \, \|)$ which is not a E-algebra semi-norm. Let $\mathcal{T}$ be a constructible set admitting $\| \, . \, \|$ as first element and ϕ as last element. Since $\mathcal{T}$ is well ordered, the subset S of the $\psi \in \mathcal{T}$ which are not E-algebra semi-norms admits a maximum element θ. If θ admits a precedent ξ, then there exists a subsemigroup S such that ξ is S-multiplicative and satisfies $\theta = \xi^S$. But by hypothesis ξ is a E-algebra semi-norm, and by Lemma 5.9 so is θ, a contradiction. Consequently, θ has no precedent, and then we have $\theta = \inf\{\psi \in \mathcal{T} \mid \theta < \psi\}$. But $\mathcal{T}$ is all ordered. Therefore, by Proposition 5.10 θ is a E-algebra semi-norm because so are all $\psi \in \mathcal{T}$ such that $\theta < \psi$. □

Corollary 5.12: *Let $\|\ .\ \|$ be a semi-multiplicative E-algebra semi-norm of A. Then $Min(A, \|\ .\ \|)$ is included in $Mult(A, \|\ .\ \|)$.*

Lemma 5.13: *Let S be a subsemi-group of B and let $\theta \in \mathcal{F}(B, \mathbb{R}_+)$ be S-multiplicative. There exists $f \in Min(B, \theta)$ such that $f(x) = \theta(x)\ \forall x \in S$.*

Proof: Let $\mathcal{H}$ be the set of constructible subsets of $Z(B, S, \theta)$. Clearly, $Z(B, S, \theta)$ is not empty, hence $\mathcal{H}$ isn't either. On $\mathcal{H}$ we denote by $\preceq$ the order relation defined as $N \preceq N'$ if N is a beginning section of N'. Then $\mathcal{H}$ is inductive for its order.

Let N_0 be a maximal element of $\mathcal{H}$ and let $f_0 = \inf\{\sigma \mid \sigma \in N_0\}$. Clearly f_0 lies in $Z(B, S, \theta)$. The family $\mathcal{G}$ of subsemi-groups $J \supset S$ such that f_0 is J-multiplicative and satisfies $f(s) \neq 0\ \forall s \in J$ is an inductive family with respect to inclusion. Let T_0 be a maximal element of $\mathcal{G}$. Suppose $T_0 \neq B \setminus Ker(f_0)$ and let $x \in B \setminus (T_0 \cup Ker(f_0)$. Let T_1 be the subsemi-group generated by x and T_0. Then f_0 is not T_1-multiplicative. But since f_0 is submultiplicative and since T_0 is maximal, there exists $t \in T_0$ such that $f_0(tx) < f_0(t)f_0(x)$, hence $f_0^{T_0} \neq f_0$. On the other hand, since $f_0^{T_0}$ obviously is S-multiplicative, $f_0^{T_0}$ belongs to $Z(B, S, \theta)$.

Hence, we can check that $N_0 \cup \{f_0^{T_0}\}$ is a constructible subset of $Z(B, S, \theta)$, a contradiction to the hypothesis "N_0 is maximal". Consequently, $T_0 = B \setminus Ker(f_0)$. Thus, f_0 is T_0-multiplicative. On the other hand, f_0 trivially satisfies $f(xy) \leq f(x)f(y) = 0\ \forall x \in Ker(f_0), y \in B$, hence $f(xy) = f(x)f(y)\ \forall x \in Ker(f_0), y \in B$. This finishes proving that f_0 is B-multiplicative, and therefore belongs to $Min(B, \theta)$. Finally, by definition, all elements of $Z(B, S, \theta)$ satisfies $f(x) = \theta(x)\ \forall x \in S$, so f_0 is the f we looked for. □

Theorem 5.14: *Let $\|\ .\ \|$ be a semi-multiplicative E-algebra semi-norm of A. Then $Min(A, \|\ .\ \|)$ is not empty and is a boundary for $(A, \|\ .\ \|)$.*

Proof: Let $x \in A$ and let S_x be the subsemigroup generated by x in A. Since $\|\ .\ \|$ is semi-multiplicative, $\|\ .\ \|$ is obviously S_x-multiplicative. Hence by Lemma 5.13 there exists $f \in Min(A, \|\ .\ \|)$ such that $f(x) = \|x\|$, and by Corollary 5.12 $Min(A, \| \, . \, \|)$ is included in $Mult(A, \| \, . \, \|)$, hence it is a boundary for $(A, \| \, . \, \|)$. □

Notation: Let γ be a semi-group homomorphism from B into B'. We will denote by $\overline{\gamma}$ the mapping from $\mathcal{F}(B', \mathbb{R})$ into $\mathcal{F}(B, \mathbb{R})$ defined by $\overline{\gamma}(\phi) = \phi \circ \gamma$, $\forall \phi \subset \mathcal{F}(B', \mathbb{R})$.

Lemma 5.15: *Let γ be a semi-group homomorphism from B into B' and let θ (resp. θ') be semi-multiplicative functions such that $\theta = \theta' \circ \gamma$. The restriction of $\overline{\gamma}$ to $Z(B', \theta')$ is a surjection onto $Z(B, \theta)$. Moreover, $Min(B, \theta) \subset \overline{\gamma}(Min(B', \theta'))$.*

Proof: Suppose Proposition 5.15 is not true. Then, there exists $\nu \in Z(B, \theta)$ such that $\nu \neq \phi \circ \gamma \ \forall \phi \in Z(B', \theta')$. Let $\mathcal{T}$ be a constructible ordered set admitting θ as first element and ν as last element. Let $\mathcal{S}$ be the set of $\phi \in \mathcal{T}$ such that $\phi \neq \nu' \circ \gamma \ \forall \nu' \in Z(B', \theta)$, let ψ be the maximum element of $\mathcal{S}$, and let $\mathcal{N}$ be the set of $\phi \in \mathcal{T}$ which are of the form $\phi' \circ \gamma$, with $\phi' \in Z(B', \theta')$ and satisfy $\phi \geq \psi$. Let $\mathcal{Q}$ be the family of constructible subsets of $Z(B', \theta')$ admitting θ' as a first element, consisting of ϕ' such that $\phi' \circ \gamma \in \mathcal{N}$. Then $\mathcal{Q}$ is inductive. Let X be a maximal element of $\mathcal{Q}$ and let $\zeta = \inf\{\phi \in X\}$. By Lemma 5.8 $X \cup \{\zeta\}$ is a constructible subset of $Z(B', \theta')$. In $\mathcal{T}$, we notice that $\inf\{\xi \circ \gamma \in \mathcal{T} \mid \xi \in X\} = \zeta \circ \gamma$, hence of course $\psi \leq \zeta \circ \gamma$. Then, since $\psi \in \mathcal{T}$, by Lemma 5.8 $\zeta \circ \gamma$ belongs to $\mathcal{T}$, and therefore belongs to $\mathcal{N}$. But then, by Lemma 5.8 $X \cup \{\zeta\}$ belongs to $\mathcal{Q}$, hence $\zeta \in X$ because X is maximal. Consequently, $\zeta \circ \gamma > \psi$, and hence, $\zeta \circ \gamma$ has follower ν such that $\psi \leq \nu$. Let S be a subsemigroup of B such that $\nu = (\zeta \circ \gamma)^S$ and let $\nu' = \zeta^{\gamma(S)}$. Then $\nu = \nu' \circ \gamma$. On the other hand, ν' is the follower of ζ in the set $X' = X \cup \{\nu'\}$, hence X' is a constructible set, so X' clearly belongs to $\mathcal{Q}$, a contradiction with the hypothesis X maximal. This proves that for every $\nu \in Z(B, \theta)$ there exists $\nu' \in Z(B', \theta')$ such that $\nu = \nu' \circ \gamma$.

Now, we can show that $Min(B, \theta) \subset \overline{\gamma}(Min(B', \theta'))$. Let $\zeta \in Min(B, \theta)$. As we just showed, there exists $\xi \in Z(B', \theta')$ such that $\zeta = \xi \circ \gamma$. Let $Y = \gamma(B) \setminus \xi^{-1}(0)$. Then Y is a subsemi-group of $(B', .)$ and we can check that ξ is Y-multiplicative. By Lemma 5.13 there exists $\zeta' \in Min(B', \xi)$ such that $\zeta'(x) = \zeta(x) \ \forall x \in Y$, hence $\zeta = \zeta' \circ \gamma$. But of course $Min(B', \xi) \subset Min(B', \theta')$, hence f belongs to $\overline{\gamma}(Min(B', \theta'))$. □

We must recall two classical results [4,5]:

Theorem 5.16 (Urysohn): *D be a compact and let α, β be two different points of D. There exists a continuous mapping f from D into $\mathbb{R}_+$ such that $f(\alpha) = 0$ and $f(\beta) = 1$.*

Lemma 5.17: *Let D be a compact. The Gelgand mapping ω from D into $\mathcal{X}(\mathcal{C}(D, \mathbb{R}), \mathbb{R})$ defined as $\omega(a) = \phi_a$, with $\phi_a(f) = f(a)$, $f \in \mathcal{C}(D, \mathbb{R})$ actually is a bijection from D onto $\mathcal{X}(\mathcal{C}(D, \mathbb{R}), \mathbb{C})$ and the mapping $\overline{\omega}$ defined as $\overline{\omega}(a) = |\phi_a|$, with $\phi_a(f) = |f(a)|$, $f \in \mathcal{C}(D, \mathbb{R})$ is a bijection from D onto $Mult(\mathcal{C}(D, \mathbb{R}), \| \,.\, \|_D)$.*

Lemma 5.18: *Let D be a compact, and let ϕ be the mapping from D into $Mult(\mathcal{C}(D, \mathbb{R}), \| \,.\, \|_D)$ defined by $\phi(\alpha)(f) = |f(\alpha)|_\infty$. Then ϕ is a bijection from D onto $Mult(\mathcal{C}(D, \mathbb{R}), \| \,.\, \|_D)$ which is equal to $Min(\mathcal{C}(D, \mathbb{R}), \| \,.\, \|_D)$.*

Proof: By Lemma 5.17 the mapping ϕ from D into $Mult(\mathcal{C}(D, \mathbb{R}), \| \,.\, \|_D)$ defined as $\phi(a) = \psi_a$, with $\psi_a(f) = |f(a)|$ $f \in \mathcal{C}(D, \mathbb{R})$, is a bijection from D onto $Mult(\mathcal{C}(D, \mathbb{R}), \| \,.\, \|_D)$. By Corollary 5.12 $Min(\mathcal{C}(D, \mathbb{R}), \| \,.\, \|_D)$ is included in $Mult(\mathcal{C}(D, \mathbb{R}), \| \,.\, \|_D)$, so we just have to show that

$$Mult(\mathcal{C}(D, \mathbb{R}), \| \,.\, \|_D) \subset Min(\mathcal{C}(D, \mathbb{R}), \| \,.\, \|_D).$$

Let $\alpha \in D$, and let us show that $\phi(\alpha)$ belongs to $Min(\mathcal{C}(D, \mathbb{R}), \| \,.\, \|_D)$. Let $S = \{f \in \mathcal{C}(D, \mathbb{R}) \mid |f(\alpha)|_\infty = \|f\|_D = 1\}$. Then S is a subsemi-group of $\mathcal{C}(D, \mathbb{R})$ and the norm $\| \,.\, \|_D$ is S-multiplicative, hence by Lemma 5.13 there exists $\zeta \in Min(\mathcal{C}(D, \mathbb{R}), \| \,.\, \|_D)$ such that $\zeta(f) = 1 \; \forall f \in S$. But there exists $\beta \in D$ such that $\phi(\beta) = \zeta$, hence $|f(\beta)|_\infty = 1$. Consequently, for every $f \in \mathcal{C}(D, \mathbb{R})$ such that $|f(\alpha)|_\infty = 1$, f must also satisfy $|f(\beta)|_\infty = 1$. Since D is compact, by Theorem 5.16 this implies $\alpha = \beta$ hence $\phi(\alpha)$ belongs to $Min(\mathcal{C}(D, \mathbb{R}), \| \,.\, \|_D)$. □

Theorem 5.19: *Let $\| \,.\, \|$ be a semi-multiplicative E-algebra semi-norm of A. Then there exists a Shilov boundary for $(A, \| \,.\, \|)$ which is the closure of $Min(A, \| \,.\, \|)$ in $Mult(A, \| \,.\, \|)$.*

Proof: By Corollary 5.12 and Theorem 5.14, $Min(A, \| \,.\, \|)$ is not empty, is included in $Mult(A, \| \,.\, \|)$ and is a boundary for $(A, \| \,.\, \|)$. Now, let F be a closed boundary for $(A, \| \,.\, \|)$ and let H be the multiplicative semi-group $(\mathcal{C}(F, \mathbb{R}), .)$. Since F is a closed subset of $Mult(A, \| \,.\, \|)$, by Lemma 5.1 it is a compact. Let γ be the mapping defined on A, taking values in $\mathcal{C}(F, \mathbb{R})$, as $\gamma(x)(\alpha) = \alpha(x)$, $\alpha \in F$. Let $x \in A$.

On one hand, by Lemma 5.6 every element ψ of $Mult(A, \| \,.\, \|)$ by definition satisfies $\|x\| \geq \sup\{\psi(x) \mid \psi \in F\}$. On the other hand, since F is a boundary for $(A, \| \,.\, \|)$ there exists $\varphi \in F$ such that $\|x\| = \sup\{\psi(x) \mid \psi \in F\}$, and consequently, $\|\gamma(x)\|_F = \|x\|$.

Now, since $\| \,.\, \|_F \circ \gamma = \| \,.\, \|$ and since γ is a semi-group homomorphism from $(A, .)$ into $(\mathcal{C}(F, \mathbb{R}), .)$, we can apply Lemma 5.15 and we have

$$Min(A, \| \,.\, \|) \subset \overline{\gamma}(Min(\mathcal{C}(F, \mathbb{R}), \| \,.\, \|_F)).$$

Let $\nu \in Min(A, \| \,.\, \|)$. So, there exists $\psi \in Min(\mathcal{C}(F, \mathbb{R}), \| \,.\, \|_F)$ such that $\nu = \psi \circ \gamma$. Now, let ϕ be the mapping from F into $Mult(\mathcal{C}(F, \mathbb{R}), \| \,.\, \|_F)$ defined as $\phi(\alpha)(f) = |f(\alpha)|_\infty$, $f \in \mathcal{C}(F, \mathbb{R})$. By Lemma 5.18, ϕ is a bijection from F onto $Min(\mathcal{C}(F, \mathbb{R}), \| \,.\, \|_F)$. Consequently, $\psi \circ \gamma$ is of the form $\phi(\alpha)$. And finally, for every $x \in A$ we have

$$\nu(x) = \overline{\gamma}(\phi(\alpha)(x) = (\phi(\alpha) \circ \gamma)(x) = |\gamma(x)(\alpha)|_\infty = |\alpha(x)|_\infty = \alpha(x),$$

because by definition α lies in $Mult(A, \| \,.\, \|)$. Thus ν belongs to F. Therefore $Min(A, \| \,.\, \|)$ is included in F and then the closure of $Min(A, \| \,.\, \|)$ is the smallest of all closed boundaries for $(A, \| \,.\, \|)$. □

Corollary 5.20: *Let $(E, | \,.\, |)$ be ultrametric and let $\| \,.\, \|$ be a semi-multiplicative E-algebra semi-norm. For every $x \in A$, there exists $\varphi \in Mult(A, \| \,.\, \|)$ such that $\varphi(x) = \|x\|$.*

Remark: In Chapter 22 we shall see that a minimal boundary for the spectral semi-norm of an ultrametric algebra is not necessarily closed.

Chapter 6

Spectral Semi-Norm

We shall recall the basic properties of continuous multiplicative semi-norms in an ultrametric normed algebra: the set of continuous multiplicative semi-norms is compact for the topology of simple convergence, and its superior envelope is a semi-multiplicative semi-norm called the spectral semi-norm. Applying Chapter 5, we will show that the set of continuous multiplicative semi-norms admits a Shilov boundary.

Notation: Throughout the chapter, L will denote a field provided with a non trivial ultrametric absolute value, and we will denote by $(A, \| \, . \, \|)$, $(A', \| \, . \, \|')$ commutative L-algebras with unity. We will show the existence of a spectral semi-norm equal to the supremum of continuous multiplicative semi-norms.

ϕ will be a L-algebra homomorphism from A into A' and we will denote by ϕ^* the mapping from $Mult(A')$ into $Mult(A)$ defined as $\phi^*(\varphi) = \varphi \circ \phi$.

Lemma 6.1: *Let $\varphi \in SM(A)$. For every subset W of $Mult(A, \| \, . \, \|)$ the mapping ϕ defined on A by $\phi(x) = \sup_{\varphi \in W} \varphi(x)$ belongs to $SM(A, \| \, . \, \|)$. Moreover, if A has a unity u, and if φ is not identically 0, then $\varphi(\lambda u) = |\lambda|$ whenever $\lambda \in L$.*

Proof: Given $W \subset SM(A, \| \, . \, \|)$, clearly we have $\sup_{\varphi \in W} \varphi(x) \leq \|x\|$, and therefore $\sup_{\varphi \in W} \varphi(x)$ belongs to $SM(A, \| \, . \, \|)$. Finally, let u be the unity in A. Either $\varphi(u) = 0$ and then $\varphi(x) = 0$ whenever $x \in A$, or $\varphi(u) = 1$ and then we have $\varphi(\lambda u) = |\lambda|\varphi(u) = |\lambda|$ whenever $\lambda \in L$. $\square$

According to Lemma 1.11 in [30], we have Lemma 6.2:

Lemma 6.2: *$Mult(A, \| \, . \, \|)$ is compact with respect to the topology of simple convergence.*

Corollary 6.3: *Let $t \in A$. Then $\{\phi(t) \mid \phi \in Mult(A, \| \, . \, \|)\})$ is a closed bounded subset of $\mathbb{R}$.*

Lemma 6.4 is obvious:

Lemma 6.4: *Let ϕ be a L-algebra homomorphism from A into another commutative L-algebra with unity A'. For each $\varphi \in Mult(A')$, $\varphi \circ \phi$ belongs to $Mult(A)$.*

Lemma 6.5: *If ϕ is surjective then ϕ^* is injective. If ϕ is an isomorphism, then ϕ^* is a homeomorphism from $Mult(A')$ onto $Mult(A)$ with respect to topologies of simple convergence on both sets.*

Proof: Suppose ϕ is surjective and let φ_1, $\varphi_2 \in Mult(A')$ be such that $\phi^*(\varphi_1) = \phi^*(\varphi_2)$. Then $\varphi_1(\phi(f)) = \varphi_2(\phi(f)) \ \forall f \in A$. But since ϕ is surjective we have $\varphi_1(g) = \varphi_2(g) \ \forall g \in A'$, hence $\varphi_1 = \varphi_2$.

Now suppose that ϕ is an isomorphism. It is obviously seen that $(\phi^*)^{-1} = (\phi^{-1})*$ and that ϕ^* transforms the filter of neighborhoods of any $\varphi \in Mult(A')$ into this of $\varphi \circ \phi$ in $Mult(A)$. □

Lemma 6.6: *Let A be a direct product $A_1 \times \ldots \times A_q$ whose norm is the product norm of the norm $\| \, . \, \|_j$ of algebras A_j, respectively. Then the mapping Ξ defined on $\bigcup_{j=1}^q Mult(A_j)$ as $\Xi(\psi_j)(x_1, \ldots x_q) = \psi_j(x_j)$ is a bijection from $\bigcup_{j=1}^q Mult(A_j)$ onto $Mult(A, \| \, . \, \|)$.*

Proof: By Lemma 6.5 it is an injection. Putting

$$u_1 = (1, 0, \ldots, 0), \ \ldots \ , u_q = (0, , \ldots, 0, 1), \text{ for each } j = 1, \ldots, q,$$

we check that for every $\varphi \in Mult(A, \| \, . \, \|)$ we have $\varphi(u_k) = 1$ for a certain k and $\varphi(u_j) = 0 \ \forall j \neq k$. Consequently such a φ is of the form $\Xi(\psi_k)$, which shows that this mapping is surjective onto $Mult(A, \| \, . \, \|)$. □

Lemma 6.7 is immediate:

Lemma 6.7: *Let ϕ be continuous. The restriction of ϕ^* to $Mult(A', \| \, . \, \|')$ takes values in $Mult(A, \| \, . \, \|)$.*

Notation: In the hypothesis of Lemma 6.7 we will denote by $\widehat{\phi}$ the restriction of ϕ^* to $Mult(A', \| \, . \, \|')$.

Theorem 6.8: *Suppose that A' is a dense L-subalgebra of A and assume that ϕ is continuous. Then ϕ^* is injective. Moreover if ϕ is injective and bicontinuous, then ϕ^* is a homeomorphism from $Mult(A', \| \, . \, \|')$ onto $Mult(A, \| \, . \, \|)$.*

Proof: Let $A'' = \phi(A)$. Let φ_1, $\varphi_2 \in Mult(A, \| \, . \, \|)$ satisfy $\varphi_1 \circ \phi = \varphi_2 \circ \phi$. Then $\varphi_1(\phi(f)) = \varphi_2(\phi(f))$, $\forall f \in A$, hence $\varphi_1(g) = \varphi_2(g)$ $\forall g \in A''$. But since A'' is dense in A, and since φ_1, φ_2 are continuous with respect to the norm of A', we conclude that $\varphi_1(g) = \varphi_2(g)$ $\forall g \in A'$, hence $\varphi_1 = \varphi_2$.

Now, suppose ϕ is a bicontinuous injection from A into A'. Let $\psi \in Mult(A, \| \, . \, \|)$. We can define $\varphi \in Mult(A'', \| \, . \, \|')$ as $\varphi(\phi(f)) = \psi(f)$ and extend by continuity the definition of φ to A'. Since ϕ is bicontinuous from $(A, \| \, . \, \|)$ onto $(A'', \| \, . \, \|')$, φ belongs to $Mult(A'', \| \, . \, \|')$ and therefore its expansion to A' belongs to $Mult(A', \| \, . \, \|')$. Thus, ϕ^* is a bijection from $Mult(A', \| \, . \, \|')$ onto $Mult(A, \| \, . \, \|)$. We will check that ϕ^* is bicontinuous. Let $\varphi \in Mult(A', \| \, . \, \|')$ and let $\psi = \varphi \circ \phi$. Let $\mathcal{V}(\varphi \circ \phi, f_1, \ldots, f_n, \epsilon)$ be a neighborhood of $\varphi \circ \phi$. In $Mult(A', \| \, . \, \|')$, $\mathcal{V}(\psi, \phi \circ f_1, \ldots, \phi \circ f_n, \epsilon)$ is a neighborhood of ψ whose image by ϕ^* is just $\mathcal{V}(\varphi \circ \phi, f_1, \ldots, f_n, \epsilon)$. Consequently ϕ^* is continuous.

Conversely, let $\mathcal{V}(\varphi, g_1, \ldots, g_n, \epsilon)$ be a neighborhood of φ. Since the multiplicative semi-norms are ultrametric and since A'' is dense in A' we can find $g'_1, \ldots, g'_n \in A''$ such that $\mathcal{V}(\varphi, g_1, \ldots, g_n, \epsilon) = \mathcal{V}(\varphi, g'_1, \ldots, g'_n, \epsilon)$. Each g'_i is of the form $\phi(f_i)$ $(1 \leq i \leq n)$, so

$$\mathcal{V}(\varphi \circ \phi, f_1, \ldots, f_n, \epsilon) = (\phi^*)^{-1}(\mathcal{V}(\varphi, g'_1, \ldots, g'_n, \epsilon)),$$

which finishes proving that $(\phi^*)^{-1}$ also is bicontinuous. □

Notation: When A' is a dense L-subalgebra of A henceforth, we will consider that $Mult(A, \| \, . \, \|)$ and $Mult(A', \| \, . \, \|)$ are equal.

Theorem 6.9: *For every $x \in A$ the sequence $\left(\| \, x^n \, \|^{\frac{1}{n}}\right)_{n \in \mathbb{N}}$ has a limit denoted by $\| \, x \, \|_{si}$, satisfying $\|x\|_{si} \leq \|x\|$ $\forall x \in A$ and the mapping f defined in A as $f(x) = \| \, x \, \|_{si}$ belongs to $SM(A, \| \, . \, \|)$ and is ultrametric. Moreover $\| \, x \, \|_{si} < 1$ if and only if $\lim_{n \to \infty} \|x^n\| = 0$.*

Proof: The statement is well known and proven, for instance, in [30] as Theorem 1.13 when the semi-norm $\| \, . \, \|$ is a norm. Actually the proof holds all the same when it is just a semi-norm. □

By Theorem 5.19 we have Corollary 6.10:

Corollary 6.10: *There exists a Shilov boundary for $(A, \| \,.\, \|_{si})$ which is included in $Mult(A, \| \,.\, \|)$ and is equal to the closure of $Min(A, \| \,.\, \|_{si})$ in $Mult(A, \| \,.\, \|)$.*

Theorem 6.11: *Let $f \in A$. Then $\|f\|_{si} = 0$ if and only if $\lim_{n\to\infty} \|(\lambda f)^n\| = 0\ \forall \lambda \in L$.*

Proof: If $\|f\|_{si} = 0$, then of course $\|\lambda f\|_{si} = 0$, hence $\lim_{n\to\infty} \|(\lambda f)^n\| = 0\ \forall \lambda \in L$. Now, suppose $\lim_{n\to\infty} \|(\lambda f)^n\| = 0\ \forall \lambda \in L$. Suppose $\|f\|_{si} \neq 0$, and let $\lambda \in L$ be such that $|\lambda| > \frac{1}{\|f\|_{si}}$. Then $\lim_{n\to\infty} \|(\lambda f)^n\|_{si} = +\infty$, which contradicts the hypothesis $\lim_{n\to\infty} \|(\lambda f)^n\| = 0$. □

Definition: $(A, \| \,.\, \|)$ will be said to be *uniform* if the norm $\| \,.\, \|$ is equivalent to $\| \,.\, \|_{si}$.

In the same way as in archimedean analysis, we must also mention this proposition:

Proposition 6.12: *$\| \,.\, \|$ and $\| \,.\, \|_{si}$ are two equivalent norms on A if and only if there exists $C \in \mathbb{R}_+$ such that $\|x^2\| \leq C\|x\|^2\ \forall x \in A$. And they are equal if and only if $\|x^n\| \leq \|x\|^n\ \forall n \in \mathbb{N}^*$, $\forall x \in A$.*

Theorem 6.13: *Let $\| \,.\, \|$ be a norm of L-algebra on A. Then $\| \,.\, \|$ is semi-multiplicative if and only if the semi-norm $\| \,.\, \|_{si}$ associated to $\| \,.\, \|$ is equal to $\| \,.\, \|$.*

On the other hand, if A is complete for its norm, then for every $x \in A$ satisfying $\|x\|_{si} < 1$, $1 - x$ is invertible in A, the set of invertible elements in A is open, each maximal ideal of A is closed and for every $\psi \in \mathcal{X}(A, L)$, the mapping $|\psi|$ from A into $\mathbb{R}_+$ defined as $|\psi|(x) = |\psi(x)|$ belongs to $Mult(A, \| \,.\, \|_{si})$.

Proof: By Theorem 6.9, if $\| \,.\, \|_{si}$ is the L-algebra norm, then it is semi-multiplicative. Conversely, if the L-algebra norm is semi-multiplicative, it is equal to $\| \,.\, \|_{si}$ because $\| x^n \|^{\frac{1}{n}} = \|x\|$ for every $x \in A$.

Now assume that A is complete. Let $x \in A$ satisfy $\|x\|_{si} < 1$. By Theorem 6.9 we have $\lim_{n\to\infty} x^n = 0$ and then the series $\sum_{n=0}^{\infty} x^n$ converges in A to a limit S that obviously satisfies $S(1-x) = 1$. As a consequence, the set of invertible elements in A is open. Since a maximal ideal is either closed or dense, here each maximal ideal of A is closed.

Moreover, for every $\lambda \in L$ satisfying $\|x\|_{si} < |\lambda|$, $1 - \frac{x}{\lambda}$ is invertible in A. Now, $|\psi|$ obviously belongs to $Mult(A)$. Assume that $|\psi(x)| > \| x \|_{si}$

for some $x \in A$. Let $\lambda = \psi(x) \in L$ and let $y = 1 - \frac{x}{\lambda}$. Then $\psi(y) = 0$, hence $1 - \frac{x}{\lambda} \in Ker(\psi)$. But $|\lambda| > \| x \|_{si}$ hence $\|\frac{x}{\lambda}\|_{si} < 1$, so $1 - \frac{x}{\lambda}$ is invertible. This just contradicts $1 - \frac{x}{\lambda} \in Ker(\psi)$. Finally $|\psi(x)| \leq \| x \|_{si}$ for every $x \in A$, hence $|\psi| \in Mult(A, \| \cdot \|_{si})$. □

Theorem 6.14 was given in several works: [33], [44] and [30, Theorem 1.16].

Theorem 6.14: *Let F be a field extension of L provided with a L-algebra semi-norm of L-algebra $\| \cdot \|$. Then $\| \cdot \|$ is a L-algebra norm and there exists an ultrametric absolute value φ on F extending that of L, such that $\varphi(x) \leq \|x\|$ whenever $x \in A$.*

Theorem 6.15 is an immediate consequence [30, Theorem 1.16].

Theorem 6.15: *Let $(A, \| \cdot \|)$ be complete. For every maximal ideal $\mathcal{M}$ of A, there exists $\varphi \in Mult(A, \| \cdot \|)$ such that $Ker(\varphi) = \mathcal{M}$ and φ is of the form $|\chi|$, with $\chi \in \mathcal{A}$, and $| \cdot |$ an absolute value on the field $\chi(A)$ expanding that of L.*

Remark: As shown in [7] and in [30, Theorem 40.5] there exist uniform ultrametric Banach commutative algebras without non zero divisors of zero, with unity, admitting no continuous absolute values, and therefore, there exist Banach ultrametric commutative K-algebras with unity admitting closed prime ideal which are not the kernel of any continuous multiplicative semi-norm. Thus, as far as ultrametric Banach algebras concerned, Theorem 6.15 is specific to maximal ideals, in the general case.

Corollary 6.16: *Given $x \in A$, then $|sp(x)| \subset \{\phi(x) \mid \phi \in Mult_m(A, \| \cdot \|)\}$.*

Theorem 6.17: *Let F, G be a partition of $Mult_m(A, \| \cdot \|)$ and suppose that there exists an idempotent u such that $\phi(u) = 1\ \forall \phi \in F$, and $\phi(u) = 0\ \forall \phi \in G$. Then u is unique satisfying these relations.*

Proof: Let $\phi \in F$. By Theorem 6.15 ϕ is of the form $|\chi|$ with $\chi \in \mathcal{X}(A)$, the field image of A by χ being provided with an absolute value extending that of L. Since $\chi(u)$ must be equal to 0 or 1, we have $\chi(u) = 1$. And similarly, if $\phi \in G$, it is of the form $|\chi|$ with $\chi \in \mathcal{X}(A)$, and of course $\chi(u) = 0$. Consequently, by Lemma 1.14, u is unique to satisfy $\chi(u) = 1\ \forall \chi \in F$, $\chi(u) = 0\ \forall \chi \in G$. □

Proposition 6.18 is classical (in particular, we can obtain it by Proposition 1.17, and Proposition 4.1 in [30]):

Proposition 6.18: *Let A be a normed L-algebra whose norm $\| \cdot \|$ is ultrametric, let $r \in]0, +\infty[$ and let B be the set of the series $\sum_{n=0}^{\infty} a_n x^n$ such that $\lim_{n\to\infty} \|a_n\| r^n = 0$. Then B, provided with the multiplication of series, is a commutative algebra which contains A and admits 1 for unity. Let $\| \cdot \|_r$ be defined on B by $\| \sum_{n=0}^{\infty} a_n x^n \|_r = \sup_{n\in\mathbb{N}} \|a_n\| r^n$. Then $\| \cdot \|_r$ is a norm of L-algebra on B. Moreover, if the norm of E is multiplicative, then so is $\| \cdot \|_r$. Further, if A is complete, then so is B.*

Theorem 6.19: *Given $\varphi \in Mult(A)$, φ belongs to $Mult(A, \| \cdot \|)$ if and only if it satisfies $\varphi(t) \le \|t\|_{si}$ whenever $t \in A$, and we have $Mult(A, \| \cdot \|) = Mult(A, \| \cdot \|_{si})$. Further, for every $t \in A$ we have*

$$\|t\|_{si} = \sup\{\varphi(t) \mid \varphi \in Mult(A, \| \cdot \|)\}$$

and there exists $\varphi \in Mult(A, \| \cdot \|)$ such that $\varphi(t) = \|t\|_{si}$.

Proof: Since $\|t\|_{si} \le \|t\|$ for all $t \in A$, each $\varphi \in Mult(A)$ satisfying $\varphi(t) \le \|t\|_{si}$ whenever $t \in A$, obviously belongs to $Mult(A, \| \cdot \|)$. Conversely, suppose that $\varphi \in Mult(A)$ satisfies $\varphi(t) > \|t\|_{si}$ for certain $t \in A$. As in the proof of Lemma 6.1, there exists $q \in \mathbb{N}^*$ and $\lambda \in L$ such that $\varphi(t)^q > |\lambda| > \|t\|_{si}^q$. Let $u = \frac{t}{\lambda}$. Then we have $\|u\|_{si} < 1 < \varphi(u)$, hence by Theorem 6.9, $\lim_{n\to\infty} u^n = 0$, whereas $\lim_{n\to\infty} \varphi(u^n) = +\infty$. This shows $\varphi \notin Mult(A, \| \cdot \|)$. As a consequence, we have $Mult(A, \| \cdot \|) = Mult(A, \| \cdot \|_{si})$. The last statements come from Corollary 6.10. □

Remark: In Theorem 1.18 of [30] by inadvertence we wrote that there exists $\varphi \in Mult_m(A, \| \cdot \|)$ such that $\varphi(t) = \|t\|_{si}$, which is obviously wrong as show many basic examples (see the remark following Corollary 17.9).

Chapter 7

Hensel Lemma

Definition and notation: Recall that L denotes a field provided with a non trivial ultrametric absolute value $| \,.\, |$.

Let E be a commutative ultrametric normed L-algebra with unity. Let $n \in \mathbb{N}$ and let $\| \,.\, \|_0$ be the norm of E. The E-algebra of polynomials in n variables $E[X_1, \ldots, X_n]$ is provided with the *Gauss norm* $\| \,.\, \|$ defined as

$$\left\| \sum_{i_1,,i_n} a_{i_1,\ldots,i_n} X_1^{i_1} \ldots X_n^{i_n} \right\| = \sup_{i_1,\ldots,i_n} \|a_{i_1,\ldots,i_n}\|_0.$$

By Proposition 6.18 this norm is a L-algebra norm. In particular, if the norm of E is multiplicative, so is the norm $\| \,.\, \|$.

We denote by $E\{X_1, \ldots, X_n\}$ the set of power series in n variables

$$\sum_{i_1,,i_n} a_{i_1,\ldots,i_n} X_1^{i_1} \ldots X_n^{i_n} \text{ such that } \lim_{i_1+\ldots+i_n \to \infty} a_{i_1,\ldots,i_n} = 0.$$

The elements of such an algebra are called *the restricted power series in n variables, with coefficients in E*. Hence by definition $E[X_1, \ldots, X_n]$ is dense in $E\{X_1, \ldots, X_n\}$. By Proposition 6.18, if E is complete, $E\{X_1, \ldots, X_n\}$ is just the completion of $E[X_1, \ldots, X_n]$.

Throughout the chapter, A will denote a commutative L-algebra with provided with an absolute value $| \,.\, |$ which extends that of L and satisfies $\{ \, |t| \mid x \in A\} = |L|$. We will denote by $v(.)$ the valuation associated to the absolute value, by V the subring $\{t \in A \mid |x| \leq 1\}$ of A and by $\mathcal{M}$ the prime ideal $\{t \in A \mid |t| < 1\}$ of V. Then it is seen that $\mathcal{M} = M_L V$. Moreover, the absolute value $\| \,.\, \|$ defined on $A[x]$ obviously can be extended to $A\{x\}$. The valuation defined on $A\{x\}$ by the Gauss norm $\| \,.\, \|$, which is an absolute value extending that of A, will be denoted by $v(.)$ again. Let $W = \{f \in A\{X\} \mid \|f\| \leq 1\}$ and let $\mathcal{N} = \{f \in A\{X\} \mid \|f\| < 1\}$. Given $f \in W$ we will denote by $\overline{a}$ the residue class of a in $\frac{W}{\mathcal{N}}$.

Given $f = \sum_{n=0}^{\infty} a_n x^n \in A\{x\}$, here we will denote by $N(f)$ the greatest of the integers m such that $|a_m| = \|f\|$.

Lemmas 7.1 and 7.2 are immediate:

Lemma 7.1: $\mathcal{M} = M_L V$, $\mathcal{N} = M_L W$, $\frac{W}{\mathcal{N}} = \frac{V}{\mathcal{M}}[\overline{x}]$.

Lemma 7.2: *Let $f \in A\{x\}$ be such that $\|f\| = 1$. Then $N(f) = \deg(\overline{f})$.*

Lemma 7.3: *Let $f,\ g \in A\{x\}$. Then $N(fg) = N(f) + N(g)$.*

Proof: Suppose first that $v(f) = v(g) = 0$. Let $q = N(f)$, $t = N(g)$. By Lemma 7.1, in $\frac{V}{\mathcal{M}}[\overline{x}]$, $\overline{f}$ is a polynomial of degree q, $\overline{g}$ is a polynomial of degree t, hence $\overline{fg}$ is a polynomial of degree $q + t$, which shows the statement. Now consider the general case. Since $\{\ |t| \mid t \in A\} = |L|$ we can find $\lambda \in L$ such that $v(\lambda f) = v(\lambda g) = 0$, so we are led to the same conclusion. □

Lemma 7.4: *Let $P \in A[x]$ and let $D \in V[x]$ be monic. Let $Q, R \in A[x]$ satisfy $P = DQ + R$ and $\deg(R) < \deg(D)$. Then we have $v(Q) \geq v(P)$ and $v(R) \geq v(P)$.*

Proof: We can clearly assume $P \neq 0$. Then, by multiplying P by a suitable constant $\lambda \in L$, we can also assume $v(P) = 0$. Since D is monic, the Euclidean division of P by D is clearly possible in $V[x]$, and therefore Q is the quotient, R is the rest of this division, due to the fact that $\deg(R) < \deg(D)$. So we have $v(Q) \geq 0$, $v(R) \geq 0$ because both Q, R belong to $V[x]$. □

Theorem 7.5: *Let A be complete with respect to $|\,.\,|$. Let $f \in A\{x\}$ and let $D \in V[x]$ be monic. There exists a unique pair $(g, R) \in A\{x\} \times A[x]$ such that $f = Dg + R$, and $\deg(R) < \deg(D)$. Moreover, we have $v(g) \geq v(f)$ and $v(R) \geq v(f)$.*

Proof: Since A is complete, then so is $A\{x\}$ with respect to the norm $\| \,.\, \|$. Consider the mapping ϕ, defined on $A[x]$, which associates to each $P \in A[x]$ the quotient Q in the division by D. By Lemma 7.4 ϕ is continuous with respect to the absolute value $\| \,.\, \|$. Since $A\{x\}$ is complete and since $A[x]$ is dense in $A\{x\}$, then ϕ has continuation to a mapping $\widehat{\phi}$ from $A\{x\}$ to $A\{x\}$ such that $v(\widehat{\phi}(f)) \geq v(f)$. In the same way, consider the mappings ψ

defined on $A[x]$ which associates to each $P \in A[x]$ the rest R in the division of P by D. By Lemma 7.4 ψ is continuous, therefore it can be continuously extended to a mapping $\widehat{\psi}$ from $A\{x\}$ to $A[x]$ such that $v(\widehat{\psi}(f)) \geq v(f)$ and $\deg(\widehat{\psi}(f)) < \deg(D)$. Thus, by putting $g = \widehat{\phi}(f)$ and $R = \widehat{\psi}(f)$ we have proven the existence of the pair (g, R).

Now, suppose we have another pair $(h, S) \in A\{x\} \times A[x]$ such that $f = Dh + S$, and $\deg(S) < \deg(D)$. Then $D(h - g) = R - S$. Since D is monic, by Lemma 7.3 we have $N(D(h - g)) \geq \deg(D)$, while $N(R - S) \leq \deg(D) - 1$, a contradiction. Hence $h = g,\ S = R$. □

Corollary 7.6: *Let A be complete with respect to $|\ .\ |$. Let $f \in A\{x\}$ and let $D \in V[x]$ be monic. If D divides f in $A\{x\}$ then it divides f in $A[x]$.*

Notation: Given a ring B and $g, h \in B[x]$, $\mathcal{I}(g, h)$ will denote the ideal generated by g and h in $B[x]$.

Lemma 7.7: *Let $g, h \in V[x]$ be monic, such that $\mathcal{I}(\overline{g}, \overline{h}) = \frac{V}{\mathcal{M}}[\overline{x}]$ and let $q \in \mathbb{N}$ be such that $q < \deg(g) + \deg(h)$. There exist $S, W \in L[x]$ satisfying*

$$v(Sg + Wh - x^q) > 0,\ v(S) \geq 0,\ v(W) \geq 0,$$
$$\deg(S) < \deg(h),\ \deg(W) < \deg(g).$$

Proof: Since $\mathcal{I}(\overline{g}, \overline{h}) = \frac{V}{\mathcal{M}}[\overline{x}]$, there exists φ and $\psi \in \mathcal{L}[x]$ such that $\varphi\overline{g} + \psi\overline{h} = \overline{1}$, $\deg(\varphi) < \deg(\overline{h})$, $\deg(\psi) < \deg(\overline{g})$. Let $B, E \in V[x]$ satisfy $\overline{B} = \varphi$, $\overline{E} = \psi$, $\deg(B) = \deg(\varphi)$, $\deg(E) = \deg(\psi)$. Thus we have $\overline{Bg + Eh - 1} = \overline{0}$ i.e.

$$v(Bg + Eg - 1) > 0. \tag{7.1}$$

Moreover, $v(B) = 0$. We now consider the Euclidean division of Bx^q by h and Ex^q by g, respectively. We obtain $Bx^q = B_0h + S$ and $Ex^q = E_0g + W$. By Lemma 7.4 we have $\min(v(S), v(B_0)) \geq v(Bx^q) = 0$. Next, by hypothesis we have

$$\deg(S) < \deg(h) \tag{7.2}$$

and

$$\deg(W) < \deg(g) \tag{7.3}$$

Let $F = Bg + Eh - 1$. Then we have $Bx^q = (B_0 + E_0)gh + Sg + Wh - x^q$. Since $q < \deg(g) + \deg(h)$, by (7.2) and (7.3) we see that

$$\deg(Sg + Wh - x^q) < \deg(g) + \deg(h)$$

and therefore $Sg+Wh-x^q$ is just the remainder of the Euclidean division of Fx^q by gh. But then, by Lemma 7.4, we have $v(Sg+Wh-x^q) \geq v(Fx^q) = v(F)$. And by (7.1) and by definition of F it is seen that $v(F) > 0$. This finishes proving that $v(Sg + Wh - x^q) > v(x^q)$. □

Notation: Let $g, h \in V[x]$ be monic and satisfy $\mathcal{I}(\overline{g}, \overline{h}) = \frac{V}{\mathcal{M}}[\overline{x}]$. We will denote by $E(g, h)$ the set of constants $\lambda \in \mathbb{R}_+$ such that, for every polynomial $Q \in L[x]$ satisfying $\deg(Q) < \deg(g) + \deg(h)$, there exist $F, G \in L[x]$ satisfying $v(Fg + Gh - Q) \geq v(Q) + \lambda$, $v(F) \geq v(Q)$, $v(G) \geq v(Q)$, $\deg(F) < \deg(h)$, $\deg(G) < \deg(g)$.

Lemma 7.8: *Let $g, h \in V[x]$, be monic and satisfy $\mathcal{I}(\overline{g}, \overline{h}) = \frac{V}{\mathcal{M}}[\overline{x}]$, and let $d = \deg(g) + \deg(h)$. Then $E(g, h)$ is a non empty interval whose lower bound is 0. Moreover, given $\lambda \in E(g, h)$, and monic polynomials ϕ, $\psi \in V[x]$ such that $v(g - \phi) \geq \lambda$, $v(h - \psi) \geq \lambda$, then $E(\phi, \psi) = E(g, h)$.*

Proof: We can apply Lemma 7.7 to each polynomial $Q_n = x^n$ for every $n = 0, \ldots, d-1$. Thus, we have polynomials S_n, W_n satisfying

$$v(S_n g + W_n h - x^n) > 0,\ v(S_n) \geq 0,\ v(W_n) \geq 0,$$
$$\deg(S_n) < \deg(h),\ \deg(T_n) < \deg(g).$$

We put

$$\lambda_n = v(S_n g + W_n h - x^n),\ (0 \leq n \leq d-1).$$

Now let $Q = \sum_{n=0}^{d-1} a_n x^n$, let $S = \sum_{n=0}^{d-1} a_n S_n$, $W = \sum_{n=0}^{d-1} a_n W_n$ and let $\lambda = \min_{0 \leq n \leq d-1} \lambda_n$. Clearly we have

$$\begin{aligned} v(Sg + Wh - Q) &\geq \min_{0\leq n\leq d-1} (v(a_n) + \lambda_n) \\ &\geq \min_{0\leq n\leq d-1} v(a_n) + \min_{0\leq n\leq d-1} \lambda_n = v(Q) + \lambda. \end{aligned}$$

But trivially: $v(S) \geq \min_{0\leq n\leq d-1} v(a_n)$, $\quad v(S) \geq v(Q)$,

$$\begin{aligned} \deg(W) &\leq \max_{0\leq n\leq d-1} (\deg(S_n)) < \deg(h), \deg(W) \\ &\leq \max_{0\leq n\leq d-1} (\deg(W_n)) < \deg(h). \end{aligned}$$

So, λ lies in $E(g, h)$. Then it is obviously seen that $E(g, h)$ is a non empty interval, and that its lower bound is 0.

Now, let $y \in E(g, h)$ and let ϕ, $\psi \in V[x]$ be monic and satisfy $v(g-\phi) > y$, $v(h - \psi) > y$. Since $v(S) \geq v(Q)$, $v(W) \geq v(Q)$, it is easy to see that

$v(S(g-\phi)+W(h-\psi)>y+v(Q)$, and therefore $v(S\phi+W\psi-Q)>y+v(Q)$. This shows that $y\in E(\phi,\psi)$, and therefore $E(g,h)\subset E(\phi,\psi)$. But similarly we have $E(\phi,\psi)\subset E(g,h)$. □

Lemma 7.9: *Let A be complete and let $f(x)=\sum_{j=0}^{\infty}a_jx^j\in A\{x\}$ be such that $\|f\|=|a_q|=1$, with $N(f)=q$. Let $g,h\in V[x]$ be monic and satisfy $\mathcal{I}(\overline{g},\overline{h})=\frac{V}{\mathcal{M}}[\overline{x}]$. Let $\lambda\in E(g,h)$. There exist monic polynomials S, $W\in L[x]$ satisfying*

$$\begin{aligned}&v(Sg+Wh-f)\geq\lambda+v(f),\quad \deg(S)\ <\deg(g),\\ &\quad\deg(W)\leq\max(\deg(h),q-\deg(g)),\\ &v(S)\geq v(f),\quad v(W)\geq v(f).\end{aligned}$$

Proof: By Theorem 7.5 we consider the Euclidean division of f by gh : $f=\ell gh+Q_1$. Hence $\deg(Q_1)\ <\ \deg(g)+\deg(h)$. By Theorem 7.5 we have

$$v(Q_1)\geq v(f), \tag{7.4}$$

$$v(\ell)\geq v(f) \tag{7.5}$$

By Lemma 7.8 there exist $S_1,W_1\in L[x]$ satisfying

$$v(S_1g+W_1\ h-Q_1)\geq v(Q_1)+\lambda(g,h), \tag{7.6}$$

$$v(S_1)\geq v(Q_1), \tag{7.7}$$

$$v(W_1)\geq v(Q_1), \tag{7.8}$$

$$\deg(S_1)<\deg(h), \tag{7.9}$$

$$\deg(W_1)<\deg(g), \tag{7.10}$$

Now we put $S=S_1+\ell h$, $W=W_1$. So we have $Sg+Wh-f=S_1+\ell gh+W_1h-\ell gh-Q_1$ and therefore by (7.6) we obtain $v(Sg+Wh-f)\geq v(Q_1)+\lambda$. Hence by (7.4) we obtain

$$v(Sg+Wh-f)\geq v(f)+\lambda. \tag{7.11}$$

By (7.4), (7.8) it is seen that

$$v(W)\geq v(f). \tag{7.12}$$

Next, we have $v(h)=0$, hence by (7.5) we see that

$$v(\ell h)\geq v(f). \tag{7.13}$$

But by (7.7) we have $v(S_1) \geq v(f)$ and therefore by (7.13) we obtain

$$v(S) \geq v(f). \tag{7.14}$$

Finally, by definition we have $\deg(\ell) = \deg(f) - \deg(gh)$ and therefore

$$\deg(W) \leq \max(\deg(S_1), \deg(\ell h)) \leq \max(\deg(h), q - \deg(g)). \tag{7.15}$$

Thanks to (7.10), (7.11), (7.12), (7.14), (7.15). □

Theorem 7.10 (Hensel Lemma): *A is supposed to be complete. Let $f \in A\{x\}$ be such that $\|f\| = 1$ and such that $\overline{f}$ splits in $\frac{V}{\mathcal{M}}[\overline{x}]$ in the form $\gamma\eta$ with $\mathcal{I}(\gamma, \eta) = \frac{V}{\mathcal{M}}[\overline{x}]$. There exists a unique pair $(g, h) \in A[x] \times A\{x\}$ such that g is monic and satisfies $f = gh$, $\overline{g} = \gamma$, $\overline{h} = \eta$, $\deg(g) = \deg(\gamma)$.*

Proof: We can obviously take monic polynomials $g_0, h_0 \in V[x]$ such that $\overline{g_0} = \gamma$, $\overline{h_0} = \eta$. We put $\nu = v(f - g_0h_0)$, and take $\tau \in E(g_0, h_0)$ satisfying $\tau \leq \nu$. We will construct sequences $(g_n)_{n\in\mathbb{N}}, (h_n)_{n\in\mathbb{N}}$ in $L[x]$ satisfying for all $n \geq 0$:

(i_n) $v(f - g_nh_n) \geq (n+1)\tau$,
(ii_n) $v(g_n - g_{n-1}) \geq n\tau, v(h_n - h_{n-1}) \geq n\tau$,
(iii_n) $\deg(h_n) \leq \deg(f) - \deg(g_0), \deg(g_n) = \deg(g_0)$,
(iv_n) $\overline{g_n} = \gamma, \overline{h_n} = \eta$,
(v_n) $\tau \in E(g_n, h_n)$.
(vi_n) g_n is monic.

First we put $f_1 = f - g_0h_0$. We notice that $\deg(f_1) = \deg(f)$. We now apply Lemma 7.9, to the case when $(Q, g, h) = (f_1, g_0, h_0)$: there exist $S_1, W_1 \in A[x]$ satisfying

$$\deg(W_1) < \deg(g_0),$$
$$\deg(S_1) < \deg(f) - \deg(g_0), \tag{7.16}$$
$$v(S_1) \geq \tau, \tag{7.17}$$
$$v(W_1) \geq \tau, \tag{7.18}$$
$$v(S_1g_0 + W_1h_0 - P_1) \geq \tau + v(P_1). \tag{7.19}$$

Next we put $g_1 = g_0 + W_1$, $h_1 = h_0 + S_1$. We check that $(i_1), (ii_1), (iii_1), (iv_1)$ are satisfied. Moreover, by (7.17) and (7.18), and by Lemma 7.8, τ lies in $E(g_1, h_1)$, hence v_1 is satisfied.

Now we suppose we have already constructed pairs (g_m, h_m) satisfying (i_m), (ii_m), (iii_m), (iv_m), (v_m) for every $m = 0, \ldots, n$. Then we put

$f_{n+1} = f - g_n h_n$. We can apply Lemma 7.9 to the case when (Q, g, h) is equal to (f_{n+1}, g_n, h_n). So, we can obtain $S_{n+1}, W_{n+1} \in A[x]$ satisfying

$$v(S_{n+1}g_n + W_{n+1}h_n - f_{n+1}) \geq \tau + v(f_{n+1}), \tag{7.20}$$

$$\deg(W_{n+1}) < \deg(g_n), \tag{7.21}$$

$$\deg(S_{n+1}) \leq \max(\deg(h_n), \deg(f_{n+1}) - \deg(g_n)), \tag{7.22}$$

$$v(S_{n+1}) \geq v(f_{n+1}), v(W_{n+1}) \geq v(f_{n+1}). \tag{7.23}$$

By (7.20), and by v_n) we obtain

$$v(S_{n+1}g_n + W_{n+1}h_n - f_{n+1}) \geq (n+2)\tau. \tag{7.24}$$

Now we put $g_{n+1} = g_n + W_{n+1},\ h_{n+1} = h_n + S_{n+1}$. We check that

$$\begin{aligned} f - g_{n+1}h_{n+1} &= (f_{n+1} - h_n W_{n+1} - g_n S_{n+1}) - S_{n+1}W_{n+1} \\ &= f_{n+1} - hW_{n+1} - g_{n+1}S_{n+1} + (h_{n+1} - h_n)W_{n+1} \\ &\quad + (g_{n+1} - g_n)S_{n+1} + S_{n+1}W_{n+1}. \end{aligned}$$

By (7.21) we notice that G_{m+1} is monic, hence vi_{n+1}) is satisfied. By (ii_m) true for $m \leq n$, we notice that

$$v(g_n - g_{n+1}) \geq (n+1)\tau, v(h_n - h_{n+1}) \geq (n+1)\tau \tag{7.25}$$

which gives (ii_{n+0}), and by (7.23) and (7.24), we obtain (i_{n+1}) $v(f - g_{n+1}h_{n+1}) \geq (n+2)\tau$. Relations (iii_{n+1}), (iv_{n+1}) are easily checked. By (7.25) and Lemma 7.8 Relation (v_{n+1}) is also clear. Therefore the sequences $(g_n)_{n\in\mathbb{N}}$, $(h_n)_{n\in\mathbb{N}}$ satisfying (i_n), (ii_n), (iii_n), (iv_n), (v_n) are now constructed. Since A is complete, the A-module $F_q[x]$ of the polynomials of degree $m \leq q$ is obviously complete with respect to the Gauss norm. Then by Relations ii_n) both sequences $(g_n)_{n\in\mathbb{N}}, (h_n)_{n\in\mathbb{N}}$ converge in $F_q[x]$. We put $g = \lim_{n\to\infty} g_n,\ h = \lim_{n\to\infty} h_n$. Clearly by (iii_n) we have $\deg(g) = \deg(g_0) = \deg(\gamma)$. By iv_n) we have $\overline{g} = \gamma,\ \overline{h} = \eta$, and finally by i_n) we have $v(f - gh) = +\infty$ hence $f = gh$.

Thus we have shown the existence of the pair (g, h). Now, since g is a monic polynomial, the equality $f = gh$ appears to be the Euclidean division of f by g. Consequently, such a pair is unique. □

Corollary 7.11: *A is supposed to be complete. Let $f(x) = \sum_{j=0}^{\infty} a_j x^j \in A\{x\}$ and let $q = N(f)$. If a_q is invertible in A then f is associated to a monic polynomial $P \in V[x]$ of degree q, in $A\{x\}$.*

$\ln - \deg E = g_n b_n$. We can apply Lemma 7.9 to the case when [illegible] is equal to $(\bar{f}_{n+1}, g_n, t_{n+1})$. So, we can obtain S_{n+1}, W_{n+1} [illegible] satisfying

$$t(S_{n+1}, g_n) \geq g_n b_n \ldots (\bar{f}_n) \geq \ldots (\bar{f}_n - 1) \qquad (7.20)$$

$$\deg(W_{n+1}) \leq \deg(t_{n+1}) \qquad (7.21)$$

$$\deg(S_{n+1}) \leq \max(\deg(t_n), \deg(W_{n+1}) - \deg(g_n)) \qquad (7.22)$$

$$t(S_{n+1}) \geq \ldots (t_{n+1}) \ldots (W_{n+1}) \ldots \qquad (7.23)$$

By (7.20), and $\ldots$ we obtain

$$t(S_{n+1} \ldots) \geq \ldots (b_n - \ldots) \geq (r+2)\ldots \qquad (7.24)$$

Now we put $\ldots$ We check that [illegible]

[illegible]

[illegible]

[illegible]

By 7.11 we conclude that C_{n+1} is monic, hence $\ldots$ is satisfied. [illegible] Let $m \leq n$, we need to [illegible]

$$t(\ldots) \geq (r+1)\ldots(b_n - \ldots) \geq (r+1)\ldots \qquad (7.25)$$

which gives [illegible] and by (7.23) and (7.24) we obtain [illegible] Solution [illegible]

[illegible] The [illegible] Since $\ldots$ is [illegible] of the [illegible] with respect to the [illegible] Then [illegible] We put [illegible] Clearly [illegible] we have [illegible] and finally by [illegible] we have [illegible]

Thus we have shown the existence of the pair (g, h). Now, since g is a monic polynomial, the quantity [illegible] so by the Euclidean division of [illegible] and g is unique.

Corollary 7.13. [illegible]

Chapter 8

Infraconnected Sets

Infraconnected sets were introduced in order to provide a class of sets as wide as possible, aimed at playing the same role as connected sets do in complex analysis towards holomorphic functions. We will see that for many things, this class of sets is quite satisfactory. Many proofs of results given in this chapter can be found in [30, Chapter 2].

Notation and definitions: Throughout this chapter the valuation of L is supposed to be dense. Let $a \in L$ and let $r \in \mathbb{R}_+$. We will denote by $d(a,r)$ the disk $\{x \in L | \; |x-a| \leq r\}$, by $d(a,r^-)$ the disk $\{x \in L | \; |x-a| < r\}$ and we will denote by $C(a,r)$ the set $= \{x \in L \mid |x-a| = r\}$.

Given r_1 and r_2 such that $0 < r_1 < r_2$ we will denote by $\Gamma(a,r_1,r_2)$ the annulus $\{x \in L | \; r_1 < |x-a| < r_2\}$ and by $\Delta(a,r_1,r_2)$ the annulus $\{x \in K | \; r_1 \leq |x-a| \leq r_2\}$.

We know that if $b \in d(a,r)$ then $d(b,r) = d(a,r)$. In the same way if $b \in d(a,r^-)$ then $d(b,r^-) = d(a,r^-)$. Moreover given two disks T and T' such that $T \cap T' \neq \emptyset$ then either $T \subset T'$ or $T' \subset T$.

Of course the following three statements are equivalent:

(i) $d(a,r) = d(a,r^-)$;
(ii) $C(a,r) = \emptyset$;
(iii) $r \notin |\, L\,|$.

Moreover the disks $d(b,r^-)$ included in $C(a,r)$ (resp. in $d(a,r)$) are the disks $d(b,r^-)$ such that $b \in C(a,r)$ (resp. in $d(a,r)$). They are called *the classes* of $C(a,r)$ (resp. of $d(a,r)$).

Henceforth D will denote a subset of L.

The closure of D is denoted by $\overline{D}$ and the interior of D is denoted by $\overset{\circ}{D}$.

We put $diam(D) = \sup\{|x-y| \mid x \in D, y \in D\}$ and $diam(D)$ is named the *diameter* of D.

If D is bounded of diameter R we denote by $\widetilde{D}$ the disk $d(a, R)$ for any $a \in D$. If D is not bounded we put $\widetilde{D} = L$.

Given a point $a \in L$ we put $\delta(a, D) = \inf\{|x-a| \mid x \in D\}$. Then $\delta(a, D)$ is named *the distance of a to D*.

Lemma 8.1: *$\widetilde{D} \setminus \overline{D}$ admits a unique partition of the form $(T_i)_{i\in I}$, where each T_i is a disk of the form $d(a_i, r_i^-)$ with $r_i = \delta(a_i, D)$.*

Definition: Such disks $d(a_i, r_i^-)$ are called *the holes* of D.

Examples:

(1) The holes of a disk $d(a, r^-)$ with $r \in |\ L\ |$ are the classes of $C(a, r)$.
(2) The only one hole of $L \setminus d(0, 1^-)$ is $d(0, 1^-)$.
(3) The holes of $L \setminus d(0, 1)$ are the disks $d(a, 1^-)$ with $a \in d(0, 1)$.

Definition: *A* set D is said to be *infraconnected* if for every $a \in D$, the mapping I_a from D to $\mathbb{R}_+$ defined by $I_a(x) = |x - a|$ has an image whose closure in $\mathbb{R}_+$ is an interval. In other words, a set D is not infraconnected if and only if there exist a and $b \in D$ and an annulus $\Gamma(a, r_1, r_2)$ with $0 < r_1 < r_2 < |a - b|$ such that $\Gamma(a, r_1, r_2) \cap D = \emptyset$.

D is said to be *strongly infraconnected* if for every hole $T = d(a, r^-)$ with $r \in |\ L\ |$, there exists a sequence $(x_n)_{n\in\mathbb{N}}$ in D such that $|x_n - a| = |x_n - x_m| = r$ whenever $n \neq m$.

Lemma 8.2: *If D is infraconnected of diameter $R \in \mathbb{R}$ (resp. $+\infty$) then $\overline{I_a(D)} = [0, R]$ (resp. $\overline{I_a(D)} = [0, +\infty[$).*

Lemma 8.3: *Let D be strongly infraconnected. Let $a \in D$ and let $r \in |\ L\ |$ be such that $r < diam(D)$. There exists a sequence $(x_n)_{n\in\mathbb{N}}$ in D such that $|x_n - a| = |x_n - x_m| = r$ whenever $n \neq m$.*

Corollary 8.4: *A strongly infraconnected set is infraconnected.*

Lemma 8.5: *Let D be infraconnected and let α belong to a hole T of diameter ρ. The closure of the set $\{|x - \alpha| \mid x \in D\}$ is an interval whose lower bound is ρ.*

Theorem 8.6: *Let A and B be two infraconnected subsets of L such that $A \cap B \neq \emptyset$. Then $A \cup B$ is infraconnected.*

Notation: Let $\mathcal{R}_D$ be the relation defined in D as $x\mathcal{R}_D y$ if there exists an infraconnected subset of D containing both x, y.

Corollary 8.7: *The relation $\mathcal{R}_D$ is an equivalence relation.*

Definition: The equivalence classes with respect to $\mathcal{R}_D$ are called *the infraconnected components.*

Examples: (1) $d(0,1^-) \cup d(1,1^-)$ is infraconnected. Its holes are the disks $d(\alpha,1^-)$ with $|\alpha| = |\alpha - 1| = 1$.

(2) Let $r \in]0,1[$ and let $D = d(0,1^-) \cup d(1,r)$. Then D is not infraconnected, its infraconnected components are $d(0,1^-)$ and $d(1,r)$. The holes of D are the disks $d(\alpha,1^-)$ with $|\alpha| = |\alpha-1| = 1$ and the disks $d(\alpha,|\alpha-1|^-)$ with $r < |\alpha - 1| < 1$.

Theorem 8.8: *Let A and B be infraconnected subsets such that $\widetilde{A} = \widetilde{B}$. Then $A \cup B$ is infraconnected.*

Definition: We will call *an empty annulus of D* an annulus $\Gamma(a, r_1, r_2)$ such that

(i) $r_1 = \sup\{|x-a| \quad |x \in D, |x-a| < r_2\}$,
(ii) $r_2 = \inf\{\, |x-a| \quad |x \in D, |x-a| > r_1\}$.

The set $d(a,r_1) \cap D$ will be denoted by $\mathcal{I}_D(\Gamma(a,r_1,r_2))$ while the set $(L \setminus d(a,r_2^-)) \cap D$ will be denoted by $\mathcal{E}_D(\Gamma(a,r_1,r_2))$. When there is no risk of confusion about the set D we will just write $\mathcal{I}(\Gamma(a,r_1,r_2))$, (resp. $\mathcal{E}(\Gamma(a,r_1,r_2))$) instead of $\mathcal{I}_D(\Gamma(a,r_1,r_2))$, (resp. $\mathcal{E}_D(\Gamma(a,r_1,r_2))$).

Remarks: (1) By definition, D is not infraconnected if and only if it admits an empty annulus.

(2) If a set D admits an empty annulus $\Gamma(a,r_1,r_2)$, then $\{\mathcal{I}(\Gamma(a,r_1,r_2)), \mathcal{E}(\Gamma(a,r_1,r_2))\}$ is a partition of D.

Examples: Let $r \in]0,1[$, let $D = d(0,r) \cup d(1,1^-)$ and let $D' = d(0,r^-) \cup d(1,r)$. Then $\Gamma(0,r,1)$ is an empty annulus of D and also of D'. In the same way $\Gamma(1,r,1)$ is also an empty annulus of D'.

Notation: Let $\mathcal{O}(D)$ be the set of empty annuli of D. Given Λ_1 and $\Lambda_2 \in \mathcal{O}(D)$, it is easily seen that $\mathcal{I}(\Lambda_1) \subset \mathcal{I}(\Lambda_2)$ is equivalent to $\mathcal{E}(\Lambda_1) \supset \mathcal{E}(\Lambda_2)$. We will denote by $\leq$ the relation defined on $\mathcal{O}(D)$ by $\Lambda_1 \leq \Lambda_2$ if $\mathcal{I}(\Lambda_1) \subset \mathcal{I}(\Lambda_2)$. It is easily seen that $\leq$ is an order relation on $\mathcal{O}(D)$.

We will denote by $<$ the relation defined by $\Lambda_1 < \Lambda_2$ if $\Lambda_1 \leq \Lambda_2$ and $\Lambda_1 \neq \Lambda_2$.

The following Lemmas 8.9 and 8.10 are easily seen.

Lemma 8.9: *Let Λ_1 and Λ_2 be two empty annuli of D. The following assertions are equivalent:*

(*i*) Λ_1 *and* Λ_2 *are not comparable with respect to the order* $\leq$;
(*ii*) $\mathcal{I}(\Lambda_1) \subset \mathcal{E}(\Lambda_2)$;
(*iii*) $\mathcal{I}(\Lambda_2) \subset \mathcal{E}(\Lambda_1)$;
(*iv*) $\mathcal{I}(\Lambda_1) \cap \mathcal{I}(\Lambda_2) = \emptyset$.

Lemma 8.10: *Let $\Lambda \in \mathcal{O}(D)$ and let $x \in \mathcal{I}(\Lambda)$ (resp. $x \in \mathcal{E}(\Lambda)$). The infraconnected component of x is included in $\mathcal{I}(\Lambda)$ (resp. in $\mathcal{E}(\Lambda)$). If $\Lambda' \in \mathcal{O}(D)$ is such that $\Lambda < \Lambda'$ then $\mathcal{I}(\Lambda') \cap \mathcal{E}(\Lambda) \neq \emptyset$.*

Corollary 8.11: *Let Σ be an empty annulus of D. The family of the empty annuli Λ of D such that $\Lambda \geq \Sigma$ is totally ordered.*

We will also need Lemma 8.12 (Lemma 3.12 in [30]). The most classical example of an ultrametric complete algebraically closed field is the field $\mathbb{C}_p$ (one can find a description in [30, Chapter 8]).

Lemma 8.12: *Let $a \in \widetilde{D}$, let ρ be the distance from a to D and let R be such that $\rho \leq R \leq diam(D)$. For $j = 1, \dots, q$ let $\alpha_j \in d(a, R)$ and let $r'_j, r''_j \in \mathbb{R}_+$ be such that $r'_j < R < r''_j$. Then*

$$\bigcap_{j=0}^{q} (\Gamma(\alpha_j, r'_j, r''_j) \cap D) \neq \emptyset.$$

Sets having finitely many infraconnected components are characterized in [30]:

Theorem 8.13: *D has finitely many infraconnected components if and only if it has finitely many empty annuli. Moreover if so does D then one of the infraconnected components is*

$$A_0 = \bigcap_{\Sigma \in \mathcal{X}(D)} \mathcal{E}(\Sigma)$$

while the others are of the form

$$A_i = \mathcal{I}(\Lambda_i) \bigcap \left(\bigcap_{\Sigma < \Lambda_i} \mathcal{E}(\Sigma) \right), \quad \text{with } \Lambda_i \in \mathcal{O}(D).$$

Definition: A bounded closed infraconnected subset A of L is said to be *affinoid* if it has finitely many holes, and if $diam(A)$ and each diameter of hole lies in $|L|$. More generally, a set A is said to be *affinoid* if it is a finite union of infraconnected affinoid sets.

From Theorem 8.13, Lemma 8.14 is easily proven:

Lemma 8.14: *A finite union and a finite intersection of affinoid subsets of L is affinoid.*

Theorem 8.15: *Let A be a commutative ultrametric normed K-algebra with unity. Suppose that A has an element u such that $sp(u)$ admits an empty annulus $\Gamma(a, r_1, r_2)$. Then there exist ψ_1, $\psi_2 \in Mult(A, \| \, . \, \|)$ such that $\psi_1(t) = r_1$, $\psi_2(t) = r_2$.*

Proof: By definition, $r_1 = \sup\{|\lambda| \mid \lambda \in sp(u) \; |\lambda| < r_2\}$, and $r_2 = \inf\{|\lambda| \mid \lambda \in sp(u) \; |\lambda| > r_1\}$. Consequently, both r_1, r_2 lie in the closure of $\{\psi(u) \mid \psi \in Mult(A, \| \, . \, \|)\}$, which, by Lemma 6.3, is a closed subset of $\mathbb{R}$. hence there exist ψ_1, $\psi_2 \in Mult(A, \| \, . \, \|)$ such that $\psi_1(t) = r_1$, $\psi_2(t) = r_2$. □

Chapter 9

Monotonous Filters

Monotonous filters are indispensable to understand the behaviour of rational functions, and all kind of analytic functions. Circular filters which are closely to monotonous filters, will be introduced in the next chapter. Most of the proofs of results given in this chapter can be found in [30, Chapter 3].

Definitions and notation: In all this chapter L is a complete ultrametric field whose valuation is dense and D is a subset of L. According to classical definitions, we call *base* of a filter $\mathcal{F}$ a subset $\mathcal{B}$ of $\mathcal{F}$ such that every element of $\mathcal{F}$ contains an element of $\mathcal{B}$. A filter $\mathcal{F}$ is said to be *secant with* D if the family of sets $\{B \cap D \mid B \in \mathcal{F}\}$ is a filter. When a filter $\mathcal{F}$ is secant with a subset D of L, the filter $\{B \cap D \mid B \in \mathcal{F}\}$ is called *intersection of* $\mathcal{F}$ *with* D, and denoted by $\mathcal{F} \cap D$.

A filter $\mathcal{F}$ on L will be said to be *thinner* than a filter $\mathcal{G}$ if every element of $\mathcal{G}$ belongs to $\mathcal{F}$. In such a case, $\mathcal{G}$ will also said to be *less thin than* $\mathcal{F}$.

A sequence $(u_n)_{n\in\mathbb{N}}$ in L will be said to be *thinner* than a filter $\mathcal{G}$ if so is the filter defined by the sets $A_q = \{u_n | n \geq q\}$ $(q \in \mathbb{N})$. In such a case, $\mathcal{G}$ will also said to be *less thin than* the sequence $(u_n)_{n\in\mathbb{N}}$.

A sequence $(u_n)_{n\in\mathbb{N}}$ in L will be said to be *an increasing distances sequence* (resp. *a decreasing distances sequence*) if the sequence $|u_{n+1} - u_n|$ is strictly increasing (resp. decreasing) and has a limit $\ell \in \mathbb{R}_+^*$.

The sequence $(u_n)_{n\in\mathbb{N}}$ will be said to be *a monotonous distances sequence* if it is either an increasing distances sequence or a decreasing distances sequence. A sequence $(u_n)_{n\in\mathbb{N}}$ in L will be said to be *an equal distances sequence* if $|u_n - u_m| = |u_m - u_q|$ whenever $n, m, q \in \mathbb{N}$ such that $n \neq m \neq q$.

Let $a \in \widetilde{D}$ and $R \in \mathbb{R}_+^*$ be such that $\Gamma(a, r, R) \cap D \neq \emptyset$ whenever $r \in]0, R[$ (resp. $\Gamma(a, R, r) \cap D \neq \emptyset$ whenever $r > R$). We call *an increasing* (resp. *a decreasing*) *filter of center a and diameter R, on D* the filter $\mathcal{F}$ on D that admits for base the family of sets $\Gamma(a, r, R) \cap D$ (resp. $\Gamma(a, R, r) \cap D$). For every sequence $(r_n)_{n \in \mathbb{N}}$ such that $r_n < r_{n+1}$ (resp. $r_n > r_{n+1}$) and $\lim_{n \to \infty} r_n = R$, it is seen that the sequence $\Gamma(a, r_n, R) \cap D$ (resp. $\Gamma(a, R, r_n) \cap D$) is a base of $\mathcal{F}$. Such a base will be called *a canonical base.*

We call *a decreasing filter with no center, of canonical base* $(D_n)_{n \in \mathbb{N}}$ *and diameter* $R > 0$, *on D* a filter $\mathcal{F}$ on D that admits for base a sequence $(D_n)_n \in \mathbb{N}$ in the form $D_n = d(a_n, r_n) \cap D$ with $D_{n+1} \subset D_n$, $r_{n+1} < r_n$, $\lim_{n \to \infty} r_n = R$, and $\bigcap_{n \in \mathbb{N}} d(a_n, r_n) = \emptyset$.

Given an increasing (resp. a decreasing) filter $\mathcal{F}$ on D of center a and diameter r we will denote by $\mathcal{P}_D(\mathcal{F})$ the set $\{x \in D | \ |x - a| \geq r\}$ (resp. the set $\{x \in D | \ |x - a| \leq r\}$) and by $\mathcal{C}_D(\mathcal{F})$ the set $\{x \in D | \ |x - a| < r\}$ (resp. the set $\{x \in D | \ |x - a| > r\}$). When there is no risk of confusion we will only write $\mathcal{P}(\mathcal{F})$ instead of $\mathcal{P}_D(\mathcal{F})$, and $\mathcal{C}(\mathcal{F})$ instead of $\mathcal{C}_D(\mathcal{F})$. Besides $\mathcal{C}_D(\mathcal{F})$ will be named *the body of* $\mathcal{F}$ and $\mathcal{P}_D(\mathcal{F})$ will be named *the beach of* $\mathcal{F}$.

We call *a monotonous filter on D* a filter which is either an increasing filter or a decreasing filter (with or without a center). Given a monotonous filter $\mathcal{F}$ we will denote by $diam(\mathcal{F})$ its diameter.

The field L is said to be *spherically complete* if every decreasing filter on L has a center in L. The field $\mathbb{C}_p$ for example is known not to be spherically complete [10]. However, every algebraically closed complete ultrametric field admits a spherically complete algebraically closed extension (see for example [30, Chapter 7]).

Theorem 9.1: *Let $(u_n)_{n \in \mathbb{N}}$ be a bounded sequence in L. Either we may extract a convergent subsequence or we may extract a monotonous distances subsequence or we may extract an equal distances subsequence from the sequence $(u_n)_{n \in \mathbb{N}}$.*

Lemma 9.2: *Let $(a_n)_n \in \mathbb{N}$ be an increasing distances (resp. a decreasing distances) sequence in D. There exists a unique increasing (resp. decreasing) filter $\mathcal{F}$ on D such that the sequence $(a_n)_n \in \mathbb{N}$ is thinner than $\mathcal{F}$. Moreover, given a monotonous filter $\mathcal{F}$ on D, there exists a unique monotonous filter $\mathcal{G}$ on L inducing $\mathcal{F}$.*

Lemma 9.3: *Let D be infraconnected. Let $\mathcal{F}$ be an increasing filter (resp. a decreasing filter) on L, of center $\alpha \in \widetilde{D}$ and diameter $R \leq diam(D)$*

(resp. $R < diam(D)$) such that α does not belong to a hole of diameter $\rho \geq R$ (resp. $\rho > R$). Then $\mathcal{F}$ is secant with D and induces on D an increasing filter (resp. a decreasing filter) of center α and diameter R, on D.

Definition: Let $\mathcal{F}$ be an increasing (resp. a decreasing) filter of center a and diameter R on D. Then $\mathcal{F}$ is said *to be pierced* if for every $r \in]0, R[$, (resp. $r < R$), $\Gamma(a, r, R)$ (resp. $\Gamma(a, R, r)$) contains some hole T_m of D.

A decreasing filter with no center $\mathcal{F}$ on D is said *to be pierced* if for every $m \in \mathbb{N}$, $\widetilde{D}_m \setminus \widetilde{D}_{m+1}$ contains some hole T_m of D.

Remarks: The definition of a pierced filter with no center also applies to a decreasing filter with a center and then is equivalent to the one given just above for such a filter.

If $\mathcal{F}$ is an increasing (resp. a decreasing) filter of center a, of diameter R, then $\mathcal{F}$ is pierced if and only if there exists a sequence of holes $(T_n)_{n\in\mathbb{N}}$ of D such that $\delta(a, T_n) < \delta(a, T_{n+1})$ (resp. $\delta(a, T_n) > \delta(a, T_{n+1})$), $\lim_{n\to\infty}\delta(a, T_n) = R$.

Given a Cauchy filter $\mathcal{F}$ on D, of limit a in L, we will call *a canonical base of $\mathcal{F}$* a sequence D_m in the form $d(a, r_m) \cap D$ with $0 < r_m < r_{m+1}$ and $\lim_{m\to\infty} r_m = 0$. The filter $\mathcal{F}$ is said *to be pierced* if for every $m \in \mathbb{N}, \widetilde{D}_m$ contains some hole of D.

Let $a \in \widetilde{D}$. Let $(T_{m,i})_{1\leq i\leq s(m), m\in\mathbb{N}}$ be a sequence of holes of D which satisfies $\delta(a, T_{m,i}) = d_m$ $(1 \leq i \leq h_m)$, $d_m < d_{m+1}$ (resp. $d_m > d_{m+1}$), $\lim_{m\to\infty} d_m = S > 0$.

The sequence $(T_{m,i})_{1\leq i\leq s(m), m\in\mathbb{N}}$ is called *an increasing (resp. a decreasing) distances holes sequence that runs the increasing (resp. decreasing) filter of center a, of diameter S.*

Now let $(T_{m,i})_{1\leq i\leq s(m), m\in\mathbb{N}}$ be a sequence of holes of D that satisfies

$$\delta(T_{m+1,j}, T_{m,i}) = d_m \ (1 \leq i \leq s(m),\ 1 \leq j \leq s(m+1)),\ d_m > d_{m+1},$$
$$\lim_{m\to\infty} d_m = R > 0,$$

where the filter $\mathcal{F}$ of base $D_m = d(a_m, d_m) \cap D$ is a decreasing filter with no center. The sequence $(T_{m,i})_{1\leq i\leq s(m), m\in\mathbb{N}}$ is called *a decreasing distances holes sequence that runs $\mathcal{F}$.*

In each case, the sequence $(d_m)_{m\in\mathbb{N}}$ is called *monotony* of the sequence $(T_{m,i})_{1\leq j\leq s(m+1),\ m\in\mathbb{N}}$.

Summarizing these definitions, an increasing (resp. decreasing) distances holes sequence that runs an increasing (resp. decreasing) filter $\mathcal{F}$ will be just named *an increasing* (resp. *decreasing*) *distances holes sequence* and the filter $\mathcal{F}$ will be named *the increasing* (resp. *decreasing*) *filter associated to the sequence* $(T_{m,i})_{1\leq i\leq s(m),m\in\mathbb{N}}$. The diameter of $\mathcal{F}$ will be called *the diameter of the sequence* $(T_{m,i})_{1\leq i\leq s(m),m\in\mathbb{N}}$. If $\mathcal{F}$ has a center a , a will be named *the center of the sequence* $(T_{m,i})_{1\leq i\leq s(m),m\in\mathbb{N}}$. If $\mathcal{F}$ has no center, the sequence $(T_{m,i})$ will be called *a decreasing distances holes sequence with no center.*

Let $(T_{m,i})_{1\leq i\leq s(m),m\in\mathbb{N}}$ be a monotonous distances holes sequences and for every $(m,i)_{1\leq i\leq s(m),m\in\mathbb{N}}$ let $\rho_{m,i} = diam(T_{m,i})$. The number $\inf_{1\leq i\leq s(m),m\in\mathbb{N}}\rho_{m,i}$ will be called *piercing of the sequence* $(T_{m,i})_{1\leq i\leq s(m),m\in\mathbb{N}}$.

If a monotonous distances holes sequence has a piercing $\rho > 0$, it will be said to be *well pierced.* If a monotonous filter $\mathcal{F}$ is run by a well pierced monotonous holes sequence, $\mathcal{F}$ will be said to be well pierced.

In each case the sequence of circles $C(a, d_m)$ when $\mathcal{F}$ has center a (resp. $C(a_{m+1}, d_m)$ when $\mathcal{F}$ has no center) will be said *to run the filter* $\mathcal{F}$, and *to carry* the monotonous distances holes sequence $(T_{m,i})_{1\leq i\leq s(m),m\in\mathbb{N}}$.

A monotonous distances holes sequences $(T_{m,i})_{1\leq i\leq s(m),m\in\mathbb{N}}$ will be said to be *simple* if $s(m) = 1$ for all $m \in \mathbb{N}$.

Moreover, a sequence of holes $(T_m)_{m\in\mathbb{N}}$ of D will be called *a Cauchy sequence of holes of limit* $a \in L$ if $\lim_{m\to\infty}\delta(a, T_m) = 0$. Such a sequence will be said *to run* the Cauchy filter of base $\{d(a,r)\cap D | r > 0\}$.

Let $(T_m)_{m\in\mathbb{N}}$ be monotonous distances holes sequence or an equal distances holes sequence. We will call *superior gauge* of the sequence the number $\limsup_{m\to} diam(T_m)$.

As an obvious corollary of Theorem 9.1, we have Theorem 9.4:

Theorem 9.4: *Let $(T_n)_{n\in\mathbb{N}}$ be a bounded sequence of holes of D. Either we may extract from the sequence $(T_n)_{n\in\mathbb{N}}$ a subsequence which converges in L, or we may extract a monotonous distances holes sequence or we may extract an equal distances holes sequence.*

Lemma 9.5 is obvious:

Lemma 9.5: *Let $(T_m)_{m\in\mathbb{N}}$ be monotonous distances holes sequence. Then the superior gauge s of the sequence satisfies $s \leq r$.*

Definition: Given a monotonous filter $\mathcal{F}$, we will call *superior gauge of* $\mathcal{F}$ the upper bound of all superior gauges of monotonous holes sequences running $\mathcal{F}$.

From the definition, Lemma 9.6 is obvious again:

Lemma 9.6: *Let $\mathcal{F}$ be an increasing (resp. decreasing) filter on D of center a, of diameter r, of superior gauge s. For every $\epsilon > 0$, there exists a hole T of D included in $\Gamma(a, r, r + \epsilon)$ (resp. $\Gamma(a, r - \epsilon, r)$) such that $diam(T) > s - \epsilon$.*

Let $\mathcal{F}$ be a decreasing filter on D with no center, of diameter r, of superior gauge s. For every $\epsilon > 0$, there exists a hole T of D included in a disk $d(a, r + \epsilon)$ such that $diam(T) > s - \epsilon$.

Notation: From now on, to the end of the chapter, γ denotes the homographic function $b + \frac{1}{x-a}$ with $a, b \in L$.

Proposition 9.7: *Let $\alpha \in L$, $r > 0$ and let $a \in L$ be such that $|a - \alpha| < r$. Then $\gamma(C(\alpha, r)) = C(b, \frac{1}{r})$.*

Proof: We may assume $b = 0$ and then the proof is immediate. □

Corollary 9.8: *Let $\alpha \in L, r_1, r_2 \in]0, +\infty[$ with $|a - \alpha| < r_1 < r_2$. Then*

$$\gamma(\Gamma(\alpha, r_1, r_2)) = \Gamma\left(b, \frac{1}{r_2}, \frac{1}{r_1}\right).$$

Corollary 9.9: *Let $\mathcal{F}$ be the increasing (resp. decreasing) filter of center α and diameter $R > |a - \alpha|$, on $L \setminus \{a\}$. Then $\gamma(\mathcal{F})$ is the decreasing (resp. increasing) filter of center b and diameter $\frac{1}{R}$.*

Lemma 9.10: *Let $\alpha \in L$ be such that $|\alpha - a| \neq r$. Then*

$$\gamma(C(\alpha, r)) = C\Big(\gamma(\alpha), \frac{r}{|a - \alpha|^2}\Big).$$

Corollary 9.11: *Let $\alpha \in L$ and $r, r' \in]0, +\infty[$ be such that $0 < r < r' < |a - \alpha|$. Then we have*

$$\gamma\Big(\Gamma(\alpha, r, r')\Big) = \Gamma\Big(\gamma(\alpha), \frac{r}{|a - \alpha|^2}, \frac{r'}{|a - \alpha|^2}\Big),$$

$$\gamma(d(\alpha, r)) = d\Big(\gamma(\alpha), \frac{r}{|a - \alpha|^2}\Big), \ \gamma(d(\alpha, r^-)) = d\Big(\gamma(\alpha), \Big(\frac{r}{|a - \alpha|^2}\Big)^-\Big).$$

Corollary 9.12: *Let $\mathcal{F}$ be the increasing (resp. decreasing) filter of center α and diameter R on $L \setminus \{a\}$ with $|a - \alpha| > R$. Then $\gamma(\mathcal{F})$ is an increasing (resp. a decreasing) filter of center $\gamma(\alpha)$, of diameter $\frac{R}{|a - \alpha|^2}$ on $L \setminus \{b\}$.*

Chapter 10

Circular Filters

We are going to define circular filters, which roughly characterize the absolute values on $K(x)$. We shall construct a tree structure and a distance on the set of circular filters.

Definitions and notation: Throughout the chapter, D will denote a subset of L. Let $a \in \widetilde{D}$, let $\rho = \delta(a, D)$ and let $R \in]0, +\infty[$ be such that $\rho \leq R \leq diam(D)$. We call *circular filter of center a and diameter R on L* the filter $\mathcal{F}$ which admits as a generating system the family of sets $\Gamma(\alpha, r', r'')$ with $\alpha \in d(a, R), r' < R < r''$, i.e. $\mathcal{F}$ is the filter which admits for base the family of sets of the form $\bigcap_{i=1}^{q} \Gamma(\alpha_i, r'_i, r''_i))$ with $\alpha_i \in d(a, R), r'_i < R < r''_i$ $(1 \leq i \leq q \ , \ q \in \mathbb{N})$.

For reasons that will appear when characterizing the absolute values of $K(x)$, a decreasing filter with no center, of canonical base $(D_n)_{n\in\mathbb{N}}$ is also be called *a circular filter on L with no center, of canonical base* $(D_n)_{n\in\mathbb{N}}$.

Finally the filter of neighbourhoods of a point $a \in L$ is called *circular filter of the neighbourhoods of a on L*. It will also be named *circular filter of center a and diameter* 0 or *Cauchy circular filter of limit a*.

A circular filter on L will be said to be *large* if it has diameter different from 0, and to be punctual if it is a cauchy circular filter.

Given a circular filter $\mathcal{F}$, its diameter will be denoted by $diam(\mathcal{F})$ and the set of its centers is denoted by $\mathcal{Q}(\mathcal{F})$. So, if $\mathcal{F}$ is a circular filter of center a and diameter r, then $\mathcal{Q}(\mathcal{F}) = d(a, r)$.

Given a circular filter $\mathcal{F}$ on L, an infraconnected affinoid subset B of L will be called *$\mathcal{F}$-affinoid* if it belongs to $\mathcal{F}$. Thus, in particular, if $\mathcal{F}$ has center a and diameter r, a $\mathcal{F}$ affinoid either is a disk $d(a, s)$, with $s \in |L|$, or is of the form $d(a,r'') \setminus \cup_{i=1}^{q} d(a_i, r'^-)$, with r', $r'' \in |L|$, $r' < r < r''$, and $|a_i - a_j| = r \ \forall i \neq j$.

Lemma 10.1 is easily checked:

Lemma 10.1: *Let $\mathcal{F}$, $\mathcal{G}$ be two different circular filters. Then $\mathcal{F}$ is not secant with $\mathcal{G}$. Moreover, if $\mathcal{Q}(\mathcal{F}) \neq \emptyset$, then $\mathcal{Q}(\mathcal{F}) \neq \mathcal{Q}(\mathcal{G})$.*

Definitions and Notation: We call *circular filter on D* the filter induced on D by a circular filter on L secant with D.

The set of circular filters on L secant with D will be denoted by $\Phi(D)$. In the same way, the set of large circular filters on L secant with D will be denoted by $\Phi'(D)$.

Thus by definition, every circular filter on D is induced by a unique circular filter on L. Given a circular filter $\mathcal{F}$ on D induced by a circular filter $\mathcal{G}$ on L, we will denote by $\mathcal{Q}(\mathcal{F})$ the set of its centers in L, i.e. $\mathcal{Q}(\mathcal{G})$.

Given a circular filter $\mathcal{F}$ on D, induced by a circular filter $\mathcal{G}$ on L, we shall call *$\mathcal{F}$-affinoid* any $\mathcal{G}$-affinoid.

Lemma 10.2: *Let $\mathcal{F}$ be a circular filter. Then $\mathcal{F}$ admits a base consisting of the family of all $\mathcal{F}$-affinoids. If $\mathcal{F}$ does not admit a countable base, it has a center and its diameter belongs to $|L|$. If $\mathcal{F}$ has no center and is secant with an affinoid subset E of L then E lies in $\mathcal{F}$. If $\mathcal{F}$ has center a and diameter r, then an affinoid set E lies in $\mathcal{F}$ if and only if satisfy $E \cap (L \setminus d(a,r)) \neq \emptyset$, $E \cap d(b, r^-) \neq \emptyset \ \forall b \in d(a,r)$.*

Proof: By definition, a circular filter with no center has a countable base, and of course so does a Cauchy circular filter. In both cases, it admits a base consisting of a family of diks which are $\mathcal{F}$-affinoid sets. Now, consider a circular filter of center a and diameter r. Therefore, $\mathcal{F}$ admits for base the family of annuli $d(a, r + \frac{1}{n} \setminus (\cup_{i=1}^{q} d(a_i, (r - \frac{1}{n})^-)$ where the a_i are centers of $\mathcal{F}$ satisfying $|a_i - a_j| = r$. In particular, if $r \notin |L|$ we have $q = 1$ and we obtain a base of the form $\Gamma(a, r - \frac{1}{n}, r + \frac{1}{n})$ which is countable.

Now, suppose that $\mathcal{F}$ is secant with an affinoid subset E of L. Let $(A_n)_{n \in \mathbb{N}}$ be a canonical base of $\mathcal{F}$. Since each A_n admits common points with E, each is included in $\widetilde{E}$, and therefore it is included in E if and only if it contains no hole of E. But since $\mathcal{F}$ has no center, $\bigcap_{n=0}^{\infty} A_n = \emptyset$, hence there exists $q \in \mathbb{N}$ such that $A_n \subset E \ \forall n \geq q$, and therefore $E \in \mathcal{F}$.

Now suppose that $\mathcal{F}$ has center a and diameter r. If $E \in \mathcal{F}$, it obviously satisfies $E \cap (L \setminus d(a,r)) \neq \emptyset$, $E \cap d(b, r^-) \neq \emptyset \ \forall b \in d(a,r)$. Now, suppose that E satisfies $E \cap (L \setminus d(a,r)) \neq \emptyset$, $E \cap d(b, r^-) \neq \emptyset \ \forall b \in d(a,r)$. Since E

has finitely many holes, on one hand there exists $s > r$ such that $\Gamma(a, r, s) \subset E$, and on the other hand, all classes of $d(a, r)$ are included in E, except finitely many: $d(b_j, r^-)$, $1 \leq j \leq n$. And for each $j = 1, \ldots, n$, there exists $r_j < r$ such that $\Gamma(b_j, r_j, r) \subset E$. Finally, E contains the set $d(a, s) \setminus \cup_{j=1}^n d(b_j, r_j)$ which obviously lies in $\mathcal{F}$. □

Remark: By Lemma 10.2 it is easily seen that there exist circular filters without countable base if and only if the residue class field of L is not countable.

The following Proposition 10.3 is just a translation of the definitions.

Proposition 10.3: *Let $\mathcal{F}$ be an increasing filter (resp. a decreasing filter) of center a and diameter r, on D. Then the circular filter of center a and diameter r on L is secant with D, and is the only circular filter on D less thin than $\mathcal{F}$.*

Conversely let $\mathcal{F}$ be a circular filter of center a and diameter r on D, secant with $d(a, r^-)$ (resp. $L \setminus d(a, r)$). Then the increasing filter (resp. decreasing filter) of center a and diameter r on L is secant with D and thinner than $\mathcal{F}$.

Proposition 10.4 is immediate and is Proposition 3.14 in [30]:

Proposition 10.4: *Let D be infraconnected, let $a \in \widetilde{D}$ and let S be the closure of $\{|x - a| \; |x \in D\}$ in $\mathbb{R}$. For every $r \in S$ the circular filter $\mathcal{F}$ of center a and diameter r on L is secant with D.*

Lemma 10.5: *Let $\mathcal{F}$ be a circular filter secant with two disks $d(a, r)$ and $d(b, s)$. Then either $d(a, r) \subset d(b, s)$ or $d(b, s) \subset d(a, r)$.*

Proof: Let $R = \delta(d(a, r), d(b, s))$. Suppose that none of the two inclusions is satisfied. Then $d(a, r) \cap d(b, s) = \emptyset$. But since $\mathcal{F}$ is secant with $d(a, r)$, then $diam(\mathcal{F}) \leq r$, and therefore $\mathcal{F}$ is not secant with $C(a, R)$, in particular neither is it with $d(b, s)$. □

Corollary 10.6: *Let D be infraconnected, and let $d(a, r^-)$ be a hole of D. Then, the circular filter of center a and diameter r on L is secant with D.*

Proposition 10.7: *Let $(a_n)_{n \in \mathbb{N}}$ be a sequence in L that is either a monotonous distances sequence or a constant distances sequence. Then there exists a unique circular filter on L less thin than the sequence (a_n).*

Proof: By Proposition 3.15 in [30] we know that there exists a unique circular filter on L less thin than the sequence (a_n). By definition, this filter is secant with the set $\{a_n \mid n \in \mathbb{N}\}$, hence with D. □

Corollary 10.8: *Let $(a_n)_{n\in\mathbb{N}}$ be a bounded sequence in L. Then there exists a subsequence $(a_{n_t})_{t\in\mathbb{N}}$ and a unique circular filter $\mathcal{F}$ on L less thin than the subsequence $(a_{n_t})_{t\in\mathbb{N}}$.*

Proof: Since the sequence $(a_n)_{n\in\mathbb{N}}$ is bounded, by Theorem 9.1 we can extract either a monotonous distances subsequence or a constant distances subsequence, or a converging subsequence. In all cases, once such a subsequence is chosen, there exists a unique circular filter $\mathcal{F}$ on L less thin than the subsequence. □

Remark: If $\mathcal{F}$ is the circular filter of center a and diameter r, it is not secant with $C(a,r)$ if and only if $r \notin |L|$.

Definitions and notation: A circular filter $\mathcal{F}$ on L will be said to be *(a,r)-approaching* if it is secant with the circle $C(a,r)$, or if it is the circular filter of center a and diameter r.

A circular filter $\mathcal{F}$ on L will be said to be *D-bordering* if it is secant with both D and $L\setminus\overline{D}$. In the same way, a circular filter on D will be said to be *D-bordering* if it is induced by a D-bordering circular filter on L.

If D is bounded, $\mathcal{F}$ will be said to be *D-peripheral*, or to be *peripheral to D* if $\mathcal{Q}(\mathcal{F}) = \widetilde{D}$.

Finally, a circular filter $\mathcal{F}$ will be said to be *strictly D-bordering* if either it is D-peripheral, or it is peripheral to some hole of D.

We will denote by $\Sigma(D)$ the set of $\varphi_{\mathcal{F}} \in Mult(K[x])$ such that $\mathcal{F}$ is D-bordering, and by $\Sigma_0(D)$ the set of $\varphi_{\mathcal{F}} \in Mult(K[x])$ such that $\mathcal{F}$ is strictly D-bordering.

Proposition 10.9: *Let D be closed and infraconnected and let $\mathcal{F}$ be a circular filter on L. If every element of $\mathcal{F}$ contains infinitely many holes of D, then $\mathcal{F}$ is D-bordering. Conversely, if $\mathcal{F}$ is D-bordering but not strictly D-bordering, then every element of $\mathcal{F}$ contains infinitely many holes of D and then, either D admits a monotonous pierced filter thinner than $\mathcal{F}$, or $\mathcal{F}$ has a center a and D admits an equal distances holes sequence $(T_n)_{n\mathbb{N}}$ such that $\delta(T_n, T_m) = \delta(T_n, a) = diam(\mathcal{F})$ for all $m \neq n$.*

Proof: Suppose that every element of $\mathcal{F}$ contains infinitely many holes of D. Then $\mathcal{F}$ is obviously secant with $L \setminus D$. Let $A \in \mathcal{F}$ be affinoid and let $s = diam(\mathcal{F})$. Suppose that $A \cap D = \emptyset$. If $D \subset L \setminus \widetilde{A}$, then every hole of D has empty intersection with A, a contradiction. But then, since D is infraconnected, there exists a hole $T = d(a, r^-)$ of A such that $D \subset T$. Since A is affinoid, r belongs to $|L|$, hence there exists another affinoid

$B \in \mathcal{F}$, with a hole $d(a, s^-)$, with $s > r$, thereby $\widetilde{D} \cap B = \emptyset$, and finally every hole of D has empty intersection with B, a contradiction again. Thus, $\mathcal{F}$ is also secant with D, and therefore $\mathcal{F}$ is D-bordering.

Conversely, consider a D-bordering filter $\mathcal{F}$ which is not strictly D-bordering. Let $s = diam(\mathcal{F})$. Let $A \in \mathcal{F}$ be affinoid and let $a \in A \cap (L \setminus D)$. Since D is closed, a belongs to a hole $T = d(a, r^-)$ of D. Since $A \cap D \neq \emptyset$, A is not included in $d(a, r^-)$. We will show that there exists $B \in \mathcal{F}$ such that $B \subset A$ and $T \cap B = \emptyset$. For any $r' \in]0, r[$ the set $B = A \setminus d(a, r')$ is an affinoid set that also belongs to $\mathcal{F}$. Since T is not included in B, then B must contain another hole of D. Thus, by induction we can construct a strictly decreasing sequence $(A_m)_{m \in \mathbb{N}}$ of affinoid sets such that each A_m contains a hole S_m which is not included in A_{m+1}. Thus, every element of $\mathcal{F}$ contains infinitely many holes of D. More precisely, suppose that D does not admit a monotonous pierced filter thinner than $\mathcal{F}$. If $\mathcal{F}$ had no center, it would be a not pierced decreasing filter, and therefore would admit a base consisting of a sequence of disks $(D_n)_{n \in \mathbb{N}}$ included in D, a contradiction to the hypothesis: every element of $\mathcal{F}$ contains infinitely many holes of D. Consequently, $\mathcal{F}$ admits a center a. Let $r = diam(\mathcal{F})$. By Proposition 9.4 we can extract from the sequence $(S_m)_{m \in \mathbb{N}}$ a subsequence $(T_n)_{n \in \mathbb{N}}$ which is either a monotonous distances holes sequence, or an equal distances holes sequence. Consider the sequence $(B_n)_{n \in \mathbb{N}}$ defined as $B_n = \cup_{j=n}^{\infty} T_j$. This sequence is the base of a filter thinner than $\mathcal{F}$. If the sequence $(T_n)_{n \in \mathbb{N}}$ is a monotonous distances holes sequence, then the sequence $(B_n)_{n \in \mathbb{N}}$ defined as $B_n = \cup_{j=n}^{\infty} T_j$ is the base of a filter thinner than a monotonous filter $\mathcal{G}$ thinner than $\mathcal{F}$, and such that the filter induced by $\mathcal{G}$ on D is pierced, a contradiction to the hypothesis. Consequently, the sequence $(T_n)_{n \in \mathbb{N}}$ is an equal distances holes sequence of center b: $\delta(T_n, T_m) = l = \delta(b, T_n)\ \forall n \neq m$. If $l < r$, (resp. $l > r$), there exists an annulus $\Gamma(b, r', r'') \in \mathcal{F}$ such that $T_n \subset d(b, r')$ (resp. $T_n \subset K \setminus d(b, r'')$), and therefore $T_n \cap \Gamma(b, r', r'') = \emptyset$. But since the sequence (B_n) is a base of a filter thinner than $\mathcal{F}$, each B_n has a non empty intersection with $\Gamma(b, r', r'')$. Consequently, $l = r = \delta(a, T_n)\ \forall n \in \mathbb{N}$. □

Lemma 10.10: *Let D be an affinoid subset of K. Then the set of D-bordering filters is finite.*

Proof: If D is infraconnected, it has finitely many holes. And if it is not infraconnected, it has finitely many infraconnected components $D_1, \ldots, D_q$, hence the set of its D-bordering filters is the union of the sets of D_j-bordering filters, $1 \leq j \leq q$. □

Lemma 10.11: *Let $\mathcal{F}$ be a circular filter on L and let $a \in L$. There exists a unique $r > 0$ such that $\mathcal{F}$ is (a,r)-approaching. Moreover, for every $s > diam(\mathcal{F})$ there exists a unique disk E of the form $d(b,s)$ which belongs to $\mathcal{F}$.*

Proof: Let $\rho = diam(\mathcal{F})$. If a is a center of $\mathcal{F}$, then $\mathcal{F}$ is (a,ρ)-approaching. Suppose now a is not a center of $\mathcal{F}$. Then we can find a disk $D \in \mathcal{F}$ which does not contain a: let $r = \delta(a,D)$. Then $\mathcal{F}$ is secant with $C(a,r)$. Let $s > diam(\mathcal{F})$. If $\mathcal{F}$ has a center a, then $\mathcal{F}$ is secant with $d(a,s)$. Now, suppose that $\mathcal{F}$ has no center. By definition of the circular filters with no center, there exists a disk $d(b,r) \in \mathcal{F}$ such that $r < s$. Then $d(b,s)$ obviously belongs to $\mathcal{F}$. Moreover, in both cases, any disk $d(b,s)$ other than $d(a,s)$ has empty intersection with $d(a,s)$, hence $\mathcal{F}$ is not secant with $d(b,s)$. □

Lemma 10.12: *Let $\mathcal{F}$ be a filter admitting a base $(A_i)_{i\in J}$ consisting of infraconnected affinoid sets. There exists at least one circular filter $\mathcal{G}$ secant with $\mathcal{F}$.*

Proof: Suppose first that $\bigcap_{i\in J} \widetilde{A}_i = \emptyset$. Then $(\widetilde{A}_i)_{i\in J}$ is a base of a circular filter with no center $\mathcal{G}$ which actually is equal to $\mathcal{F}$. Suppose now that $\bigcap_{i\in J} \widetilde{A}_i \neq \emptyset$ and let $a \in \bigcap_{i\in J} \widetilde{A}_i$. Let $r = \inf_{i\in J} diam(A_i)$. Then the circular filter of center a and diameter r is clearly secant with $\mathcal{F}$, which ends the proof. □

Lemma 10.13: *Let $\mathcal{G}$ be a circular filter on K of center a and diameter r and let $\mathcal{F}$ be a filter secant with $\mathcal{G}$, admitting a base $(A_i)_{i\in J}$ consisting of infraconnected affinoid sets. Either $\mathcal{F}$ is less thin than $\mathcal{G}$, or $\mathcal{F}$ is secant with another circular filter $\mathcal{G}'$.*

Proof: Since $\mathcal{F}$ is secant with $\mathcal{G}$, for each $i \in J$ we have $A_i \cap C(a,r) \neq \emptyset$ and $diam(A_i) \geq r$. Suppose that $\mathcal{F}$ is not less thin than $\mathcal{G}$. Then there exists an affinoid infraconnected set $B \in \mathcal{G}$ which does not contain A_i, whenever $i \in J$. Let $s = \inf_{i\in J} diam(A_i)$.

Suppose first that $\mathcal{G}$ has no center. Then we can assume that B is a disk $d(b,\lambda)$, with $\lambda \in]r,s]$. Let $t \in]r,s[\cap|L|$. Since the A_i are infraconnected and affinoid, for each $i \in J$ and for every class D of $C(b,t)$ except at most for finitely many, D is included in A_i. Consequently, $\mathcal{F}$ is secant with the circular filter $\mathcal{G}'$ of center b and diameter t.

Suppose now that $\mathcal{G}$ has center a. Since B has finitely many holes, either $s > r$, or there exists a hole $T = d(b,\rho)$ of B such that $T \cap A_i \neq \emptyset\ \forall i \in J$. Suppose first that $s > r$. Let $t \in]r,s[\cap|L|$. Similarly to the previous case, since the A_i are infraconnected and affinoid, for each $i \in J$ and for every class D of $C(a,t)$ except at most for finitely many, D is included in A_i. Consequently, $\mathcal{F}$ is secant with the circular filter $\mathcal{G}'$ of center a and diameter t. Similarly, suppose now that there exists a hole $T = d(b,\rho)$ of B such that $T \cap A_i \neq \emptyset\ \forall i \in J$. We can find $t \in]\rho, r[\cap|L|$ and then all classes of $C(b,t)$ except at most finitely many are included in A_i whenever $i \in J$. Consequently, $\mathcal{F}$ is secant with the circular filter $\mathcal{G}'$ of center b and diameter t, which ends the proof. $\square$

Lemma 10.14: *Let $\mathcal{T}$ be a circular filter secant with B, of diameter r and let $B \in \mathcal{T}$ be $d(a,r'') \setminus \cup_{i=1}^{q} d(a_i,r')$ where the a_i are centers of $\mathcal{T}$ satisfying $|a_i - a_j| = r\ \forall i \neq j$. Let $\epsilon = \frac{1}{2}\min(r''-r, r-r')$, let $s'' =]r, r+\epsilon[\cap|L|$, $s' \in]r-\epsilon, r[\cap|L|$. Let $A = d(a,s'') \setminus \cup_{i=1}^{q} d(a_i,s')$. Let $\mathcal{F}$ be a circular filter secant with A and let $\mathcal{G}$ be a circular filter satisfying $\delta(\mathcal{F},\mathcal{G}) < \epsilon$. Then $A \in \mathcal{T}$, $A \subset B$ and $\mathcal{G}$ is secant with B.*

Proof: It is obvious that $A \in \mathcal{T}$, and we check that $A \subset B$ because $\widetilde{A} \subset \widetilde{B}$, and each hole of B is included in a hole of A. Now, let us check that $\mathcal{G}$ is secant with $\widetilde{B}$. Indeed, if $\mathcal{G}$ is not secant with $\widetilde{A}$, then due to the hypothesis $\delta(\mathcal{F},\mathcal{G}) < \epsilon$ we can check that it is secant with a disk containing A, of diameter r'', hence it is secant with $\widetilde{B}$. Similarly, if $\mathcal{G}$ is secant with a hole $d(a,r'^-)$ of A, due to the hypothesis $\delta(\mathcal{F},\mathcal{G}) < \epsilon$, then $\mathcal{G}$ is secant with $\Gamma(a,r',r) \subset B$. $\square$

Chapter 11

Tree Structure and Metric on Circular Filters

In this chapter, we will show that the set of circular filters is provided with a tree structure as defined in Chapter 2, and that the diameter here is an increasing function with values in $\mathbb{R}$, defining distances associated to this structure, as shown in [9, Chapter 2]. The first remarks on that tree structures are due to E. Motzkin [41]. Finally, J.C. Yoccoz and J. Rivera-Letellier considered a similar distance [42].

Definitions and notation: Given two circular filters $\mathcal{F}$ and $\mathcal{G}$, $\mathcal{F}$ is said *to surround* $\mathcal{G}$ if either $\mathcal{G}$ is secant with $\mathcal{Q}(\mathcal{F})$, or if $\mathcal{F} = \mathcal{G}$. Similarly, a circular filter $\mathcal{F}$ is said *to surround a monotonous filter* $\mathcal{G}$ if it surrounds the circular filter associated to $\mathcal{G}$. A monotonous filter $\mathcal{F}$ is said *to surround a circular filter* $\mathcal{G}$ if its associated circular filter surrounds $\mathcal{G}$, and $\mathcal{F}$ is said *to surround a monotonous filter* $\mathcal{G}$ if the circular filter associated to $\mathcal{F}$ surrounds the circular filter associated to $\mathcal{G}$.

We will denote by $\preceq$ the relation on the set of circular filters defined as $\mathcal{F} \preceq \mathcal{G}$ if $\mathcal{G}$ surrounds $\mathcal{F}$ and by $\prec$ the relation defined as $\mathcal{F} \prec \mathcal{G}$ if $\mathcal{F} \preceq \mathcal{G}$ and $\mathcal{F} \neq \mathcal{G}$.

By definition of the relation $\preceq$, Lemma 10.14 is then immediate:

Lemma 11.1: *Let $\mathcal{F}, \mathcal{G}$ be two circular filters such that $\mathcal{Q}(\mathcal{F}) \neq \emptyset$ and $\mathcal{Q}(\mathcal{G}) \neq \emptyset$. Then $\mathcal{F} \preceq \mathcal{G}$ if and only if $\mathcal{Q}(\mathcal{F}) \subset \mathcal{Q}(\mathcal{G})$.*

Corollary 11.2: *Let $\mathcal{F}, \mathcal{G}$ be two circular filters such that $\mathcal{F} \preceq \mathcal{G}$. Then $diam(\mathcal{F}) \leq diam(\mathcal{G})$. Moreover, $\mathcal{F} = \mathcal{G}$ if and only if $diam(\mathcal{F}) = diam(\mathcal{F})$.*

Theorem 11.3: *The relation $\preceq$ is an order relation on $\Phi(L)$ and $\prec$ is the strict order associated to this order relation.*

Proof: Indeed, the relation is reflexive by definition. Let $\mathcal{F}, \mathcal{G}, \mathcal{H}$ be 3 circular filters. Suppose that $\mathcal{F} \preceq \mathcal{G}$ and $\mathcal{G} \preceq \mathcal{F}$. If $\mathcal{F} \neq \mathcal{G}$, then by definition $\mathcal{Q}(\mathcal{F}) \neq \emptyset$, $\mathcal{Q}(\mathcal{G}) \neq \emptyset$, $\mathcal{F}$ is secant with $\mathcal{Q}(\mathcal{G})$ and $\mathcal{G}$ is secant with $\mathcal{Q}(\mathcal{F})$. Consequently, $\mathcal{Q}(\mathcal{F}) = \mathcal{Q}(\mathcal{G})$ and therefore by Corollary 11.2, $\mathcal{F} = \mathcal{G}$. Finally, suppose that $\mathcal{F} \preceq \mathcal{G}$ and $\mathcal{G} \preceq \mathcal{H}$, with $\mathcal{F} \neq \mathcal{G} \neq \mathcal{H}$. Then $\mathcal{F}$ is secant with $\mathcal{Q}(\mathcal{G})$, hence with $\mathcal{Q}(\mathcal{H})$ and therefore $\mathcal{F} \preceq \mathcal{H}$. □

Theorem 11.4: *Let D be infraconnected and let $\mathcal{F} \in \Phi(D)$. Then $\mathcal{F}$ is a minimal element in $\Phi(D)$ if and only if:*

(1) it is punctual, or
(2) it has no center, or
(3) $\mathcal{Q}(\mathcal{F}) \cap D = \emptyset$.

Proof: On one hand it is easily seen that if $\mathcal{F}$ is punctual, or has no center, then it is minimal in $\Phi(L)$, and therefore in $\Phi(D)$. Suppose now that $\mathcal{Q}(\mathcal{F}) \cap D = \emptyset$. Suppose that $\mathcal{G} \in \Phi(D)$ satisfies $\mathcal{G} \prec \mathcal{F}$. Then $\mathcal{G}$ is secant with $\mathcal{Q}(\mathcal{F})$, and has a diameter strictly inferior to the diameter r of $\mathcal{F}$, hence there exists a disk $d(b, s) \in \mathcal{G}$, with $s < r$. Consequently, $d(b, s) \subset \mathcal{Q}(\mathcal{F})$. Since $d(b, s) \cap D = \emptyset$, we see that $\mathcal{G}$ is not secant with D.

Now, let $\mathcal{F} \in \Phi(D)$ be minimal in $\Phi(D)$. Suppose that $\mathcal{F}$ is not punctual and has a center a, and let $r = diam(\mathcal{F})$. If $\mathcal{Q}(\mathcal{F}) \cap D \neq \emptyset$, then for any $a \in \mathcal{Q}(\mathcal{F}) \cap D$, of course $\mathcal{F}$ surrounds the filter of neighborhoods of a, a contradiction to the hypothesis "$\mathcal{F}$ minimal in $\Phi(D)$". Thus, $\mathcal{Q}(\mathcal{F}) \cap D = \emptyset$, which completes the proof. □

Lemma 11.5: *Let $\mathcal{F}, \mathcal{G}$ be two circular filters such that $\mathcal{Q}(\mathcal{F}) \cap \mathcal{Q}(\mathcal{G}) \neq \emptyset$. Then $\mathcal{F}$ and $\mathcal{G}$ are comparable for $\preceq$.*

Proof: Since $\mathcal{Q}(\mathcal{F}) \cap \mathcal{Q}(\mathcal{G}) \neq \emptyset$, and since both $\mathcal{Q}(\mathcal{F})$, $\mathcal{Q}(\mathcal{G})$ are disks, we can suppose for instance $\mathcal{Q}(\mathcal{F}) \subset \mathcal{Q}(\mathcal{G})$. Then $\mathcal{F}$ is secant with $\mathcal{Q}(\mathcal{G})$, hence $\mathcal{F} \preceq \mathcal{G}$. □

Theorem 11.6: *Let $\mathcal{F}$ be a circular filter on L and let $s > diam(\mathcal{F})$. There exists a unique circular filter on L, of diameter s, surrounding $\mathcal{F}$.*

Proof: By Lemma 10.11 there exists a unique disk $d(b, s)$ such that $\mathcal{F}$ is secant with this disk. Then the circular filter $\mathcal{G}$ of center b and diameter s obviously surrounds $\mathcal{F}$. Conversely, let $\mathcal{H}$ be another circular filter of diameter s, surrounding $\mathcal{H}$. Since $s > diam(\mathcal{F})$, $\mathcal{F}$ is secant with $\mathcal{Q}(\mathcal{H})$ which is of the form $d(b, s)$, hence $\mathcal{H} = \mathcal{G}$. □

Proposition 11.7: *Let $\mathcal{F}$, $\mathcal{G}$ be two circular filters surrounding a certain circular filter. Then $\mathcal{F}$ and $\mathcal{G}$ are comparable for $\preceq$.*

Proof: Suppose that both $\mathcal{F}$, $\mathcal{G}$ surrounds $\mathcal{H}$. If $\mathcal{F} = \mathcal{H}$, then $\mathcal{G}$ just surrounds $\mathcal{F}$. Now suppose that both $\mathcal{F}$, $\mathcal{G}$ strictly surrounds $\mathcal{H}$. Then $\mathcal{H}$ is secant with both $\mathcal{Q}(\mathcal{F})$ and $\mathcal{Q}(\mathcal{G})$. But by Lemma 10.5 either $\mathcal{Q}(\mathcal{F}) \subset \mathcal{Q}(\mathcal{G})$ or $\mathcal{Q}(\mathcal{G}) \subset \mathcal{Q}(\mathcal{F})$, hence by Lemma 11.5 $\mathcal{F}$ and $\mathcal{G}$ are comparable. □

By Proposition 10.4, we obtain Theorem 11.8:

Theorem 11.8: *Let D be infraconnected, let $\mathcal{F}$ be a circular filter on L secant with D and let $r \in]diam(\mathcal{F}), diam(D)[$. The unique circular filter of diameter r surrounding $\mathcal{F}$ is secant with D.*

Proof: By Lemma 10.11 there exists a unique disk $d(a, r)$ $\mathcal{F}$ is secant with. By Proposition 10.4 the circular filter of center a and diameter r is secant with D, and on the other hand, by Theorem 11.6 this is the unique circular filter of diameter r surrounding $\mathcal{F}$. □

Proposition 11.9: *Let $\mathcal{F}$, $\mathcal{G}$ be two circular filters on L which are not comparable for the relation $\preceq$. There exist disks $F \in \mathcal{F}$ and $G \in \mathcal{G}$ such that $F \cap G = \emptyset$. Moreover, given $F' \in \mathcal{F}$, $G' \in \mathcal{G}$ such that $F' \cap G' = \emptyset$, we have $\delta(F, G) = \delta(F', G') > \max(diam(\mathcal{F}), diam(\mathcal{G}))$.*

Proof: Suppose first that both $\mathcal{F}$, $\mathcal{G}$ have no center. Then $\mathcal{F}$ (resp. $\mathcal{G}$) admits a canonical base $\mathcal{D}$ (resp. $\mathcal{E}$) consisting of a family of disks. But since the two filters are not secant, we can obviously find $F \in \mathcal{D}$ and $G \in \mathcal{E}$ such that $F \cap G = \emptyset$.

Suppose now that $\mathcal{F}$ has centers and let $d(a, r) = \mathcal{Q}(\mathcal{F})$. If all disks $G \in \mathcal{G}$ contain $d(a, r)$, then $\mathcal{G}$ surrounds $\mathcal{F}$, hence we can find a disk $G = d(b, u) \in \mathcal{G}$ which does not contain $d(a, r)$. Of course, if $G \subset d(a, r)$, then $\mathcal{F}$ surrounds $\mathcal{G}$, hence $G \cap d(a, r) = \emptyset$. Consequently, we have $|a - b| > r$ and $u < |a - b|$. Let $s \in]r, |a - b|[$. Then the disk $F = d(a, s)$ belongs to $\mathcal{F}$ and is such that $F \cap G = \emptyset$.

Since $\mathcal{F}$ is a filter, we have $F \cap F' \neq \emptyset$, hence either $F \subset F'$, or $F' \subset F$. In the same way, $G \cap G' \neq \emptyset$, hence either $G \subset G'$, or $G' \subset G$. Without loss of generality, we can assume that $F \subset F'$. If $G \subset G'$, then our claim is obvious. Suppose now that $G' \subset G$. If $F' \cap G \neq \emptyset$, then either $G \subset F'$, therefore $G \subset G'$, a contradiction, or $F' \subset G$, hence $F \subset G$, a contradiction again. Consequently, we have $F' \cap G = \emptyset$. Putting $l = \delta(F', G)$ we have $|a - b| = l \ \forall a \in F', \ b \in G$, hence in particular, $\delta(F, G) = \delta(F', G') = l$. Then we notice that $l > diam(F) > diam(\mathcal{F})$, and $l > diam(G) > diam(\mathcal{G})$. □

Notation: Let $\mathcal{F}$, $\mathcal{G}$ be two circular filters which are not comparable for $\preceq$, we notice that, by Proposition 11.9, $\delta(F,G)$ does not depend on the choice of disks $F \in \mathcal{F}$ and $G \in \mathcal{G}$, so we can put $\lambda(\mathcal{F},\mathcal{G}) = \delta(F,G)$ with F, G disks such that $F \in \mathcal{F}, G \in \mathcal{G},\ F \cap G = \emptyset$.

Theorem 11.10: *Let D be infraconnected and let $\mathcal{F},\mathcal{G} \in \Phi(D)$. There exists* $\sup(\mathcal{F},\ \mathcal{G}) \in \Phi(D)$ *and it is the unique circular filter of diameter $\lambda(\mathcal{F},\mathcal{G})$ which surrounds both $\mathcal{F}$, $\mathcal{G}$.*

Proof: The claim is trivial when the two filters are comparable for $\preceq$. So we assume they are not. Let $l = \lambda(\mathcal{F},\mathcal{G})$. By Theorem 11.6 there exists a unique circular filter $\mathcal{S} \in \Phi(L)$ of diameter l surrounding $\mathcal{F}$. Then $\mathcal{S}$ has centers. Let $\mathcal{Q}(\mathcal{S}) = d(a,l)$. Then by Proposition 11.9 $\mathcal{G}$ contains disks included in $d(a,l)$ and therefore is secant with $d(a,l)$, hence $\mathcal{S}$ surrounds $\mathcal{G}$. We will check that $\mathcal{S}$ is the smallest element of the set of filters on L surrounding both $\mathcal{F}$ and $\mathcal{G}$. Indeed, let $\mathcal{H}$ be a circular filter surrounding $\mathcal{F}$ and $\mathcal{G}$. Then both $\mathcal{F}$ and $\mathcal{G}$ are secant with $\mathcal{Q}(\mathcal{H})$. Let $d(b,s) = \mathcal{Q}(\mathcal{H})$. Consider disks $F = d(\alpha,\rho) \in \mathcal{F}$ and $G = d(\beta,\nu) \in \mathcal{G}$ such that $F \cap G = \emptyset$. Since $F \cap d(b,s) \neq \emptyset$ and $G \cap d(b,s) \neq \emptyset$, and since $F \cap G = \emptyset$, it is easily seen that both F, G are included in $d(b,s)$, hence $s \geq l$. But since both $\mathcal{S}$, $\mathcal{H}$ surround $\mathcal{F}$, by Proposition 11.7 they are comparable for $\preceq$. Then, since $l \leq s$, $\mathcal{H}$ surrounds $\mathcal{S}$, and therefore $\mathcal{S}$ is the smallest element of the set of filters on L surrounding both $\mathcal{F}$ and $\mathcal{G}$.

Now, suppose that both $\mathcal{F}$, $\mathcal{G}$ are secant with D. Then $d(\alpha,\rho) \cap D \neq \emptyset$, and $d(\beta,\nu) \cap D \neq \emptyset$, hence $\mathcal{S}$ which is the circular filter of center α and diameter $l = |\alpha - \beta|$ is secant with D because D is infraconnected. □

Theorem 11.11: *If D is infraconnected, $\Phi(D)$ is a tree with respect to the order $\preceq$ and the mapping diam is strictly increasing from $\Phi(D)$ to $\mathbb{R}_+$.*

Proof: This is an obvious consequence of Theorems 11.3, 11.4, 11.10 and Proposition 11.7. □

Notation: Let $\mathcal{F}$, $\mathcal{G}$ be two circular filters and let $\mathcal{S} = \sup(\mathcal{F},\mathcal{G})$. We put $\delta(\mathcal{F},\mathcal{G}) = \max(diam(\mathcal{S}) - diam(\mathcal{F}),\ diam(\mathcal{S}) - diam(\mathcal{G}))$ and $\delta'(\mathcal{F},\mathcal{G}) = 2diam(\mathcal{S}) - diam(\mathcal{F}) - diam(\mathcal{G})$.

Remark: Particularly, if $\mathcal{F} \preceq \mathcal{G}$, we have $\delta(\mathcal{F},\mathcal{G}) = diam(\mathcal{G}) - diam(\mathcal{F})$.

According to Chapter 2, we can state Corollary 11.12

Corollary 11.12: *δ is the supremum distance associated to the mapping diam defined on the tree $\Phi(K)$ and δ' is the whole distance associated to the mapping diam defined on the tree $\Phi(K)$.*

Lemma 11.13 is immediate and will be useful:

Lemma 11.13: *Let $\mathcal{F}$ be a circular filter and let $B \in \mathcal{F}$ be infraconnected and affinoid. Let $\mathcal{G}_1, \dots, \mathcal{G}_n$ be the B-bordering filters and let $\lambda = \min_{1 \le j \le n}(\delta(\mathcal{G}, \mathcal{G}_j))$. For every $\mathcal{G} \in \Phi(K)$ such that $\delta(\mathcal{G}, \mathcal{F}) < \lambda$, B lies in $\mathcal{G}$.*

Proof: Every disk $d(a, r)$ with $a \in B$ and $r < \lambda$ is clearly included in B. Consequently, $\mathcal{G}'$ is secant with B. Now, suppose that $B \notin \mathcal{G}'$. Then $\mathcal{G}'$ is B-bordering, and therefore is one the $\mathcal{G}_j$ $(1 \le j \le n)$, a contradiction to the hypothesis $\delta(\mathcal{G}, \mathcal{F}) < \lambda$. Hence B lies in $\mathcal{G}'$. □

Theorem 11.14: *Every bounded monotonous sequence of $\Phi(L)$ has a limit with respect to the metric topology.*

Proof: Let $(\mathcal{F}_n)_{n \in \mathbb{N}}$ be a bounded monotonous sequence of circular filters, and for each $n \in \mathbb{N}$, let $r_n = diam(\mathcal{F}_n)$. Without loss of generality, we can obviously assume that the sequence is strictly monotonous. Consequently, for each $n \in \mathbb{N}$, each filter $\mathcal{F}_n$ has a center. Let $l = \lim_{n \to \infty} r_n$. Suppose first that it is a decreasing sequence. For each $n \in \mathbb{N}$ we can find a center $a_n \notin d(a_{n+1}, r_{n+1})$. Thus, the sequence $(a_n)_{n \in \mathbb{N}}$ is such that the sequence $(|a_{n+1} - a_n|)_{n \in \mathbb{N}}$ is strictly decreasing. Hence by Proposition 10.7 there exists a unique circular filter $\mathcal{F}$ less thin than the sequence $(a_n)_{n \in \mathbb{N}}$ of radius l. And then, we check that the sequence $(\mathcal{F}_n)_{n \in \mathbb{N}}$ converges to $\mathcal{F}$ with respect to the metric topology. If the sequence is increasing, it is easily seen that the sequence converges to the circular filter of center a_1 and diameter l. □

Theorem 11.15 ([11]): *$\Phi(L)$ is complete with respect to the metric topology.*

Proof: Let $(\mathcal{F}_n)_{n \in \mathbb{N}}$ be a Cauchy sequence with respect to the metric topology. For every $m, n \in \mathbb{N}$, let $\mathcal{S}_{m,n}$ denote $\sup(\mathcal{F}_m, \mathcal{F}_n)$. Next, for each $n \in \mathbb{N}$, let $r_n = diam(\mathcal{F}_n)$, and $s_n = \sup_{m \ge n}\{diam(\mathcal{S}_{n,m})\}$. We will show that the sequence $d(a_n, s_n)_{n \in \mathbb{N}}$ is decreasing with respect to inclusion.

Indeed, let $m, t > n$. Both $\mathcal{S}_{n,t}, \mathcal{S}_{n,m}$ are secant with $d(a_n, s_n)$ and surround $\mathcal{F}_n$, hence are comparable with respect to $\preceq$. Then, $\sup(\mathcal{S}_{n,t}, \mathcal{S}_{n,m})$ is secant with $d(a_n, s_n)$, and surrounds $\mathcal{F}_n$, $\mathcal{F}_m$, $\mathcal{F}_t$, hence surround $\mathcal{S}_{m,t}$. Consequently, $\mathcal{S}_{m,t}$ is secant with $d(a_n, s_n)$. Now, let us fix $t > n$. This true for all $m \geq t$, hence $diam(\mathcal{S}_{m,t}) \leq s_n \; \forall m \geq t$, and therefore $s_m \leq s_n$. Thus, for every $n \in \mathbb{N}$, let $\mathcal{G}_n$ be the circular filter of center a_n and diameter s_n. The sequence $(\mathcal{G}_n)_{n\in\mathbb{N}}$ is then decreasing with respect to the order $\preceq$ and therefore, by Theorem 11.14, has a limit $\mathcal{G}$ with respect to the metric topologies. Let $s = \lim\limits_{n\to\infty} s_n$. Since the sequence $(\mathcal{F}_n)_{n\in\mathbb{N}}$ is a Cauchy sequence, and since by definition of the distance δ, we have $\delta(\mathcal{F}_n, \mathcal{S}_{n,m}) \leq \delta(\mathcal{F}_n, \mathcal{F}_m)$, it is clear that $\lim\limits_{n\to\infty} s_n - r_n = 0$. Consequently, $\lim\limits_{n\to\infty} \delta(\mathcal{F}_n, \mathcal{G}) = 0$ and therefore the Cauchy sequence has limit $\mathcal{G}$. □

Remark: In [42] another distance is defined on the set of large circular filters, as $2\log(diam(\sup(\mathcal{F},\mathcal{G})-\log(diam(siam\mathcal{F})-\log(diam(\mathcal{G})$. This last distance defines the same topology on the set of large circular filters, but does not admit the same Cauchy sequences and does not expand the usual distance on the field L defined by the absolute value. As a consequence, the set of large circular filters appears to be complete for such a distance and therefore a sequence converging to a point $a \in L$ in the usual sense, actually has no limit for this distance.

Theorem 11.16: *Let B be a totally ordered subset of $\Phi(K)$. Then B admits an infimum $\mathcal{T}$ with respect to the other $\preceq$. Further, subset A of $\Phi(K)$ admits a supremum with respect to the order $\preceq$ if and only if it is bounded with respect to δ.*

Proof: Let B be a totally ordered subset of $\Phi(K)$ and let $r = \inf\{diam\mathcal{G} | \mathcal{G} \in B\}$. We can see that all sequences $(\mathcal{G}_n)_{n\in\mathbb{N}}$ of B satisfying $\lim_{n\to\infty} diam(\mathcal{G}_n) = r$ are Cauchy sequences and admit the same limit $\mathcal{T}$. Therefore $\mathcal{T}$ is clearly the infimum of B with respect to $\preceq$. Now, let A be a subset of $\Phi(K)$. If it admits a supremum with respect to the order $\preceq$, it is obviously bounded. Conversely, assume that A is bounded. We can clearly find a disk $d(b, s)$ such that all elements of A is secant with $d(b, s)$. Consequently, the circular filter of center b and diameter s surrounds all elements of A. Now, by Proposition 11.7, the set $A*$ of circular filters surrounding all elements of A is totally ordered, and therefore admits an infimum $\mathcal{S}$, and we have $\mathcal{G} \preceq \mathcal{S}$ for every $\mathcal{G} \in A$. Consequently, $\mathcal{S}$ is the supremum of A. □

Chapter 12

Rational Functions and Algebras $R(D)$

Notation: Throughout this chapter and in the following, K will denote an algebraically closed complete ultrametric field and D will be an infinite subset of K.

Rational without poles in a set D of the field K are the only material we handle to define a kind of holomorphic functions in D. But first, we have to know perfectly the properties of rational functions, with regards to the ultrametric structure of K.

Given a function f from D to K, $\|f\|_D$ denotes $\sup\{|f(x)| \mid x \in D\} \in [0, +\infty]$. And we denote by $R(D)$ (resp. $R_b(D)$) the K-algebra of all rational functions $h \in K(x)$ with no pole in D (resp. the K-algebra of all rational functions $h \in K(x)$ with no pole in D which are bounded in D).

Given $P(x) = \sum_{n=0}^{q} a_n x^n \in K[x]$, and $\mu \in \mathbb{R}$, we put

$$v(P, \mu) = \inf_{0 \leq j \leq q} v(a_n) + n\mu,$$

we denote by $N^+(P, \mu)$ the greatest of the integers j such that $v(a_j) + j\mu = \inf_{0 \leq j \leq q} v(a_n) + n\mu$, and by $N^-(P, \mu)$ the least of the integers j such that $v(a_j) + j\mu = \inf_{0 \leq j \leq q} v(a_n) + n\mu$.

By results of Chapters 4 in [30] (Lemmas 4.4, 4.6 and 4.8, Theorem 4.11 and Corollaries 4.12 and 4.13), we have the following Lemmas:

Lemma 12.1: *Let $P(x) \in K[x]$. Let $r \in \mathbb{R}_+$ and let $a \in K$ be such that $|a| \leq r$. Then $|P(x)|$ has a limit $\varphi_{a,r}(P)$ when $|x - a|$ approaches r but remains different from r.*

Moreover, if $P(x) = \sum_{j=0}^{n} \alpha_j (x-a)^j$, *then*

$$\varphi_{a,r}(P) = \max_{0 \le j \le n} |\alpha_j| r^j = \|P\|_{d(a,r)}.$$

Thus, $\varphi_{a,r}$ *belongs to* $Mult(K[x])$ *and has continuation to* $K(x)$*. Moreover, given* $h \in K(x)$, $\varphi_{a,r}$ *lies in* K *if and only if so does* r.

Corollary 12.2: *Let* $P(x) = \sum_{j=0}^{n} \alpha_j x^j \in K[x]$*. Then*

$$v(P, -\log r) = -\log(\varphi_{0,r}(P)).$$

Lemma 12.3: *Let* $h \in K(x)$ *and let* $r \in \mathbb{R}_+$*. For every* $a \in d(0,r)$ *we have* $\lim_{|x-a| \to r, |x-a| \neq 0} |h(x)| = \varphi_{a,r}(h)$*. Let* $x \in C(0,r)$*. If* h *has no zeros (resp. no poles) in the class of* x *in* $d(0,r)$ *then* $|h(x)| \ge \varphi_{0,r}(h)$ *(resp.* $|h(x)| \le \varphi_{0,r}(h)$*). If* h *has neither any zeros nor any poles in the class of* x *inside* $d(0,r)$*, then* $|h(x)| = \varphi_{0,r}(h)$.

Corollary 12.4: *Let* $h \in K(x) \setminus \{0\}$*. We have* $v(h(x)) = v(h, v(x))$ *for every* $x \in K$ *such that* h *has no zero* α *satisfying* $v(x-\alpha) > v(x)$ *and no pole* β *satisfying* $v(x-\beta) > v(x)$.

Theorem 12.5 (G. Garandel): *For every large circular filter* $\mathcal{F}$ *on* K*, for every rational function* $h(x) \in K(x)$, $|h(x)|$ *has a limit* $\varphi_{\mathcal{F}}(h)$ *along the filter* $\mathcal{F}$*. If* $\mathcal{F}$ *has center* 0 *and diameter* r*, then* $\varphi_{\mathcal{F}}(h) = |h|(r)$.

Lemma 12.6: *Let* $h \in K(x) \setminus \{0\}$*. The function in* μ $v(h,.)$ *is continuous and piecewise linear and has at each point* μ *a left-side derivative* $v'^l(h,\mu)$ *and a right-side derivative* $v'^r(h,\mu)$*. Moreover, if* $h \in K[x]$*, then* $v'^l(h,\mu) = N^-(h,\mu)$*, and* $v'^r(h,\mu) = N^+(h,\mu)$*. Let* $\nu \in \mathbb{R}$.

If $d(0,\omega^{-\nu})$ *contains* s *zeros and* t *poles of* h*, (taking multiplicities into account), and if* $C(0,\omega^{-\nu})$ *contains neither zeros nor poles, then* $v(h,.)$ *has a derivative equal to* $s-t$ *at* ν.

If $C(0,\omega^{-\nu})$ *contains* s *zeros and* t *poles of* h *(taking multiplicities into account), then we have* $v'^l(h,\nu) - v'^r(h,\nu) = s - t$*. Moreover, if the function* $v(f,\mu)$ *is not derivable at* ν*, then* ν *lies in* $v(K)$.

Lemma 12.7: *Let* $h \in K(x) \setminus \{0\}$ *have* s *zeros and* t *poles in* $d(0,r)$ *(taking multiplicities into account), and have neither zeros nor poles in* $\Gamma(0,r,r')$*. Then in* $\Gamma(0,r,r')$, $v(h(x))$ *is of the form* $A + (s-t)v(x)$*, with* $A = -\log(\varphi_{0,r}(h))$.

Corollary 12.8: *Let $h \in K(x) \setminus \{0\}$. If $d(0,r_1)$ contains s zeros and t poles of h, (taking multiplicities into account), and if $\Gamma(0,r_1,r_2)$ contains neither zeros nor poles of h, then $|h(x)| = \left(\frac{|x|}{r_2}\right)^{s-t}\varphi_{0,r_2}(h) = \left(\frac{|x|}{r_1}\right)^{s-t}\varphi_{0,r_1}(h)$, $\forall x \in \Gamma(0,r_1,r_2)$.*

Theorem 12.9: *Let $h \in K(x) \setminus \{0\}$. If $r \in |K|$, then $\varphi_{0,r}(h)$ lies in $|K|$. Conversely, if h has s zeros and t poles in $d(0,r)$, if $s \neq t$ with $\varphi_{0,r}(h) \in |K|$, then r lies $|K|$.*

Proof: The direct claim is an immediate consequence of Lemma 12.7. So we will show the converse claim and suppose that $\varphi_{0,\rho}(h) \in |K|$ and that $s \neq t$. If h admits zeros or poles in $C(0,r)$ then of course r lies in $|K|$. So, we assume that h has neither zeros nor poles in $C(0,r)$. Therefore we can find $\rho \in]0,r[\cap|K|$ such that h has neither any zeros nor any poles in $\Delta(0,\rho,r)$, hence h admits exactly s zeros and t poles in $d(0,\rho)$. And by Lemmas 12.3 and 12.8 we have $\varphi_{0,r}(h) = \varphi_{0,\rho}(h)\left(\frac{r}{\rho}\right)^{s-t}$. But since $\rho \in |K|$ and since $s-t \neq 0$, by the direct claim it is clear that $\frac{\rho^{s-t}}{\varphi_{0,r}(h)}$ also lies in $|K|$, and so does r. □

Notation: Henceforth, we denote by U the disk $d(0,1)$ in K. Given a disk $A = d(a,r)$ or $A = d(a,r^-)$, we put $\varphi_A = \varphi_{a,r}$.

Theorems 12.10 and 12.11 will be indispensable in Chapters 33, 34 and 35:

Theorem 12.10: *Let $h \in K(x)$ satisfy $\deg(h) > 0$. Then the set $h^{-1}(U) = \{x \in K \mid |h(x)| \leq 1\}$ is affinoid.*

Proof: Let $D = h^{-1}(U)$ and let $h = \frac{P}{Q}$, with P, $Q \in K[x]$ relatively prime, and of course $\deg(P) > \deg(Q)$. We first notice that D is bounded. Let $\Gamma(a,l',l'')$, with $a \in D$, be an empty annulus of D. Suppose that h has no zeros and no poles in $d(a,l')$. There exists $l > l'$ such that h admits no zeros and no poles in all $d(a,l)$, hence by Lemma 12.2 $|h(x)|$ is constant in all $d(a,l)$, hence $|h(x)| = |h(a)| \leq 1$ $\forall x \in d(a,l)$, thereby $d(a,l) \subset D$, a contradiction. Thus, for each empty annulus Λ of D, $\mathcal{I}(\Lambda)$ contains at least a zero or a pole of h. Consequently, D admits finitely many empty annuli and therefore by Theorem 8.13 D has finitely many infraconnected components.

Let E be an infraconnected component of D, of diameter r, and let $a \in E$. Let t (resp. s) be the number of poles (resp. zeros) of h inside $\widetilde{E}$. Since E is infraconnected, the circular filter of center a and diameter

r is secant with E and thereby $\varphi_{a,r}(h) \leq 1$. It is also secant with $K \setminus \widetilde{E}$, hence $\varphi_{a,r}(h) \geq 1$, therefore $\varphi_{a,r}(h) = 1$. On the other hand we have $s - t > 0$ because if $s - t \leq 0$, then by Lemma 12.7 there exists $r' > r$ such that $|h(x)| = \varphi_{a,r}(h)$ $\forall x \in \Gamma(a, r, r')$, a contradiction to the hypothesis $D \subset d(a, r)$. Consequently, by Lemma 12.7 the equality $\varphi_{a,r}(h) = 1$ implies $r \in |K|$.

Similarly, consider now a hole $T = d(b, \rho^-)$ of E. Let $l = \varphi_{b,\rho}(h)$. Since E is infraconnected, the circular filter of center b and diameter ρ is secant with E. and thereby $\varphi_{b,\rho}(h) \leq 1$. It is also secant with T, hence $\varphi_{b,\rho}(h) \geq 1$, therefore $\varphi_{b,\rho}(h) = 1$. Then by Lemma 12.3, T contains at least one pole of h. Thus, E has finitely many holes. Moreover, since $|h(x)| > 1$ $\forall x \in T$, by Lemma 12.8, inside T the number of poles of h is strictly greater than the number of zeros. Consequently by Lemma 12.7 ρ lies in $|K|$. □

Theorem 12.11: *Let D be affinoid. There exists $h \in R(D)$ such that $\deg(h) > 0$, $h(D) = U$, $D = h^{-1}(U)$.*

Proof: Suppose first D is infraconnected. Then D is of the form $d(a, r_0) \setminus \bigcup_{j=1}^{n} d(b_j, r_j^-)$, with $|b_i - b_j| \geq \max(r_i, r_j)$, $\forall i \neq j$, $b_j \in d(a, r_0)$ $r_j \leq r_0$ $\forall j = 1, \ldots, n$, $r_j \in |K|$ $\forall j = 0, \ldots, n$. For each $j = 0, \ldots, n$, we can take $\lambda_j \in K$ such that $|\lambda_j| = r_j$. Let $h(x) = \frac{x-a}{\lambda_0} + \sum_{j=1}^{n} \frac{\lambda_j}{x-b_j}$. Outside $d(b-j, r_j^-)$, we have $|\frac{\lambda_j}{x-b_j}| \leq 1$, inside $d(b-j, r_j^-)$, we have $|\frac{\lambda_j}{x-b_j}| > 1$. Inside $d(a, r_0)$ we have $|\frac{x-a}{\lambda_0}| \leq 1$, outside $d(a, r_0)$ we have $|\frac{x-a}{\lambda_0}| > 1$. Consequently, we check that $|h(x)| \leq 1$ if and only if $x \in D$, so we have constructed h when D is infraconnected. Moreover, by construction, we check that $\lim_{|x| \to \infty} |h(x)| = +\infty$, so $\deg(h) > 0$.

We now consider the general case and denote by $D_1, \ldots, D_q$ the infraconnected components of D. For each $i = 1, \ldots, q$, there exists $h_i \in R(D_i)$ such that $|h(x)| \leq 1$ if and only if $x \in D_i$. We then put $g = \sum_{i=1}^{n} \frac{1}{h_i}$ and $h = \frac{1}{g}$. We can easily check that $|g(x)| \geq 1$ if and only if $x \in D$. Moreover, we notice that for each $i = 1, \ldots, n$ we have $\lim_{|x| \to \infty} |h_i(x)| = +\infty$, hence $\lim_{|x| \to \infty} |g(x)| = 0$, therefore $\lim_{|x| \to \infty} |h(x)| = +\infty$. So, $\deg(h) > 0$ and h satisfies our claim. □

According to Theorem 9.3 in [30] we have Theorem 12.12:

Theorem 12.12: *$R_b(D) = R(D)$ if and only if D is closed and bounded. Moreover, if D is closed and bounded, $\| \cdot \|_D$ is a semi-multiplicative ultrametric norm of K-algebra on $R_b(D)$.*

Chapter 13

Simple Convergence on *Mult*($K[x]$)

As we saw in Chapter 5, $Mult(K[x])$ is provided with the topology of simple convergence. On the other hand, according to Chapter 11, circular filters which characterize the elements of $Mult(K[x])$, are provided with a distance. Thus, $Mult(K[x])$ admits two topologies that we shall compare.

Notation: We denote by Ψ the mapping Ψ from $\Phi(K)$ into $Mult(K[x])$ defined as $\Psi(\mathcal{F}) = \varphi_{\mathcal{F}}$.

Throughout the chapter, D will denote a subset of K and $\mathcal{U}_D$ will denote the topology of uniform convergence on D.

The characterization of all absolute values on $K(x)$ is given in [32, 34 and 30]. According to Theorems 4.14 and 10.5 in [30], we have this theorem:

Theorem 13.1 (Garandel-Guennebaud): *The mapping Ψ from $\Phi(K)$ into $Mult(K[x])$ defined as $\Psi(\mathcal{F}) = \varphi_{\mathcal{F}}$ is a bijection. Moreover, the restriction Ψ' of Ψ to $\Phi'(K)$ is a bijection from $\Phi'(K)$ onto the set of multiplicative norms on $K[x]$, and therefore onto the set of multiplicative norms on $K(x)$. Moreover, if $\mathcal{F}$ has center a and diameter r, then $\varphi_{\mathcal{F}}(h) = \varphi_{a,r}(h)\ \forall h \in K(x)$.*

Definition: A multiplicative semi-norm on $K[x]$ defined by a punctual circular filter of limit a will be called *a punctual multiplicative semi-norm* and will be denoted by φ_a.

Remarks: The punctual multiplicative semi-norm φ_a is just of the form $\phi(P) = |P(a)|$, with $a \in K$.

In a field F which is complete but not algebraically closed, such as $\mathbb{Q}_p$, we have to consider the (algebraically closed) completion Ω of an algebraic closure. Then every element ψ of $Mult(\Omega[x])$ has a restriction to $F[x]$ which obviously belongs to $Mult(F[x])$. And conversely, every element of $Mult(F[x])$ admits extensions to $Mult\Omega[x])$. Therefore, we have a surjection from the set of circular filters on Ω onto $Mult(F[x])$. But this surjection is not injective: if a and b are conjugated over F, the circular filters of centers a and b and of same diameter r define the same absolute values on $F[x]$.

Consider a circular filter $\mathcal{F}$ of center a and diameter $r \in |K|$. The point of generic disk consists of considering a transcendental element $t_{\mathcal{F}}$ over K, laying in the circle $|x-a| = r$, such that $|t_{\mathcal{F}} - u| = r \ \forall u \in K$ such that $|u-a| \leq r$ [15]. Consequently we can check that $\varphi_{\mathcal{F}}$ defines an absolute value $| \cdot |$ on $K(t_{\mathcal{F}})$ satisfying $\varphi_{\mathcal{F}}(P) = |P(t_{\mathcal{F}})$.

Theorem 13.2: *$Mult_a(K[x]) = Mult_m(K[x])$ is dense in $Mult(K[x])$ with respect to the topology of simple convergence.*

Proof: It is well known that all maximal ideal of $K[x]$ have codimension 1, hence $Mult_a(K[x]) = Mult_m(K[x])$. Now, let $\varphi_{\mathcal{F}} \in Mult(K[x])$. Let $P_1, \ldots, P_n \in K[x]$ and let $\epsilon > 0$. For each $j = 1, \ldots, n$, there exists $E_j \in \mathcal{F}$ such that $|P_j(x) - \varphi_{\mathcal{F}}(P_j)|_\infty \leq \epsilon \ \forall x \in E_j$. Let $E = \cap_{j=1}^n E_j$. Then E lies in $\mathcal{F}$ and we have $|P_j(x) - \varphi_{\mathcal{F}}(P_j)|_\infty \leq \epsilon \ \forall x \in E$, which shows that φ_a belongs to the neighborhood $\mathcal{V}(\varphi_{\mathcal{F}}, P_1, \ldots, P_n, \epsilon)$ for every $a \in E$. □

Theorem 13.3: *Let D be infraconnected, closed, bounded. Then $Mult(R(D), \| \cdot \|_D)$ is sequentially compact with respect to the topology of simple convergence.*

Proof: Let $(\psi_n)_{n\in\mathbb{N}}$ be a sequence of $Mult(R(D), \| \cdot \|_D)$. Since $Mult(R(D), \| \cdot \|_D)$ is compact, the sequence $(\psi_n)_{n\in\mathbb{N}}$ admits a point of adherence $\varphi_{\mathcal{F}}$. Let $r = diam(\mathcal{F})$. If $\mathcal{F}$ has no center, or if it has a center and if $r \notin |K|$, by Lemma 10.2 it has a countable base, hence there does exist a subsequence of the sequence $(\psi_n)_{n\in\mathbb{N}}$ converging to $\varphi_{\mathcal{F}}$. Now suppose that $\mathcal{F}$ has a center a and that $r \in |K|$. Suppose that we can extract from the sequence (ψ_n) a subsequence $(\psi_{m_t})_{t\in\mathbb{N}}$, with $\psi_{m_t} = \varphi_{\mathcal{F}_t}$, $(t \in \mathbb{N})$, such that for a certain fixed $b \in d(a,r)$, each $\mathcal{F}_t$ is (b, r_t)-approaching with $r_t \neq r$. Consider now $x - b$. For convenience, we put $\phi_t = \psi_{m_t}$ $(t \in \mathbb{N})$. By hypothesis we can reextract from this sequence $(\phi_t)_{t\in\mathbb{N}}$ a new subsequence $(\phi_{t_q})_{q\in\mathbb{N}}$ such that $\lim_{q\to\infty} \phi_{t_q}(x-b) = \varphi_{\mathcal{F}}(x-b)$. This clearly proves that $\lim_{q\to\infty} r_{t_q} = r$, whereas $r_{t_q} \neq r$. Consequently, for every element A of $\mathcal{F}$

we can find a rank $s \in \mathbb{N}$ such that $\mathcal{F}_{t_s}$ is secant with A. Therefore, for every $\epsilon > 0$ and for every $f \in R(D)$, we can find a rank $s \in \mathbb{N}$ such that $|\phi_{t_s}(f) - \varphi_{\mathcal{F}}(f)|_\infty \leq \epsilon$, hence the sequence (ϕ_{t_s}) converges to $\varphi_{\mathcal{F}}$.

Finally, suppose that we cannot find $b \in d(a,r)$ and a subsequence $(\psi_{m_t})_{t\in\mathbb{N}}$, with $\psi_{m_t} = \varphi_{\mathcal{F}_t}$, $(t \in \mathbb{N})$, such that for a certain fixed $b \in d(a,r)$, each $\mathcal{F}_t$ be (b, r_t)-approaching with b and diameter $r_t \neq r$. We first notice that there exists $r' > r$ such that none of the $\varphi_{\mathcal{F}_t}$ is secant with $\Gamma(a, r, r')$. In the same way, for every $c \in d(a,r)$, there exists $\rho(c) \in]0, r[$ such that none of the $\phi_{\mathcal{F}_t}$ is secant with $\Gamma(a, \rho, r)$. Let $(c_t)_{t\in\mathbb{N}}$ be a sequence in $d(a,r)$ such that $|c_j - c_k| = r\ \forall j \neq k$. We can easily construct a decreasing sequence of affinoid sets $(A_t)_{t\in\mathbb{N}}$ of the form

$$A_t = d(a, r') \Big\backslash \left(\bigcup_{j=1}^{t} d(c_j, \rho(c_j)^-) \right).$$

For each $t \in \mathbb{N}$, let $\epsilon_t = \min(\frac{1}{t}, r' - r, r - \rho(c_1), \ldots, r - \rho(c_t))$. By hypothesis, for each $t \in \mathbb{N}$, there exists $n_t \in \mathbb{N}$ such that $|\psi_{n_t}(x - c_t) - \varphi_{\mathcal{F}}(x - c_t)|_\infty < \epsilon_t$, hence, putting $\psi_{n_t} = \varphi_{\mathcal{F}_t}$, we can see that $\mathcal{F}_t$ is secant with $d(a, r')$ but is not secant with $\Gamma(a, r, r')$ and is not secant with $\cup_{j=1}^{t} d(c_j, r^-)$. Consequently, $\mathcal{F}_t$ is secant with a class of $d(a,r)$ different from all $d(c_j, r)\ \forall j = 1, \ldots, t$. So, we can construct a sequence $(b_t)_{t\in\mathbb{N}}$ such that $|b_j - a| = |b_j - b_k| = r\ \forall j \neq k$ and such that $\mathcal{F}_t$ is secant with $d(b_t, r^-)$.

Now, let $f \in R(D)$. For each $t \in \mathbb{N}$, since $\mathcal{F}_t$ is secant with $d(b_t, r^-)$, we can find $a_t \in d(b_t, r^-)$ such that $|\psi_{n_t}(f) - |f(a_t)||_\infty < \epsilon_t$. Then the sequence $(a_t)_{t\in\mathbb{N}}$ is thinner than $\mathcal{F}$, and therefore $\lim_{t\to\infty} |f(a_t)| = \varphi_{\mathcal{F}}(f)$. But of course $\lim_{t\to\infty} |f(a_t)| = \lim_{t\to\infty} \psi_{n_t}(f)$, so the sequence (ψ_{n_t}) converges to $\varphi_{\mathcal{F}}$ again. □

Notation: In Chapter 11 the set of circular filters was provided with an order relation $\preceq$ that makes it a tree, and we have denoted by $\prec$ the strict order associated to $\preceq$. On the other hand, we can apply to $Mult(K[x])$ the usual order $\leq$ on functions with values in $\mathbb{R}$ as $\varphi_{\mathcal{F}} \leq \varphi_{\mathcal{G}}$ if $\varphi_{\mathcal{F}}(P) \leq \varphi_{\mathcal{G}}(P)\ \forall P \in K[x]$, and we denote by $<$ the strict order associated to $\leq$, as $\varphi_{\mathcal{F}} < \varphi_{\mathcal{G}}$ if $\varphi_{\mathcal{F}}(P) \leq \varphi_{\mathcal{G}}(P)\ \forall P \in K[x]$ and $\varphi_{\mathcal{F}} \neq \varphi_{\mathcal{G}}$.

Theorem 13.4: *Let $\varphi_{\mathcal{F}}, \varphi_{\mathcal{G}} \in Mult(K[x])$. Then $\varphi_{\mathcal{F}} < \varphi_{\mathcal{G}}$ if and only if $\mathcal{F} \prec \mathcal{G}$.*

Proof: Let $\mathcal{F}, \mathcal{G} \in \Phi(K)$. Suppose first that they satisfy $\mathcal{F} \prec \mathcal{G}$ and let $s = diam(\mathcal{G})$. By Corollary 11.2 $\mathcal{F}$ is secant with a disk $d(a,r)$ strictly included in $d(b,s)$. Let $P \in K[x]$. We have

$$\varphi_{\mathcal{F}}(P) \le \|P\|_{d(a,r)} \le \|P\|_{d(a,s)} = \varphi_{\mathcal{G}}(P)\ \forall P \in K[x].$$

Now, let $Q(x) = x - a$. Then

$$\varphi_{\mathcal{F}}(Q) \le \|Q\|_{d(a,r)} = r \text{ and } \varphi_{\mathcal{G}}(Q) = \|Q\|_{d(a,s)} = s,$$

hence $\varphi_{\mathcal{F}}(Q) < \varphi_{\mathcal{G}}(Q)$. Suppose that $\mathcal{F}$ and $\mathcal{G}$ are not comparable with respect to $\preceq$. By Proposition 11.8, there exist disks $F = d(a,r) \in \mathcal{F}$ and $G = d(b,s) \in \mathcal{G}$ such that $F \cap G = \emptyset$ and $\delta(F,G) > \max(diam(\mathcal{F}), diam(\mathcal{G}))$. Consider $P(x) = x - a$ and $Q(x) = x - b$. Then

$$\varphi_{\mathcal{F}}(P) \le \|P\|_F = r \text{ and } \varphi_{\mathcal{G}}(P) = \|P\|_G = |b-a|,$$

hence $\varphi_{\mathcal{F}}(P) < \varphi_{\mathcal{G}}(P)$. In the same way, $\varphi_{\mathcal{G}}(Q) \le \|P\|_G = s$ and $\varphi_{\mathcal{F}}(Q) = \|P\|_F = |b-a|$, hence $\varphi_{\mathcal{G}}(Q) < \varphi_{\mathcal{F}}(Q)$. Thus, neither $\varphi_{\mathcal{F}} \le \varphi_{\mathcal{G}}$ nor $\varphi_{\mathcal{G}} \le \varphi_{\mathcal{F}}$ is true whenever $\mathcal{F}$ and $\mathcal{G}$ are not comparable. This finishes proving that $\varphi_{\mathcal{F}} < \varphi_{\mathcal{G}}$ if and only if $\mathcal{F} \prec \mathcal{G}$. □

Corollary 13.5: *Ψ is a strictly increasing bijection from $\Phi(K)$ onto $Mult(K[x])$ and Ψ^{-1} is a strictly increasing bijection from $Mult(K[x])$ onto $\Phi(K)$.*

Corollary 13.6: *Let D be closed. The restriction of Ψ the set of $\mathcal{F} \in \Phi(K)$ which are not converging to a point of $K \setminus D$ is a bijection from this set onto $Mult(R(D))$.*

According to Theorem 10.4 in [30] we have Theorem 13.7:

Theorem 13.7 (Garandel): *Let $\mathcal{F}$ be a large circular filter on K. Then $\varphi_{\mathcal{F}}$ belongs to $Mult(R(D), \mathcal{U}_D)$ if and only if $\mathcal{F}$ is secant with D.*

Corollary 13.8: *The mapping from $\Phi(D)$ to $Mult(R(D))$ which associates to each circular filter $\mathcal{F}$ secant with D the multiplicative semi-norm $\varphi_{\mathcal{F}}$, is a bijection from $\Phi(D)$ onto $Mult(R(D), \mathcal{U}_D)$.*

Definitions and notation: According to Chapter 2, the function *diam* defines two equivalent distances on the set of circular filters, in particular $\delta(\mathcal{F},\mathcal{G}) = diam(\sup(\mathcal{F},\mathcal{G}) - \min(diam(\mathcal{F}), diam(\mathcal{G}))$. Henceforth, we will apply this distance to $Mult(K[x])$ and to $Mult(R(D), \mathcal{U}_D)$ as well as to $\Phi(K)$ or $\Phi(D)$ by putting $\delta(\varphi_{\mathcal{F}}, \varphi_{\mathcal{G}}) = \delta(\mathcal{F},\mathcal{G})$. Thus, $Mult(K[x])$ and $Mult(R(D), \mathcal{U}_D)$ are provided with a metric topology defined by δ.

Given $h = \frac{P}{Q} \in K(x)$, with $P,\ Q \in K[x]$, we put $\deg(h) = \deg(P) - \deg(Q)$, and $\deg(h)$ will be called *the degree* of h.

Theorem 13.9: *Let $\varphi_{\mathcal{F}} \in Mult(K(x))$. Let $f \in K(x)$, and let $\epsilon > 0$. There exists an affinoid subset E of K of diameter $l > diam(\mathcal{F})$, which belongs to $\mathcal{F}$, such that $||f(x)| - \varphi_{\mathcal{F}}(f)|_\infty \leq \epsilon$, $\forall x \in E$.*

Proof: If $\mathcal{F}$ has no center, there exists a disk $d(a,l) \in \mathcal{F}$, with $r \in |K|$, containing neither zeros nor poles of f, therefore by Lemma 12.3 $|f(x)|$ is a constant equal to $\varphi_{\mathcal{F}}(f)$ in $d(a,r)$, so our claim is obvious. Now, suppose that $\mathcal{F}$ is the circular filter of center a and diameter r. Let $\Lambda_1, \ldots, \Lambda_q$ be the classes of $d(a,r)$ containing at least one zero or one pole of f. By Lemma 12.3 $|f(x)|$ is a constant equal to $\varphi_{\mathcal{F}}(f)$ in $d(a,r) \setminus (\cup_{j=1}^q \Lambda_j)$. Consider a class $\Lambda_j = d(a_j, r^-)$, and let s_j (resp. t_j) be the number of zeros (resp. poles) of f inside Λ_j, and let s_0 (resp. t_0) be the number of zeros (resp. poles) of f in all $d(a,r)$. Let $\rho \in]0,r[\cap|K|$ be such that $\left|\left(\frac{r}{\rho}\right)^{s_j - t_j} - 1)\right|\varphi_{\mathcal{F}}(f) \leq \epsilon \ \forall j = 0, \ldots, q$. Let $l = \frac{r^2}{\rho}$ and let $E = d(a,l) \setminus \cup_{j=1}^q d(a_j, \rho^-)$. By Lemma 12.7 and Corollary 12.8 we can check that the inequality $|\ |f(x)| - \varphi_{\mathcal{F}}(f)|_\infty \leq \epsilon$ holds in all E. Since $\rho < r$, E is an affinoid set which belongs to $\mathcal{F}$. Moreover, by definition, $l > diam(\mathcal{F})$. □

Theorem 13.10: *Let D be a closed bounded set and $\varphi_{\mathcal{F}} \in Mult(R(D), \| \cdot \|_D)$. There exists a base of neighborhoods of $\varphi_{\mathcal{F}}$ in $Mult(R(D), \| \cdot \|_D)$, with respect to the topology of simple convergence, consisting of the family of sets of the form $Mult(R(E \cap D), \| \cdot \|_{E\cap D})$, where E is a $\mathcal{F}$-affinoid.*

Proof: Let $f_1, \ldots, f_q \in R(D)$ and consider a neighborhood W of $\varphi_{\mathcal{F}}$: we will show that it contains a neighborhood of the form $Mult(R(E \cap D), \| \cdot \|_{E\cap D})$, where E is a $\mathcal{F}$-affinoid. Indeed, by Lemma 13.9 there exist affinoid sets $E_j \in \mathcal{F}$, $(1 \leq j \leq q)$ such that $|\ |f_j(x)| - \varphi_{\mathcal{F}}(f_j)|_\infty \leq \epsilon \ \forall x \in E_j$ $(1 \leq j \leq q)$. Then $\bigcap_{j=1}^q E_j$ is an infraconnected affinoid set which belongs to $\mathcal{F}$. We can put $E = \bigcap_{j=1}^q E_j$. So, we have an affinoid set F such that

$$|\ |f_j(x)| - \varphi_{\mathcal{F}}(f_j)|_\infty \leq \epsilon, \ \forall x \in E_j \ (1 \leq j \leq q),$$

and therefore

$$|\ |\psi(f_j)| - \varphi_{\mathcal{F}}(f_j)|_\infty \leq \epsilon, \ \forall x \in E_j \ (1 \leq j \leq q), \quad \forall \psi \in Mult(R(E \cap D), \| \cdot \|_{E\cap D}).$$

Now, by Lemma 10.2 F contains a $\mathcal{F}$-affinoid E. Then $Mult(R(E), \| \cdot \|_E)$ is a neighborhood of $\varphi_{\mathcal{F}}$ in $Mult(K(x))$ and $Mult(R(E \cap D), \| \cdot \|_{E\cap D})$ is a neighborhood of $\varphi_{\mathcal{F}}$ in $Mult(R(D), \| \cdot \|_D)$. □

Corollary 13.11: *Let $\varphi_{\mathcal{F}} \in Mult(K[x])$. There exists a base of neighborhoods of $\varphi_{\mathcal{F}}$ in $Mult(K[x])$ consisting of the family of sets of the form $Mult(R(E), \| \cdot \|_E)$, where E is a $\mathcal{F}$-affinoid.*

Corollary 13.12: *$Mult(K[x])$ is locally compact and locally sequentially compact.*

Proof: Let $\varphi_{\mathcal{F}} \in Mult(K[x])$ and let E be an affinoid set such that $Mult(R(E), \| \cdot \|_E)$ is a neighborhood of $\varphi_{\mathcal{F}}$ in $Mult(K[x])$. Then $Mult(R(E), \| \cdot \|_E)$ is a compact neighborhood of $\varphi_{\mathcal{F}}$. Moreover, it is a sequentially compact neighborhood. □

Notation: Given a set $D \subset K$, we will denote by $\Sigma(D)$ the set of $\varphi_{\mathcal{F}} \in Mult(K[x])$ such that $\mathcal{F}$ is D-bordering, and by $\Sigma_0(D)$ the set of $\varphi_{\mathcal{F}} \in Mult(K[x])$ such that $\mathcal{F}$ is strictly D-bordering.

Chapter 14

Topologies on $Mult(K[x])$

In Chapter 11 we defined a metric δ on $\Phi(K)$ and we showed in Chapter 12 that $\Phi(K)$ is complete with respect to this metric. In Chapter 14 we saw that $\Phi(K)$ is in bijection with $Mult(K[x])$ which is provided with the topology of simple convergence. Henceforth, thanks to this one to one correspondence, we will consider that both $\Phi(K)$ and $Mult(K[x])$ are provided with both topologies.

Theorem 14.1: *Let D be closed and bounded. Then the boundary of $Mult(R(D), \| \cdot \|_D)$ inside $Mult(K[x])$, with respect to the topology of simple convergence, is equal to $\Sigma(D)$.*

Proof: Let $\mathcal{F}$ be D-bordering. Let W be a neighborhood of $\varphi_{\mathcal{F}}$ in $Mult(K[x])$. By Theorem 13.10 there exists an affinoid set $E \in \mathcal{F}$ such that $Mult(R(E), \| \cdot \|_E) \subset W$. Since $E \cap (K \setminus D) \neq \emptyset$, for all $a \in E \cap (K \setminus D)$, φ_a belongs to $Mult(R(E), \| \cdot \|_E)$, which shows that $\varphi_{\mathcal{F}}$ belongs to the closure of $Mult(K[x]) \setminus Mult(R(D), \|.\|_D)$. Hence,

$$\Sigma(D) \subset Mult(K[x]) \setminus Mult(R(D), \|.\|_D).$$

Conversely, let $\varphi_{\mathcal{F}} \in Mult(R(D), \|.\|_D)$ belong to the closure of $Mult(K[x]) \setminus Mult(R(D), \|.\|_D)$ in $Mult(K[x])$. Let $E \in \mathcal{F}$ be an affinoid set. Then $Mult(R(E), \| \cdot \|_E)$ is a neighborhood of $\varphi_{\mathcal{F}}$, hence it contains an open neighborhood W of $\varphi_{\mathcal{F}}$. Therefore

$$W \cap (Mult(K[x]) \setminus Mult(R(D), \|.\|_D)) \neq \emptyset.$$

Moreover,

$$W \cap (Mult(K[x]) \setminus Mult(R(D), \|.\|_D))$$

is open in $Mult(K[x])$ because so are W and $Mult(K[x]) \setminus Mult(R(D), \|.\|_D)$. Since $Mult_a(K[x])$ is dense inside $Mult(K[x])$, there exists

$$\varphi_a \in (Mult_a(K[x]) \setminus Mult(R(D), \|.\|_D)) \cap Mult(R(E), \| \cdot \|_E).$$

Consequently, a lies in E, which proves that $\mathcal{F}$ is secant with $K \setminus D$. Therefore the boundary of $Mult(R(D), \| \cdot \|_D)$ inside $Mult(K[x])$ is included in $\Sigma(D)$, and finally these two sets are equal. □

Remark: Since the topology of simple convergence is weaker than the δ-topology, the boundary of $Mult(R(D), \| \cdot \|_D)$ inside $Mult(K[x])$, with respect to the δ-topology is obviously included in the boundary of $Mult(R(D), \| \cdot \|_D)$ inside $Mult(K[x])$, with respect to the topology of simple convergence. The equality does not hold in the general case.

Theorem 14.2: *Let D be closed, bounded and infraconnected. Then $\Sigma_0(D)$ is included in the boundary of $Mult(R(D), \| \cdot \|_D)$ inside $Mult(K[x])$, with respect to the δ-topology. Moreover, the boundary of $Mult(R(D), \| \cdot \|_D)$ inside $Mult(K[x])$, with respect to the δ-topology, is equal to $\Sigma(D)$ if and only if for every $\varphi_{\mathcal{G}} \in \Sigma(D) \setminus \Sigma_0(D)$, there exists either a monotonous distances holes sequence or an equal distances holes sequence which is thinner than $\mathcal{G}$, whose superior gauge is equal to its diameter.*

Proof: Let $\widetilde{D} = d(a, r)$, with $a \in D$. Let $\varphi_{\mathcal{G}} \in \Sigma(D)$. Suppose first that $\mathcal{G}$ is the peripheral of D, hence $\varphi_{\mathcal{G}}$ lies in the closure of $\{\varphi_{a,s} \mid s > r\}$ for both topologies, hence it lies in the closure of $Mult(K[x]) \setminus Mult(R(D), \| \cdot \|_D)$ for both topologies. In the same way, $\varphi_{\mathcal{G}}$ lies in the closure of $\{\varphi_{a,s} \mid 0 \leq s \leq r\}$ for both topologies. But since D is infraconnected, for every $s \in [0, r]$, by Proposition 10.4 the circular filter of center a and diameter s is secant with D, and then, by Corollary 13.8, $\varphi_{a,s}$ belongs to $Mult(R(D), \| \cdot \|_D)$. Consequently, $\varphi_{\mathcal{G}}$ lies in the closure of $Mult(R(D), \| \cdot \|_D)$ for both topologies. This shows that $\varphi_{\mathcal{G}}$ belongs to the boundary of $Mult(R(D), \| \cdot \|_D)$ inside $Mult(K[x])$ with respect to the δ-topology.

Similarly, suppose now that $\mathcal{G}$ is the peripheral of a hole $T = d(b, l^-)$ of D. Then $\varphi_{\mathcal{G}}$ lies in the closure of $\{\varphi_{b,s} \mid s < l\}$ for both topologies, hence it lies in the closure of $Mult(K[x]) \setminus Mult(R(D), \| \cdot \|_D)$ for both topologies. In the same way, $\varphi_{\mathcal{G}}$ lies in the closure of $\{\varphi_{a,s} \mid l \leq s \leq r\}$ for both topologies. But since D is infraconnected, for every $s \in [l, r]$, by Proposition 10.4 the circular filter of center b and diameter s is secant with D, and then, by Corollary 13.8, $\varphi_{b,s}$ belongs to $Mult(R(D), \| \cdot \|_D)$. Consequently, $\varphi_{\mathcal{G}}$ lies in the closure of $Mult(R(D), \| \cdot \|_D)$ for both topologies and therefore belongs to the boundary of $Mult(R(D), \| \cdot \|_D)$ inside $Mult(K[x])$ with respect to the δ-topology.

Now, let $\varphi_{\mathcal{G}} \in \Sigma(D) \setminus \Sigma_0(D)$. Let $r = diam(\mathcal{G})$. Suppose first that there exists a monotonous holes sequence, or an equal distances holes sequence, thinner than $\mathcal{G}$, whose superior gauge is equal to its diameter. Let $(D_n)_{n\in\mathbb{N}}$ be a base of $\mathcal{G}$ consisting of affinoid sets. Then of course $\lim_{n\to\infty} diam(D_n) = r$. Thus, there exists a monotonous distances holes sequence or an equal distances holes sequence, $(T_n)_{n\in\mathbb{N}}$ such that $\lim_{n\to\infty} diam(T_n) = r$, and $T_n \subset D_n$ $\forall n \in \mathbb{N}$. For each $n \in \mathbb{N}$, let $T_n = d(b_n, r_n^-)$, let $s_n = diam(D_n)$ and $\lambda_n = \delta(\varphi_{\mathcal{G}}, \varphi_{b_n,r_n})$. Assume first that the monotonous sequence (T_n) is decreasing. Since the circular filter $\mathcal{F}_n$ peripheral of T_n is secant with D_n, we have $diam(\sup(\mathcal{F}_n, \mathcal{G})) \leq s_n$. Consequently, by definition of the distance δ, we can check that $\lambda_n \leq \max(s_n - r_n, s_n - r)$, and finally $\lim_{n\to\infty} \lambda_n = 0$. Hence $\mathcal{G}$ is a point of adherence of the sequence $(\mathcal{F}_n)$ with respect to the δ-topology, and therefore belongs to the closure of $Mult(K[x]) \setminus Mult(R(D), \| \cdot \|_D)$ inside $Mult(K[x])$. Since it belongs to $Mult(R(D), \| \cdot \|_D)$, it belongs the boundary of $Mult(R(D), \| \cdot \|_D)$ inside $Mult(K[x])$ with respect to the δ-topology. Assume now that the sequence (T_n) is an increasing holes sequence, or an equal distances holes sequence. Then each filter $\mathcal{F}_n$ is surrounded by $\mathcal{G}$. Consequently, $\lambda_n = r - r_n$, and therefore the conclusion is immediate.

Finally, suppose now that there exist neither any monotonous holes sequence nor any equal distances holes sequence whose superior gauge is equal to its diameter. Then, we can find $\lambda < r$ and an affinoid set $B \in \mathcal{G}$ such that every hole of D included in B has a diameter $\rho \leq \lambda$. Let $B = d(a, s_0) \setminus \cup_{j=1}^{q} d(a_j, s_j^-)$ and let $\varphi_{\mathcal{F}} \in Mult(K[x]) \setminus Mult(R(D), \| \cdot \|_D)$. Either $\mathcal{F}$ is secant with $K \setminus (d(a, r_0) \cup (\cup_{\leq j \leq q} d(a_j, r_j^-)))$, and then we check that $\delta(\mathcal{G}, \mathcal{F}) \geq \min_{0\leq j\leq q} |r - s_j|_\infty$, or it is secant with a hole T_k included in B, and then $\delta(\mathcal{G}, \mathcal{F}) \geq r - \lambda$. Thus we have proven that for every

$$\varphi_{\mathcal{F}} \in Mult(K[x]) \setminus Mult(R(D), \| \cdot \|_D),$$

we have

$$\delta(\mathcal{G}, \mathcal{F}) \geq \min\left(\min_{0\leq j\leq q} |r - s_j|_\infty, |r - \lambda|_\infty\right).$$

This finishes proving that $\varphi_{\mathcal{G}}$ does not belong to the boundary of $Mult(R(D), \| \,.\, \|_D)$ inside $Mult(K[x])$ with respect to the δ-topology. $\square$

Theorem 14.3: *Let D be a closed bounded set. The topology of simple convergence is weaker than the δ-topology on $Mult(R(D), \| \,.\, \|_D)$. If at least one large circular filter on K is secant with D, then $Mult(R(D), \| \,.\, \|_D)$ is not compact for the δ-topology. However, on a totally ordered subset of $Mult(R(D), \| \,.\, \|_D)$ the two topologies are equal.*

Proof: Let $\varphi_{\mathcal{F}} \in Mult(R(D), \| \,.\, \|_D)$. First we will show that every neighborhood of $\varphi_{\mathcal{F}}$, with respect to the simple convergence topology, which is of the form $Mult(R(E \cap D), \| \,.\, \|_{E \cap D})$, where E is an affinoid set, contains a neighborhood of $\varphi_{\mathcal{F}}$ with respect to the δ-topology. Let $r = diam(\mathcal{F})$, and let l be the maximum of the diameters of holes of E. For every $\mathcal{G} \in Mult(R(D), \| \,.\, \|_D)$ such that $\delta(\mathcal{F}, \mathcal{G}) \leq r - l$, then $\mathcal{G}$ is secant with E, and therefore lies in $Mult(R(E \cap D), \| \,.\, \|_{E \cap D})$. This shows that $Mult(R(E \cap D), \| \,.\, \|_{E \cap D})$ is a neighborhood of $\varphi_{\mathcal{F}}$ with respect to the δ-topology. But by Theorem 13.10 the family of neighborhoods of $\varphi_{\mathcal{F}}$, for the simple convergence topology inside $Mult(R(D), \| \,.\, \|_D)$, of the form $Mult(R(E \cap D), \| \,.\, \|_{E \cap D})$, with E an affinoid set, forms a base of neighborhoods of $\varphi_{\mathcal{F}}$ for the simple convergence topology: this shows that every neighborhood of $\varphi_{\mathcal{F}}$ for the simple convergence topology is a neighborhood of $\varphi_{\mathcal{F}}$ for the δ-topology, and therefore, the simple convergence topology is weaker than the δ-topology.

Suppose that at least one large circular filter $\mathcal{F}$ is secant with D. There exists a sequence (a_n) in D which is thinner than $\mathcal{F}$, and we know that the sequence φ_{a_n} converges to $\varphi_{\mathcal{F}}$ with respect to the topology of simple convergence. On the other hand, the sequence satisfies $\lim_{n \to \infty} |a_{n+1} - a_n| = diam(\mathcal{F})$. But since $\delta(\varphi_a, \varphi_b) = |a - b|$, the sequence φ_{a_n} satisfies $\lim_{n \to \infty} \delta(\varphi_{a_{n+1}}, \varphi_{a_n}) = diam(\mathcal{F})$, and therefore it has no limit in $Mult(R(D), \| \,.\, \|_D)$. Consequently, if D admits at least one large circular filter, the simple convergence topology is strictly weaker than the δ-topology.

Suppose now that a subset S of $Mult(R(D), \| \,.\, \|_D)$ is totally ordered with respect to the order $\preceq$. Let $\varphi_{\mathcal{F}} \in Mult(R(D), \| \,.\, \|_D)$ and consider a neighborhood of $\varphi_{\mathcal{F}}$ with respect to the δ-topology, of the form $B = \{\psi \in Mult(R(D), \| \,.\, \|_D) \mid \delta(\psi, \varphi_{\mathcal{F}}) \leq \epsilon\}$. Let $r = diam(\mathcal{F})$. If $\mathcal{F} = \inf(S)$, we denote by E the unique disk in K of diameter $r + \epsilon$, such that $\mathcal{F}$ is secant with E, and we put $s = r - \epsilon$. Else, since there exists $\mathcal{F}_0 \in S$ such that $\mathcal{F}_0 \prec \mathcal{F}$, we put $r_0 = diam(\mathcal{F}_0)$ and consider a disk $d(b, s)$

of diameter $s \in]\max(r_0, r-\epsilon), r[$ such that $\mathcal{F}_0$ is secant with $d(b,s)$ and we denote by E the annulus $\Gamma(b,s,r+\epsilon)$. Now, in all cases, consider $\varphi_{\mathcal{G}} \in S \cap (Mult(R(E \cap D), \| \,.\, \|_{E\cap D}))$. Thus, $\mathcal{G}$ is secant with $d(b, r+\epsilon)$, hence $diam(\mathcal{G}) \leq r+\epsilon$. If $\mathcal{F} \prec \mathcal{G}$ then $r \leq diam(\text{G}) \leq r+\epsilon$, hence of course $\mathcal{G}$ lies in B. Now suppose $\mathcal{G} \prec \mathcal{F}$, hence $\mathcal{F}_0 \prec \mathcal{G} \prec \mathcal{F}$, hence $s \leq diam(\mathcal{G}) \leq r$, and therefore $\mathcal{G}$ lies in B again. This proves that $(Mult(R(E\cap D), \| \,.\, \|_{E\cap D})) \cap S$ is included in B, which obviously shows that the two topologies are equal on S. □

Corollary 14.4: *The topology of simple convergence is weaker than the δ-topology on $Mult(K[x])$. However, on a totally ordered subset of $Mult(K[x])$ the two topologies are equal.*

Later, we will need this Proposition:

Proposition 14.5: *Let $(\varphi_{\mathcal{F}_n})_{n\in N}$, $(\varphi_{\mathcal{G}_n})_{n\in N}$ be sequences in $Mult(H(D), \| \,.\, \|_D)$ such that the sequence $(\varphi_{\mathcal{F}_n})_{n\in N}$ converges to a limit $\varphi_{\mathcal{T}} \in Mult(H(D), \| \,.\, \|_D)$ with respect to the topology of simple convergence. Suppose that $\lim_{n\to\infty} \delta(\mathcal{F}_n, \mathcal{G}_n) = 0$. Then the sequence $(\varphi_{\mathcal{G}_n})_{n\in N}$ also converges to $\mathcal{T}$ with respect to the topology of simple convergence.*

Proof: Let $r = diam(\mathcal{T})$, $B \in \mathcal{T}$ be a $\mathcal{T}$-affinoid, let $B = d(a, r'') \setminus \cup_{i=1}^{q} d(a_i, r')$ where the a_i are centers of $\mathcal{T}$ satisfying $|a_i - a_j| = r$, $\forall i \neq j$. Let $\epsilon = \frac{1}{2}\min(r''-r, r-r')$, $s'' =]r, r+\epsilon[\cap |K|$, $s' \in]r-\epsilon, r[\cap |K|$ and $A = d(a, s'') \setminus \cup_{i=1}^{q} d(a_i, s')$. According to hypotheses, we can find a rank N such that $\mathcal{F}_n$ is secant with A and $\delta(\mathcal{F}_n, \mathcal{G}_n) < \epsilon \ \forall n \geq N$. Then, by Lemma 10.14 each $\mathcal{G}_n$ is secant with B for all $n \geq N$. Thus, given any $\mathcal{F}$-affinoid B, there exists a rank $N \in \mathbb{N}$ such that all $\varphi_{\mathcal{G}_n}$ belong to $Mult(H(B), \| \,.\, \|_B)$ whenever $n \geq N$. □

Definition: A Hausdorff space E is said to be *arcwise connected* if for every $a, b \in E$, there exists a continuous mapping h from $[0,1]$ to E such that $h(0) = a$, $h(1) = b$.

Theorem 14.6 ([9]): *Let S be a locally compact subset of $Mult(K[x])$ with respect to the simple convergence topology and let T be the set of circular filters $\mathcal{H}$ such that $\varphi_{\mathcal{H}} \in S$. The following statements are equivalent:*

(i) *S is connected with respect to the topology of simple convergence;*

(ii) *S is arcwise connected with respect to the topology of simple convergence;*

(iii) *S is connected with respect to the δ-topology;*

(iv) *S is arcwise connected with respect to the δ-topology;*

(v) *There exists no annulus Ξ together with filters $\mathcal{F}$, $\mathcal{G} \in T$ such that $\mathcal{F}$ be secant with $\mathcal{I}(\Xi)$, $\mathcal{G}$ be secant with $\mathcal{E}(\Xi)$, and none of circular filters $\mathcal{H} \in T$ be secant with Ξ.*

Proof: First, (*iv*) trivially implies (*iii*) and (*ii*) trivially implies (*i*). Next, by Corollary 14.4 it is obviously seen that (*iv*) implies (*ii*) and that (*iii*) implies (*i*). Consequently, (*iv*) implies (*i*). We can also easily check that (*i*) implies (*v*). Indeed, suppose that (*v*) is not true. Thus there exists an annulus $\Xi = \Gamma(a, r, s)$ together with $\mathcal{F}$, $\mathcal{G} \in T$ such that $\mathcal{F}$ is secant with $\mathcal{I}(\Xi)$, $\mathcal{G}$ is secant with $\mathcal{E}(\Xi)$, but none of the circular filters $\mathcal{H}$ are secant with Ξ whenever $\mathcal{H} \in T$. Let E be the set of $\varphi_{\mathcal{H}}$, $\mathcal{H} \in T$ such that $\mathcal{H}$ is secant with $d(a, r)$, and let F be the set of $\varphi_{\mathcal{H}}$, $\mathcal{H} \in T$ such that $\mathcal{H}$ is secant with $K \setminus d(a, s^-)$. Then E, F are two subsets of S closed in $Mult(K[x])$ for the topology of simple convergence, and make a partition of S, therefore S is not connected with respect to the topology of simple convergence. Consequently, (*i*) implies (*v*).

Thus, it only remains us to prove that (*v*) implies (*iv*). The mapping *diam* defined on $\Phi(K)$ is strictly increasing by Corollary 11.2. Let $\mathcal{F}$, $\mathcal{G} \in T$ be such $\mathcal{G}$ surrounds $\mathcal{F}$, let $r = diam(\mathcal{F})$, $s = diam(\mathcal{G})$. Let $l \in [r, s]$. Let $\mathcal{R}$ be the unique circular filter of diameter l surrounding $\mathcal{F}$. By Theorem 13.11, every neighborhood of $\varphi_{\mathcal{R}}$, with respect to the topology of simple convergence, contains a set of the form $Mult(R(E), \| \, . \, \|_E)$, with E an affinoid subset of K, and such a set contains an annulus of the form $\Gamma(a, l - \epsilon, l)$. But by hypothesis, there exists $\mathcal{B} \in T$ secant with $\Gamma(a, l - \epsilon, l)$. Consequently, every neighborhood of $\varphi_{\mathcal{R}}$ with respect to the topology of simple convergence contains elements of S, and therefore (since S is locally compact), $\mathcal{R}$ belongs to T. Consequently, the mapping *diam* satisfies $\{diam(\mathcal{H}) \mid \mathcal{F} \preceq \mathcal{H} \preceq \mathcal{G}\} = [diam(\mathcal{F}), diam(\mathcal{G})]$. Hence we can apply Theorem 2.4 which shows that S is arcwise connected with respect to the δ-topology. This finishes proving that (*v*) implies (*iv*). □

Corollary 14.7: *Let D be a closed bounded set. The following statements are equivalent:*

(*i*) *$Mult(R(D), \| \, . \, \|_D)$ is connected with respect to the topology of simple convergence,*

(*ii*) *$Mult(R(D), \| \, . \, \|_D)$ is arcwise connected with respect to the topology of simple convergence,*

(*iii*) *$Mult(R(D), \| \, . \, \|_D)$ is connected with respect to the δ-topology,*

(*iv*) *$Mult(R(D), \| \, . \, \|_D)$ is arcwise connected with respect to the δ-topology,*

(*v*) *D is infraconnected.*

Chapter 15

Spectral Properties and Gelfand Transforms

The chapter is first devoted to present several semi-multiplicative semi-norms and compare them to the spectral semi-norm. We introduce properties (o), (p), (q), (r), (s) which will keep the same meaning throught the book. Most of results in domain chapter were stated in [23]. Next, we consider two kinds of Gelfand transform in an ultrametric K-algebra: one is similar to the definition in complex analysis, but does not show strong properties. The other is more sophisticated and shows certain specify properties linked to circular filters.

Notation: Henceforth, K will denote an algebraically closed field complete for a non trivial ultrametric absolute value. Throughout the chapter, $(A, \| \, . \, \|)$ will be a commutative normed K-algebra with unity.

We can define several usual semi-multiplicative semi-norms:
$\|x\|_{sa} = \sup\{\varphi(x) \mid \varphi \in Mult_a(A, \| \, . \, \|)\}$,
$\|x\|_{sm} = \sup\{\varphi(x) \mid \varphi \in Mult_m(A, \| \, . \, \|)\}$.
Moreover, we put $\tau(x) = \sup\{|\lambda| \mid \lambda \in sp(x) \cup \{0\}\}$.

Remarks: (1) By Corollary 6.10 we have $\|x\|_{si} = \sup\{\varphi(x) \mid \varphi \in Mult(A, \| \, . \, \|)\}$.

(2) In certain cases, $sp(x)$ might be empty (for instance in a field extension of K). This is why $\tau(x)$ involves $sp(x) \cup \{0\}$.

Proposition 15.1 is immediate from Theorems 6.13, 6.17 and Corollary 6.10:

Proposition 15.1: *Assume $Max_a(A) \neq \emptyset$. Then $\|x\|_{sa} \leq \tau(x) \leq \|x\|_{sm} \leq \|x\|_{si}$ $\forall x \in A$. Moreover, $\|x\|_{sa} = \sup\{|\chi(x)| \mid \chi \in \mathcal{X}(A, K)\}$ $\forall x \in A$.*

Remark: τ is not a semi-norm. Indeed, as we will see later, there might exist $\chi \in \mathcal{X}(A) \setminus \mathcal{X}(A, K)$ and $x, y \in A$ such that $\chi(x) = 1 + t, \ \chi(y) = -t$, with $t \notin K$, and $\max(\tau(x), \tau(y)) < 1$. However we have $\chi(x+y) = 1$, hence $\tau(x + y) \geq 1$.

Notation: In A we will denote by (o), (p), (q), (r), (s) these properties:

(o) $\|x\|_{sa} = \tau(x)$ $\forall x \in A$,
(p) $\|x\|_{sa} = \|x\|_{si}$ $\forall x \in A$,
(q) $\tau(x) = \|x\|_{si}$ $\forall x \in A$,
(r) $\|x\|_{sa} = \|x\|_{sm}$ $\forall x \in A$,
(s) $\|x\|_{sm} = \|x\|_{si}$ $\forall x \in A$.

Thus, by definition, Property (p) implies Properties (o), (q), (r), and (s), and (r) implies (o).

Proposition 15.2: *Assume that A satisfies Property (p). Then A is semi-simple if and only if $\| \, . \, \|_{si}$ is a norm.*

Theorem 15.3: *Let A be complete, let $\| \, . \, \|_{si}^A$ be its spectral semi-norm and let B be a commutative normed K-algebra with unity whose spectral semi-norm $\| \, . \, \|_{si}^B$ satisfies Property (q).*

Then every K-algebra homomorphism ϕ from A to B satisfies $\|\phi(x)\|_{si}^B \leq \|x\|_{si}^A$ $\forall x \in A$.

Proof: Suppose that for certain $x \in A$ we have $\|\phi(x)\|_{si}^B > \|x\|_{si}^A$ and let $y = \phi(x)$. Then there exists $\lambda \in sp_B(\phi(x))$ such that $|\lambda| > \|x\|_{si}^A$. Then by Theorem 6.13 $\lambda - x$ is invertible in A, hence $\phi(\lambda - x)$ is invertible in B. But $\phi(\lambda - x) = \lambda - y$, a contradiction to the hypothesis $\lambda \in sp_B(\phi(x))$. □

Corollary 15.4: *Assume A is complete, let B be a uniform commutative normed K-algebra with unity satisfying Property (q) and let ϕ be a K-algebra homomorphism from A to B. Then ϕ is continuous.*

By Theorems 6.13 and 15.3 we have Corollary 15.5:

Corollary 15.5: *Let A be complete. Let F be a set and let B be a K-subalgebra with unity of the algebra of bounded functions from F to K provided with the norm $\| \, . \, \|_F$.*

Then every K-algebra homomorphism ϕ from A to B is continuous and satisfies $\|\phi(f)\|_F \leq \|f\|_{si}$ $\forall f \in A$.

Notation: In A we can define two kinds of Gelfand transform $\mathtt{G}_A$ and $\mathtt{GM}_A$. The first one, denoted by $\mathtt{G}_A$ is similar to that in complex analysis in complex analysis, consisting of associating to each element f of A the mapping $\widehat{f}$ from $\mathcal{X}(A,K)$ to K defined as $\widehat{f}(\chi) = \chi(f)$, $(\chi \in \mathcal{X}(A,K))$. The second, denoted by $\mathtt{GM}_A$ consists of associating to each element f of A the mapping f^* from $Mult(A, \|\ .\ \|)$ to $Mult(K[x])$ defined as $f^*(\phi)(P) = \phi(P \circ f)$.

As in complex analysis, Propositions 15.6, 15.7 and 15.8 are immediate:

Proposition 15.6: *$\mathtt{G}_A$ is injective if and only if the intersection of all maximal ideals of codimension 1 is null.*

Proposition 15.7: *$\mathcal{X}(A,K)$ being provided with the topology of simple convergence, for every $f \in A$, $\widehat{f}$ belongs to $\mathcal{C}(\mathcal{X}((A,K),K)$.*

Proposition 15.8: *$\mathcal{C}(\mathcal{X}((A,K),K)$ being provided with the norm of uniform convergence, $\mathtt{G}_A$ satisfies $\|\widehat{f}\| = \|f\|_{sa}$ for every $f \in A$.*

Theorem 15.9: *Assume that A satisfies Property (p). If $\mathcal{C}(\mathcal{X}(A,K),K)$ being provided with the norm of uniform convergence, then $\mathtt{G}_A$ is an isometry if and only if $\|x^2\| = \|x\|^2$, $\forall x \in A$.*

Proof: Suppose $\|x^2\| = \|x\|^2$ is true for all $x \in A$, let $x \in A$ and let $\rho = \frac{\|x\|_{si}}{\|x\|}$. Then we check that $\frac{\|x^{2^n}\|_{si}}{\|x^{2^n}\|} = \rho^{(2^n)}$ hence $\|x\|_{si} = \rho \lim_{n\to\infty} \|x^n\|^{\frac{1}{n}}$, and therefore $\rho = 1$. So, $\mathtt{G}_A$ is an isometry. The converse is trivial. □

Theorem 15.10: *Assume that A satisfies Property (p). The following two properties are equivalent on A:*

(i) A is semi-simple and $\mathtt{G}_A(A)$ is closed in $\mathcal{C}(\mathcal{X}(A,K),K)$,
(ii) A is uniform.

Proof: First, suppose that A is uniform. Let x belong to the Jacobson radical of A. In particular, x belongs to the intersection of maximal ideals of codimension 1 of A. So, we have $\|x\|_{sa} = 0$, so, by Property (p), $\|x\|_{si} = 0$, and therefore $x = 0$, hence A is semi-simple. Let h belong to the closure of $\mathtt{G}_A(A)$ in $\mathcal{C}(\mathcal{X}(A,K),K)$. Since $\mathcal{C}(\mathcal{X}(A,K),K)$ is provided with the norm of uniform convergence, we just have to consider a sequence $\widehat{f_n}$ of $\mathtt{G}_A$ converging to h. Thus, $\|h - \widehat{f_n}\| = \sup\{|\chi(f_n) - h(\chi)|\ |\chi \in \mathcal{X}(A,K)\}$. The sequence $(f_n)_{n\in\mathbb{N}}$ is a Cauchy sequence in A, with respect to $\|x\|_{si} = 0$, and therefore with respect to the norm of A because A is uniform. Let $f = \lim_{n\to\infty} f_n$. Then we can check that $\widehat{f} = \lim_{n\to\infty} \widehat{f_n}$ in $\mathcal{C}(\mathcal{X}(A,K),K)$ and therefore $h = \widehat{f}$.

Conversely, suppose that A is semi-simple and that $\mathtt{G}_A(A)$ is closed in $\mathcal{C}(\mathcal{X}(A,K),K)$. Since A is semi-simple and satisfies Property p), $\|\ .\ \|_{sa}$

is a norm equal to $\|x\|_{si}$. Consider a Cauchy sequence (f_n) with respect to the norm $\| \,.\, \|_{si}$. Then the sequence $(\widehat{f_n})$ is a Cauchy sequence in $\mathcal{C}(\mathcal{X}(A,K),K)$, and has a limit h which actually lies in $\mathtt{G}_A(A)$ because $\mathtt{G}_A(A)$ is closed in $\mathcal{C}(\mathcal{X}(A,K),K)$. Hence, there exists $f \in A$ such that $\widehat{f} = h$. Consequently, f is the limit of the sequence (f_n) with respect to the norm $\| \,.\, \|_{si}$. Thus, A is a K-algebra complete for both $\| \,.\, \|_{si}$ and $\| \,.\, \|$. And since $\|x\|_{si} \leq \|x\| \; \forall x \in A$, by Banach's Theorem the two norms are equivalent. □

Now we will examine the mapping $\mathtt{GM}_A$.

Theorem 15.11: *Given $f \in A$, the mapping f^* from $Mult(A, \| \,.\, \|)$ to $Mult(K[x])$ is continuous with respect to the topology of simple convergence.*

Proof: Let $\mathcal{G} = f^*(\phi)$. By Theorem 14.10 the family of sets $Mult(H(E), \| \,.\, \|_E)$, where E is an infraconnected affinoid subset of K lying in $\mathcal{G}$, makes a base of neighborhoods of $\varphi_{\mathcal{G}}$. So, we take an arbitrary $\mathcal{G}$-affinoid B and will show that there exists a neighborhood W of ϕ inside $Mult(A, \| \,.\, \|)$ whose image by f^* is included in $Mult(H(B), \| \,.\, \|_B)$. If $\mathcal{G}$ has center b and diameter $r > 0$, we can take an arbitrary $\mathcal{G}$-affinoid B of the form $d(b, r+\epsilon) \setminus (\cup_{j=1}^{q} d(b_j, (r-\epsilon^-))$, and if $\mathcal{G}$ has no center and but has diameter r, or is a Cauchy filter, we can take an arbitrary $\mathcal{G}$-affinoid B of the form $d(b, r+\epsilon)$. Putting $b = b_0$, we consider now the neighborhood of ϕ

$$\mathcal{W} = \{\psi \in Mult(A, \| \,.\, \|) \mid |\psi(f - b_j) - \phi(f - b_j)| \leq \epsilon, (0 \leq j \leq q)\}.$$

Then we can check that $f^*(\mathcal{W}) \subset Mult(H(B), \| \,.\, \|_B)$, which proves the claim. □

Lemma 15.12 is immediate:

Lemma 15.12: *Let $\chi \in \mathcal{X}(A,K)$ and let $f \in A$. Then $f^*(|\chi|) = \varphi_{\chi(f)}$.*

Proof: Let $\mathcal{F}$ be the circular filter such that $\varphi_{\mathcal{F}} = f^*(|\chi|)$. Then $\varphi_{\mathcal{F}}$ clearly belongs to $Mult_m(K[x])$ and $\mathcal{F}$ is the Cauchy filter of neighborhoods of $\chi(f)$ because $\varphi_{\mathcal{F}}(P) = |P(\chi(f))| \; \forall P \in K[x]$. □

Theorem 15.13: *Assume that is uniform and that the intersection of maximal ideals of codimension 1 is null. Then $\mathtt{GM}_A$ is injective.*

Proof: Suppose $f^* = g^*$ and $f \neq g$. let $r = \|f - g\|$. Since the intersection of maximal ideals of codimension 1, there exists $(\chi \in \mathcal{X}(A,K)$ such that $\chi(g - f) \neq 0$. By Lemma 15.12 we have $f^*(|\chi|) = \varphi_{\chi(f)}$ and

$g^*(|\chi|) = \varphi_{\chi(g)}$. But since $f^* = g^*$, we have $f^*(|\chi|) = g^*(|\chi|)$. Let $\mathcal{F}$ be the circular filter such that $\varphi_{\mathcal{F}} = f^*(|\chi|) = g^*(|\chi|)$. Then $\varphi_{\mathcal{F}}$ clearly belongs to $Mult_m(K[x])$ and $\mathcal{F}$ is the Cauchy filter of neighborhoods of $\chi(f)$ because $\varphi_{\mathcal{F}}(P) = |P(\chi(f))|$. Similarly, it is the Cauchy filter of neighborhoods of $\chi(g)$. Consequently, $\chi(f) = \chi(g)$, hence $\chi(f - g) = 0$, a contradiction to the hypothesis. □

Theorem 15.14: *Let A be uniform. Let $\phi \in Mult(A, \| \,.\, \|)$, let $f,\ g \in A$ be such that $f^*(\phi) \neq g^*(\phi)$. Let $\varphi_{\mathcal{F}} = f^*(\phi)$, $\varphi_{\mathcal{G}} = g^*(\phi)$. Then $diam(\sup(\mathcal{F}, \mathcal{G})) = \phi(f - g)$.*

Proof: Suppose that $f^*(\phi) \neq g^*(\phi)$. Let $r = diam(\mathcal{F})$, $s = diam(\mathcal{G})$. Suppose first that $\mathcal{F} \prec \mathcal{G}$, hence $r < s$. Let us take $l \in]r, s[$. By Theorem 10.11 there exists a unique disk $d(a, l)$ which belongs to $\mathcal{F}$ Then a is a center of $\mathcal{G}$. Thus, we have $\phi(f - a) \leq l$, $\phi(g - a) = s$, hence, by Lemma 5.3 $\phi(g - f) = s = diam(\sup(\mathcal{F}, \mathcal{G}))$.

Now, suppose that $\mathcal{F}$ and $\mathcal{G}$ are uncomparable. Let $\mathcal{S} = \sup(\mathcal{F}, \mathcal{G})$. By Proposition 12.9, we can find disks $F = d(a, r') \in \mathcal{F}$, $G = d(b, s') \in \mathcal{G}$ such that $\delta(F, G) = \lambda(\mathcal{F}, \mathcal{G}) = |a - b| = diam(\sup(\mathcal{F}, \mathcal{G}))$. Of course we have $|b - a| > r'$ and $|b - a| > s'$. Then, $\phi(g - a) = |b - a|$ and $\phi(f - a) < |b - a|$, hence $\phi(g - f) = |b - a|$, and therefore $\phi(g - f) = diam(\sup(\mathcal{F}, \mathcal{G}))$. □

Corollary 15.15: *Let A be uniform. Let $\phi \in Mult(A, \| \,.\, \|)$, let $f,\ g \in A$. Let $\varphi_{\mathcal{F}} = f^*(\phi)$, $\varphi_{\mathcal{G}} = g^*(\phi)$. Then $\delta(f^*(\phi), g^*(\phi))) \leq \phi(f - g) \leq \|f - g\|$.*

Notation: Given $\phi \in Mult(A, \| \,.\, \|)$, we will denote by Z_ϕ the mapping from A into $Mult(K[x])$ defined as $Z_\phi(f) = f^*(\phi)$.

Corollary 15.16: *Let A be uniform. The family of functions Z_ϕ, $\phi \in Mult(A, \| \,.\, \|)$ is uniformly equicontinuous with respect to the δ-topology on $Mult(K[x])$.*

Corollary 15.17: *Let A be uniform. Each function Z_ϕ, $\phi \in Mult(A, \| \,.\, \|)$ is continuous with respect to the topology of simple convergence on $Mult(K[x])$.*

Remark: Theorem 15.13 and its Corollaries apply to reduced affinoid K-algebras, as it is shown in Chapter 35.

Chapter 16

Analytic Elements

Let us briefly recall that analytic elements where introduced by Marc Krasner in order to define a general notion of holomorphic functions in sets which are not just disks: indeed, when two disks have a non empty intersection, this which has the biggest radius contains the other. As a consequence, it is hopeless to cover a set (which is not a disk) with a chained family of disks. But, according to Runge's Theorem, complex holomorphic functions may be viewed as uniform limits of rational functions. Marc Krasner adopted this point of view to defining holomorphic functions in a set D.

Definition and notation: In this chapter D is an infinite subset of K. We denote by $H(D)$ the completion of $R(D)$ for the topology $\mathcal{U}_D$ of uniform convergence on D. The elements of $H(D)$ are called the *analytic elements on D* [30,36].

By definition, the set $H(D)$ is then provided with the topology of uniform convergence on D for which it is complete, and every $f \in H(D)$ defines a function on D which is the uniform limit (on D) of a sequence $(h_n)_{n\in\mathbb{N}}$ in $R(D)$. Thus, given another set D' containing D, the restriction to D of elements of $H(D')$ enables us to consider that $H(D')$ is included in $H(D)$. Here, we will only consider bounded analytic elements, and we will denote by $H_b(D)$ the set of the elements $f \in H(D)$ bounded on D. Then $H_b(D)$ is clearly a K-vector subspace of $H(D)$ and is closed in $H(D)$. Moreover, $\| \, . \, \|_D$ is a K-algebra norm on $R_b(D)$, hence its continuation to $H_b(D)$ is a K-algebra norm that makes $H_b(D)$ a Banach K-algebra. If D

is unbounded, we will denote by $H_0(D)$ the set of the $f \in H(D)$ such that $\lim_{|x|\to+\infty, x\in D} f(x) = 0$.

According to Theorem 6.8 and Corollary 13.8, it is obviously seen that each element $\varphi_{\mathcal{F}}$ of $Mult(R(D), \| \cdot \|_D)$ has a unique continuation to $Mult(H(D), \| \cdot \|_D)$ and will be denoted by $\varphi_{\mathcal{F}}$ again. Particularly, when $\mathcal{F}$ is the circular filter of center a and diameter r, we will also denote by $\varphi_{a,r}$ the continuation of $\varphi_{a,r}$ to $Mult(H(D), \| \cdot \|_D)$. Thus we identify $Mult(H(D), \| \cdot \|_D)$ to $Mult(R(D), \| \cdot \|_D)$, and therefore all properties already proven in $Mult(R(D), \| \cdot \|_D)$ also hold in $Mult(H(D), \| \cdot \|_D)$.

Theorem 16.1 is well known [30, Chapter 10]:

Theorem 16.1: *Let D be an infinite closed subset of K. Then $H_b(D)$ is a Banach K-subalgebra of K^D. The following three conditions are equivalent*

(*i*) $H_b(D) = H(D)$
(*ii*) *$H(D)$ is topological K-vector space,*
(*iii*) *$(H(D), \| \cdot \|_D)$ is a K-Banach algebra*
(*iv*) *D is closed and bounded.*

If these conditions are satisfied, then $\| \cdot \|_D$ is a semi-multiplicative norm.

The following two lemmas are classical and easy.

Lemma 16.2: *$R(D)$ is a full K-subalgebra of $H(D)$.*

Lemma 16.3: *Let $f \in H(D)$ be such that $\inf_{x\in D}|f(x)| > 0$. Then f^{-1} belongs to $H(D)$. Moreover if D is closed and bounded, an element $g \in h(D)$ belongs to $H(D)$ if and only if $\inf_{x\in D}|g(x)| > 0$.*

Let $g, h \in H(D)$ satisfy $|g(x)| = 1$ for all $x \in D$ and $\|h-1\|_D > \|g-1\|_D$. Then we have $\|hg - 1\|_D = \|h - 1\|_D$

Lemma 16.4: *Let $a \in K\setminus D$ and let $f \in R(D)$ be such that $|f(a)| > \|f\|_D$. Then $\frac{1}{x-a}$ belongs to the closure of $K[f, x]$ in $H(D)$.*

Proof: Let A be the closure of $K[f, x]$ in $H(D)$. Since A is a Banach K-algebra, by Theorem 16.3 $f(a) - f$ is invertible in A, hence of course in $H(D)$. But by Lemma 16.2 its inverse g actually belongs to $R(D)$. Now, since a is clearly a zero of $f(a) - f$, it is a pole of g. Consequently, by Lemma 1.8 $\frac{1}{x-a}$ belongs to $K[g, x]$ which is included in A. □

Lemma 16.5: *Let $\alpha \in \overline{D} \setminus D$ and let $f \in H(D)$ be such that there exists $n \in \mathbb{N}$ such that $(x - \alpha)^n f \in H(D \cup \{\alpha\})$. There exists a unique $q \in \mathbb{N}$ such that $(x - \alpha)^q f \in H(D \cup \{\alpha\})$ and such that the value of $(x - \alpha)^q f$ at α is not zero.*

Definitions and notation: In the hypothesis of Lemma 16.5, f is said to be *meromorphic at* α and *to admit* α *as a pole of order* q.

Let $D \subset K$ be infraconnected closed and bounded, and let T be a hole of D. Let $f \in H(D)$. Then f will be said to be *meromorphic in* T if there exist finitely many points $(a_i)_{(1\leq i\leq n)}$ in T such that f has continuation to an element of $H((D\cup T)\setminus\{a_i|\ 1\leq i\leq n\})$.

Let f be meromorphic in T, and belong to $H(D\cup T)\setminus\{a_i|\ 1\leq i\leq n\})$. For each $i=1,\dots,n$ if $f\notin H((D\cup T)\setminus\{a_h\ |h\neq i\})$ then a_i is a pole of f as an element of $H((D\cup T)\setminus\{a_j|1\leq j\leq n\})$. Let q_i be its order. Then a_i will be called *a pole of* f *of order* q_i *in* T. The polynomial $P(x)=\prod_{i=1}^{n}(x-a_i)^{q_i}$ will be called *the polynomial of the poles of* f *in* T.

Let $d(b_i,r_i^-))_{i\in I}$ be the family of holes of D. We will denote by $H'(D,(b_i)_{i\in I})$ the K-subvector space of $H(D)$ consisting of the $f\in H(\overline{D}\setminus(\cup_{i\in I}d(b_i,r_i^-))$ admitting each b_i as a pole of order ≤ 1, and having no other poles in any hole $d(b_i,r_i^-)$.

Lemma 16.6 is immediate ([30, Lemmas 34.2 and 31.3]):

Lemma 16.6: *Let* D *be closed bounded and infraconnected, let* T *be a hole of* D*, and let* $f\in H(D)$*. If* f *is meromorphic in* T*, the polynomial* P *of the poles of* f *in* T *satisfies* $Pf\in H(D\cup T)$*. Conversely, if there exists a polynomial* Q *such that* Qf *lies in* $H(D\cup T)$ *then* f *is meromorphic in* T*.*

Basic results about analytic elements on disks and annuli are mainly given in chapter 13 of [30]: Theorem 16.7 is Theorem 13.1 in [30].

Theorem 16.7: *Let* $r\in\mathbb{R}_+^*$*, let* $\mathcal{F}$ *be the circular filter of center* 0 *and diameter* r *on* K*, and let* $E=d(0,r)$*. Then* $H(E)$ *is the set of the power series* $f(x)=\sum_{n=0}^{\infty}a_nx^n$ *such that* $\lim_{n\to\infty}|a_n|r^n=0$ *and we have*

$$\|f\|_E=\max_{n\in\mathbb{N}}|a_n|r^n=\varphi_{\mathcal{F}}(f)=\|f\|_{C(0,r)}.$$

For every $\alpha\in E$*,* $H(E)$ *is also equal to the set of the series* $f(x)=\sum_{n=0}^{\infty}b_n(x-\alpha)^n$ *such that* $\lim_{n\to\infty}|b_n|r^n=0$*.*

Let $B=K\setminus d(0,r^-)$*. Then* $H(B)$ *is the set of the Laurent series* $f(x)=\sum_{n=0}^{\infty}\frac{a_n}{x^n}$ *such that* $\lim_{n\to\infty}|a_n|r^{-n}=0$*, and we have*

$$\|f\|_B=\max_{n\in\mathbb{N}}|a_n|r^{-n}=\varphi_{\mathcal{F}}(f)=\|f\|_{C(0,r)}.$$

For every $\alpha\in d(0,r^-)$*,* $H(B)$ *is also equal to the set of the series* $f(x)=\sum_{n=0}^{\infty}\frac{b_n}{(x-\alpha)^n}$ *such that* $\lim\limits_{n\to\infty}|b_n|r^{-n}=0$*.*

By Theorem 14.1 and 14.4 in [30] and by Theorem 16.7 we have Theorem 16.8

Theorem 16.8: *Let $\alpha \in D$ and $r \in \mathbb{R}_+^*$ be such that $d(\alpha, r) \subset D$. Let $f \in H(D)$. In $d(\alpha, r)$, $f(x)$ is equal to a power series of the form $\sum_{n=0}^{\infty} a_n(x-\alpha)^n$ such that $\lim_{n\to\infty} |a_n|r^n = 0$. If $f(\alpha) = 0$ and if $f(x)$ is not identically zero in $d(\alpha, r)$, then there exists a unique integer $q \in \mathbb{N}^*$ such that $a_n = 0$ for every $n < q$ and $a_q \neq 0$, and α is an isolated zero of f in $d(\alpha, r)$. Moreover, there exists $g \in H(D)$ such that $f(x) = (x-\alpha)^q g(x)$.*

As a consequence of Theorems 16.7 and 16.8, we deduce Theorem 16.9:

Theorem 16.9: *The characteristic of K is supposed different from 0. Let $f \in H(d(0, r^-))$ and for every $j = 0, \dots, n$ let $a_j = \frac{f^{(j)}(0)}{j!}$. Then f is equal to the power series $\sum_{n=0}^{\infty} a_n x^n$, and $\|f\|_{d(0,r^-)} = \sup_{n\in\mathbb{N}} |a_n|$. Moreover, for every $q \in \mathbb{N}$, f is of the form $\sum_{j=0}^{q} a_j x^j + x^n g$, with $g \in H(M)$ and $\|f\|_{d(0,r^-)} = \max(|a_0|, |a_1|, \dots |a_{n-1}|r^{n-1}, \|x^n g\|_{d(0,r^-)})$.*

We must also recall Theorem 13.12 in [30]:

Theorem 16.10: *If K has characteristic 0, then an element $f \in H(d(0, r))$ has a derivative identically equal to 0 if and only if it is equal to a constant. If K has a characteristic $p \neq 0$, then an element $f \in H(d(0, r))$ has a derivative identically equal to 0 if and only if there exists $g \in H(d(0, r))$ such that $f(x) = (g(x))^p$.*

Definitions and notation: Let $f \in H(D)$, let $\alpha \in \overset{\circ}{D}$, let $r > 0$ be such that $d(\alpha, r) \subset D$, and let $f(x) = \sum_{n=q}^{\infty} b_n(x-\alpha)^n$ whenever $x \in d(\alpha, r)$, with $b_q(\alpha) \neq 0$, and $q > 0$. Then α is called *a zero of multiplicity order* q, or more simply, *a zero of order* q. In the same way, q will be named *the multiplicity order of* α.

An element $f \in H(D)$ is said to be *quasi-invertible* if it factorizes in the form Pg with $P \in K[x]$ all zeros of which lie $\overset{\circ}{D}$, and $g \in H(D)$ invertible in $H(D)$.

Theorem 16.11: *Let D be closed and bounded and let $f \in H(D)$ be quasi-invertible. There exists $h \in R(D)$ such that $|f(x)| = |h(x)|$, $\forall x \in D$.*

Proof: Since f is quasi-invertible, there exists a polynomial P whose zeros lie in D, and an invertible element g in $H(D)$ such that $f = Pg$. Let $m = \inf_{x\in D} |g(x)|$. Since g is invertible, we have $m = \frac{1}{\|\frac{1}{g}\|_D}$, hence $m > 0$, and then we can find $l \in R(D)$ such that $\|g - l\|_D < m$. Consequently, we have $|g(x)| = |l(x)|$, $\forall x \in D$ and we obtain h by putting $h = Pl$. Moreover, if $r \in |K|$, then $\varphi_{0,r}(h) \in |K|$. Conversely, if $s \neq t$ and if $\varphi_{0,r}(h) \in |K|$, then $r \in |K|$.

The following Theorem 16.12 is Theorem 15.1 in [30].

Theorem 16.12 (M.Krasner): *Let D be bounded (resp. unbounded) and infraconnected and let $f \in H_b(D)$. There exists a unique sequence of holes $(T_n)_{n\in\mathbb{N}^*}$ of D and a unique sequence $(f_n)_{n\in\mathbb{N}}$ in $H(D)$ such that $f_0 \in H(\widetilde{D})$ (resp. $f_0 \in K$), $f_n \in H_0(K \setminus T_n)$ $(n > 0)$, $\lim_{n\to\infty} \|f_n\|_D = 0$ and*

$$(i)\quad f = \sum_{n=0}^{\infty} f_n \text{ and } \|f\|_D = \sup_{n\in\mathbb{N}} \|f_n\|_D.$$

Moreover for every hole $T_n = d(a_n, r_n^-)$, we have

$$(ii)\quad \|f_n\|_D = \|f_n\|_{K\setminus T_n} = \varphi_{a_n,r_n}(f_n) \leq \varphi_{a_n,r_n}(f) \leq \|f\|_D.$$

If D is bounded and if $\widetilde{D} = d(a,r)$ we have

$$(iii)\quad \|f_0\|_D = \|f_0\|_{\widetilde{D}} = \varphi_{a,r}(f_0) \leq \varphi_{a,r}(f) \leq \|f\|_D.$$

If D is not bounded then $|f_0| = \lim_{|x|\to\infty, x\in D} |f(x)| \leq \|f\|_D$.

Let $D' = \widetilde{D} \setminus (\cup_{n=1}^{\infty} T_n)$. Then f belongs to $H(D')$ (resp. $H_b(D')$) and its decomposition in $H(D')$ is given again by (i) *and then f satisfies $\|f\|_{D'} = \|f\|_D$.*

Corollary 16.13: *Let D be closed and infraconnected. Let $(T_i)_{i\in I}$ be the family of holes of D. Let J be a subset of D and let $L = I \setminus J$. Let $E = D \cup (\cup_{i\in J} T_i)$ and let $F = D \cup (\cup_{i\in L} T_i)$. Then we have $H(D) = H_0(E) \oplus H(F)$, and for each $g \in H_0(E)$, $h \in H(F)$, we have $\|g+h\|_D = \max(\|g\|_E, \|h\|_F)$.*

Theorem 16.14: *Let D be closed and infraconnected. Then $\Sigma_0(D)$ is a minimal boundary for $(H(D), \|\cdot\|_D)$.*

Proof: By Theorem 16.12 it is a boundary for $(H(D), \|\cdot\|_D)$. Conversely, let B be a boundary included in $\Sigma_0(D)$. Let $\widetilde{D} = d(a,r)$. Considering $(x-a)$ we can see that $\varphi_{a,r}$ must belong to B because if $\phi \in \Sigma_0(D)\setminus\{\varphi_{a,r}\}$ we have $\phi(x-a) < r = \|x-a\|_D$. In the same way, let $T = d(\alpha, \rho^-)$ and let $h = \frac{1}{x-\alpha}$. Then we can see that for every $\phi \in \Sigma_0(D) \setminus \{\varphi_{\alpha,\rho}\}$ we have $\phi(h) < \frac{1}{\rho} = \|h\|_D$. Consequently, $\Sigma_0(D) = B$ and therefore is a minimal boundary for $(H(D), \|\cdot\|_D)$. □

Lemma 16.15: *Let D be infraconnected and suppose that $H(D)$ contains a dense Luroth K-algebra $K[h,x]$. Then each hole of D contains at least one pole of h.*

Proof: Let B be the closure of $K[h,x]$ in $H(D)$ and let $T_1,\ldots,T_q$ be the holes of D containing at least one hole of D and let $D' = \widetilde{D} \setminus \cup_{j=1}^{q} T_j$. Clearly $K[h,x]$ is included in $H(D')$. But since each hole of D' is a hole of D, and since $\widetilde{D'} = \widetilde{D}$, by Corollary 16.14 $H(D')$ is a closed K-subalgebra of $H(D)$. Consequently $D = D'$. □

The Mittag-Leffler Theorem suggests some new definitions.

Definitions and notation: Let $f \in H_b(D)$. We consider the series $\sum_{n=0}^{\infty} f_n$ obtained in Theorem 16.12, whose sum is equal to f in $H(D)$, with $f_0 \in H(\widetilde{D}), f_n \in H(K \setminus T_n) \setminus \{0\}$ whereas the T_n are holes of D. Each T_n will be called a *f-hole* and f_n will be called the *Mittag-Leffler term of f associated to T_n*, whereas f_0 will be called the *principal term* of f.

For each f-hole T of D, the Mittag-Leffler term of f associated to T will be denoted by $\overline{f_T}$ whereas the principal term of f will be denoted by $\overline{f_0}$. The series $\sum_{n=0}^{\infty} f_n$ will be called *the Mittag-Leffler series of f on the infraconnected set D*.

More generally, let E be an infraconnected set and $f \in H(E)$. According to Theorem 9.5 [30], f is of the form $g + h$ with $g \in R(K \setminus (\overline{E} \setminus E))$, and $h \in H_b(\overline{E})$, and such a decomposition is unique, with respect to an additive constant. For every hole T of $\overline{E}$, we will denote by $\overline{f}_T$ the Mittag-Leffler term of h associated to T, and $\overline{f}_T$ will still be named *the Mittag-Leffler term of f associated to T*.

Corollary 16.16: *Let D be infraconnected. Let $f \in H_b(D)$ let $(T_n)_{n \in \mathbb{N}^*}$ be the sequence of the f-holes, with $T_n = d(a_n, \rho_n^-)$, let $f_0 = \overline{f}_0$, and $f_n = \overline{f}_{T_n}$ for every $n \in \mathbb{N}^*$. Let $\widetilde{D} = d(a,s)$ (resp. $\widetilde{D} = K$). There exists $q \in \mathbb{N}$ such that $\|f\|_D = \|f_q\|_D$. If $q \geq 1$ then $\|f\|_D = \varphi_{a_q,r_q}(f) = \varphi_{a_q,r_q}(f_q)$. If $q = 0$ and if D is bounded (resp. is not bounded) then $\|f\|_D = \varphi_{a,s}(f) = \varphi_{a,s}(f_0)$ (resp. $\|f\|_D = |f_0|$). Further, given a hole T of D, if f belongs to $H_b(D)$, and if g belongs to $H_0(K \setminus T)$ and satisfies $f - g \in H(D \cup T)$, then $\overline{f}_T$ is equal to g.*

Corollary 16.17: *Let D be infraconnected. Let $f \in H_b(D)$. There exists a large circular filter $\mathcal{F}$ with center $\alpha \in \widetilde{D}$ secant with D such that $\varphi_{\mathcal{F}}(f) = \|f\|_D$. If D is bounded, there exists a D-bordering filter $\mathcal{F}$ such that $\varphi_{\mathcal{F}}(f) = \|f\|_D$.*

Corollary 16.18: *Let $f \in H(d(0,1^-))$ and let $(d(\alpha_m,1^-))_{m\in\mathbb{N}^*}$ be the family of the f-holes. Then f is of the form*

$$\text{(iv)} \qquad \sum_{n=0}^{\infty} a_{n,0}x^n + \sum_{m,n\in\mathbb{N}^*} \frac{a_{n,m}}{(x-\alpha_m)^n}$$

with $\lim_{n\to\infty} a_n = 0, \lim_{n\to\infty} |a_{n,m}| = 0$ *whenever*

$$m \in \mathbb{N}^* \text{ and } \lim_{m\to\infty} (\sup_{n\in\mathbb{N}^*} |a_{n,m}|) = 0.$$

Moreover f satisfies

$$\|f\|_{d(0,1^-)} = \sup_{m,n\in\mathbb{N}^*} |a_{n,m}|.$$

Conversely, every function of the form (iv), with the α_m satisfying $|\alpha_m| = |\alpha_j - \alpha_m| = 1$ whenever $m \neq j$, belongs to $H(d(0,1^-))$. The norm $\| \cdot \|_{d(0,1^-)}$ is multiplicative and equal to $\varphi_{0,1}$.

Theorem 16.19 is proven in Chapter 15 of [30]:

Theorem 16.19: *Let $r_1, r_2 \in \mathbb{R}_+$ satisfy $0 < r_1 < r_2$. Then $H(\Delta(0,r_1,r_2))$ is equal to the set of Laurent series $\sum_{-\infty}^{+\infty} a_n x^n$ with $\lim_{n\to-\infty} |a_n| r_1^n = \lim_{n\to\infty} |a_n| r_2^n = 0$ and we have*

$$\left\| \sum_{-\infty}^{+\infty} a_n x^n \right\|_{\Delta(0,r_1,r_2)} = \max\Big(\sup_{n\geq 0} |a_n| r_1^n,\ \sup_{n<0} |a_n| r_2^n \Big).$$

About maximal ideals of codimension 1, we must also recall Theorem 16.2 in [30]:

Notations: Let $D \in \mathcal{A}$. For every $a \in D$ we will denote by $\mathcal{Y}(a)$ the ideal of the $f \in H(D)$ such that $f(a) = 0$.

Theorem 16.20: *Let D be closed bounded infraconnected. The mapping ϕ from D into the set of ideals of $H(D)$ defined by $\phi(a) = \mathcal{Y}(a)$ is a bijection from D onto $Max_a(H(D))$.*

Finally, we can notice that analytic elements define uniformly continuous functions, in a closed bounded set:

Theorem 16.21: *Let D be closed and bounded and let $f \in H(D)$. Then f is uniformly continuous on D.*

Proof: Let $h \in R(D)$. Since h has finitely many poles, we can find $r_0 > 0$ such that for all $a \in D$, $d(a, r_0)$ contains no poles of h. Let $M = \|h\|_D$. Thus, by Corollary 16.18, for all $a \in D$, we have

$$\sup_{k\geq 1}\left|\frac{h^{(k)}(a)}{k!}\right| r_0^k \leq \sup_{k\geq 0}\left|\frac{h^{(k)}(a)}{k!}\right| r_0^k \leq \|h\|_{d(a,r_0)} \leq M.$$

Consequently, for every $r \in]0, r_0]$, we have

$$\sup_{k\geq 1}\left|\frac{h^{(k)}(a)}{k!}\right| r^k \leq \sup_{k\geq 1}\left(\left|\frac{h^{(k)}(a)}{k!}\right| r_0^k (\frac{r}{r_0})^k\right) \leq M\frac{r}{r_0}.$$

Thus, we can see that $h(d(a,r)) \subset d(h(a), M\frac{r}{r_0})$, which proves that h is uniformly continuous on D. Then, so is every element of $H(D)$, as a uniform limit of uniformly continuous functions on D. □

Chapter 17

Holomorphic Properties on Infraconnected Sets

Throughout the chapter, D is a subset of K. Here we shall go back to the basic properties of analytic elements which are transmitted by rational functions.

Proposition 17.1 is classical and given in [30] as Proposition 20.3:

Proposition 17.1: *Let $a \in \widetilde{D}$, let $r \in [\delta(a, D), diam(D)]$, and let $f \in H(D)$. If $\varphi_{a,r}(f) \neq 0$ the equality $|f(x)| = \varphi_{a,r}(f)$ holds in all classes of $C(a,r)$ except maybe in finitely many ones.*

Theorems 17.2 and 17.4 are proven in Chapter 23 of [30]:

Theorem 17.2: *Let $f \in H(C(0,r))$ (resp. $f \in H(d(0,r))$). The number of zeros of f in $C(0,r)$ (resp. in $d(0,r)$) is equal to $N^+(f, -\log r) - N^-(f, -\log r)$, (resp. $N^+(f, -\log r)$), taking multiplicities into account.*

Corollary 17.3: *Let $f \in H(C(0,r))$ have t zeros in $C(0,r)$. Let $q = N^+(f, -\log r) - N^-(f, -\log r)$. Then*

$$r = \sqrt[t]{\left|\frac{a_q}{a_{q+t}}\right|}.$$

Theorem 17.4: *Let $f = \sum_{n=-\infty}^{+\infty} a_n x^n \in H(C(0,r))$ have no zero in $C(0,r)$. Let $t = N^+(f, -\log r)$. Then we have $N^+(f, -\log r) = N^-(f, -\log r)$ and $v(f(x)) = v(f, -\log r) = v(a_t) - t \log r$ whenever $x \in C(0,r)$.*

Corollary 17.5: *Let $f \in H(\Gamma(0,r,s))$ have no zero in $\Gamma(0,r,s)$ and be equal to a Laurent series $\sum_{n=-\infty}^{+\infty} a_n x^n$ $\forall x \in \Gamma(0,r,s)$. Assume that there* *exists an integer* $m \in \mathbb{Z}^*$ *such that* $|a_m|\rho^m > |a_n|\rho^n \ \forall n \neq m, \ \forall \rho \in]r, s[$. *If* $m > 0$, *we have* $f(\Gamma(0, r, s)) = \Gamma(0, |a_m|r^m, |a_m|s^m)$, *and if* $m < 0$, *we have* $f(\Gamma(0, r, s)) = \Gamma(0, |a_m|s^m, |a_m|r^m)$.

Theorem 17.6 will be useful when considering restricted power series:

Theorem 17.6: *Let* $f(x) = \sum_{n=0}^{\infty} a_n x^n \in H(U)$. *Then* f *is invertible in* $H(U)$ *if and only if* $|a_0| > \sup_{n>0} |a_n|$. *And* f *is irreducible in* $H(U)$ *if and only if* $|a_0| \leq |a_1|$, $|a_1| > \sup_{n>1} |a_n|$.

Proof: If $|a_0| > \sup_{n>0} |a_n|$, by Theorem 6.13 f is obviously invertible in $H(U)$. Else, by Theorem 23.5 in [30], f admits at least one zero, and therefore is not invertible. Precisely, let q be the unique integer such that $|a_q| = \sup_{n\in\mathbb{N}} |a_n|$. If $q = 1$, then by Theorems 23.5 and 23.6 in [30], f has exactly one zero α, and since it is quasi-invertible, it is of the form $(x - \alpha)g$, with g invertible, hence f is irreducible. If $q > 1$, since f admits q zeros, f factorizes in the form $\prod_{j=1}^{q}(x - a_j)g(x)$, whereas $g \in H(U)$ has no zeros, and therefore is invertible, and therefore is not irreducible. □

Definition: D will be said to be *calibrated* if every hole has a diameter that belongs to $| K |$ and if *diam*(D) either belongs to $| K |$ or is infinite.

Theorem 17.7 is proven in Chapter 30 of [30], (as Theorem 30.6):

Theorem 17.7: *Let* D *be closed and bounded. Then* D *is calibrated and strongly infraconnected if and only if for every* $f \in H(D)$ *there exists* $\alpha \in D$ *such that* $|f(\alpha)| = \|f\|_D$.

As far as closed bounded sets are concerned, the composition of analytic elements is an analytic element (Corollary 11.2 in [30]):

Proposition 17.8: *Let* D, D' *be closed bounded subsets of* K, *and let* $f \in H(D)$ *be such that* $f(D) \subset D'$. *Let* $g \in H(D')$. *Then* $g \circ f$ *belongs to* $H(D)$.

Homomorphisms between K-algebras $H(D)$ are characterized in [30], (Proposition 11.6 and Theorem 11.7). Here, for convenience, we will restrict the statement to the case of closed bounded sets:

Proposition 17.9: *Let* D, D', D'' *be closed bounded subsets of* K *and let* $\gamma \in H(D')$ *satisfy* $\gamma(D') \subset D$. *Let* ϕ_γ *be the mapping from* $H(D)$ *into* $H(D')$ *defined as* $\phi_\gamma(f) = f \circ \gamma$. *Then* ϕ_γ *is a* K*-algebra homomorphism from* $H(D)$ *into* $H(D')$ *continuous with respect to the topology of uniform convergence on* D *for* $H(D)$ *and on* D' *for* $H(D')$. *If* $\gamma \in H(D')$ *and* $\zeta \in H(D'')$ *satisfy* $\gamma(D') = D$ *and* $\zeta(D'') = D'$, *then* $\phi_\zeta \circ \phi_\gamma = \phi_{\gamma\circ\zeta}$.

If γ is a bijection from D' onto D and if $\gamma^{-1} \in H(D)$ then ϕ_γ is a K-algebra isomorphism from $H(D)$ onto $H(D')$ bicontinuous with respect to the topology of uniform convergence on D for $H(D)$ and on D' for $H(D')$, satisfying $\left(\phi_\gamma\right)^{-1} = \phi_{\gamma^{-1}}$.

By Propositions 17.9 we can easily deduce Corollary 17.10:

Corollary 17.10: *Let ϕ_γ be an isomorphism from $H(D)$ onto $H(D')$, with $\gamma \in H(D')$. Then γ is a bijection from D' onto D such that $\gamma^{-1} \in H(D)$.*

Definition: Let D be open. An element f of $H(D)$ is said to be *strictly injective* if it is injective and such that f' has no zero in D.

According to Theorems 27.1 and 27.2 in [30], we have these statements:

Theorem 17.11: *Let K have characteristic zero. Let $f \in H(D)$ be injective in D. Then f is strictly injective.*

Theorem 17.12: *Let $a \in K$, $r \in \mathbb{R}_+$, let $f(x) = \sum_{n=0}^{\infty} a_n(x-a)^n \in H(d(a,r))$, and let $s = \sup_{n\geq 1} |a_n| r^n$ be > 0. Then the following statements are equivalent:*

(*i*) $|a_1| > |a_n| r^{n-1}$ *whenever* $n > 1$
(*ii*) $|f(x) - f(y)| = |x-y||a_1|$ *whenever* $x, y \in d(a,r)$
(*iii*) f *is strictly injective in* $d(a,r)$.

Moreover when conditions $(i), (ii), (iii)$ are satisfied, then we have $s = |a_1| r$ and $|f'(x)| = |a_1|$ whenever $x \in d(a,r)$.

Theorem 17.13 is proven in [30] in Theorem 26.1 and Corollary 26.2:

Theorem 17.13: *Let $f(x) = \sum_{n=0}^{\infty} a_n(x-a)^n \in H(d(a,r))$ and let $s = \sup_{n\geq 1} |a_n| r^n$. Then $f(d(a,r)) = d(a_0, s)$, $f(d(a,r^-)) = d(a_0, s^-)$, and $v_a(f - a_0, -\log r) = -\log s$.*

Theorem 17.14 is proven in Chapter 27 of [30], (as Corollary 27.6):

Theorem 17.14: *Let $f \in H(d(a,r))$ be strictly injective and let $d(b,s) = f(d(a,r))$. Then f^{-1} belongs to $H(d(b,s))$.*

Theorem 17.15 is proven in Chapter 10 of [30], (as Proposition 10.14):

Theorem 17.15: *Let D have an empty annulus Λ. Let w_1, w_2 be the functions defined on D by $w_1(x) = 1, w_2(x) = 0$ if $x \in \mathcal{I}(\Lambda)$ and $w_1(x) = 0, w_2(x) = 1$ if $x \in \mathcal{E}(\Lambda)$. Then w_1 and w_2 belong to $H(D)$.*

As an immediate consequence, we have Corollary 17.16.

Corollary 17.16: *If D has infinitely many infraconnected components then $H(D)$ is not noetherian.*

Henceforth, we will also use Theorem 17.17 which was proved in [30] as Theorem 21.12:

Theorem 17.17: *Let $f \in H(D)$. Then $f(D)$ is infraconnected.*

Chapter 18

$\mathcal{T}$-Filters and $\mathcal{T}$-Sequences

The behaviour of analytic elements is linked to the existence of certain pierced filters, called T-filters [20,30]. This has a strong implication on ultrametric spectral theory.

Notation: In all of this chapter D denotes an infraconnected closed set of diameter $S \in \overline{\mathbb{R}}_+$. Let $a \in \widetilde{D}$, let $r \in]\delta(a, D), S] \cap |K|$ and let $(T_i)_{i \in I}$ be the set of the holes of D included in $C(a, r)$. We will denote by $\mathcal{T}(D, a, r)$ the set $\cup_{i \in I} T_i$.

For every $q \in \mathbb{N}$ we will denote by $\mathcal{S}(D, a, r, q)$ the set of the monic polynomials P of degree q whose zeros lie in $\mathcal{T}(D, a, r)$.

Let $a \in \widetilde{D}$. Let $r \in [\delta(a, D), diam(D)] \cap |K|$ and let $q \in \mathbb{N}$. We put

$$\gamma_D(a, r, q) = r^q \inf_{P \in \mathcal{S}(D,a,r,q)} \left\| \frac{1}{P} \right\|_D.$$

Lemma 18.1: *Let $D = d(a, r^-) \setminus d(a, \rho^-)$, with $0 < \rho < r$. Then $\gamma_D(a, r, q) = \left(\frac{r}{\rho}\right)^q$.*

Proof: Indeed, for each monic polynomial P of degree q, having all its zeros in $d(a, \rho^-)$ we have $\left\| \frac{1}{P} \right\|_D = \frac{1}{\rho^q}$. □

In [30] we have shown this proposition, as Corollary 34.3.

Proposition 18.2: *Let $a \in \widetilde{D}$, let $r \in [\delta(a, D), diam(D)] \cap |K|$ and let $q \in \mathbb{N}$. Let $(T_i)_{i \in I}$ be the set of the holes of D included in $C(a, r)$ with*

$T_i = d(a_i, r_i^-)$. *Then we have*

$$\gamma_D(a, r, q)$$
$$= \frac{r^q}{\sup\left\{\inf_{1\le j\le \ell}\left(r_{i_j}^{t_j}\prod_{\substack{1\le m\le \ell\\ m\ne i_j}}|a_m - a_{i_j}|^{t_m}\right)\ \middle|\ 1\le \ell\le q,\ (i_1,\dots,i_\ell)\in I^\ell,\ \sum_{j=1}^{\ell} t_j = q\right\}}.$$

Definitions: Let $\mathcal{F}$ be an increasing filter on D of center $a \in \widetilde{D}$ and diameter S. The filter $\mathcal{F}$ will be called *an increasing T- filter* if there exists a sequence of circles $(\Sigma_m)_{m\in\mathbb{N}}$ with $\Sigma_m = C(a, d_m)$ $(m \in \mathbb{N})$ such that $\lim_{m\to\infty} d_m = S$ and $d_m < d_{m+1}$ together with a sequence of natural integers $(q_m)_{m\in\mathbb{N}}$ satisfying

$$\lim_{m\to\infty} \gamma_D(a, d_m, q_m)\prod_{j=1}^{m-1}\left(\frac{d_j}{d_m}\right)^{q_j} = 0. \tag{18.1}$$

A decreasing filter $\mathcal{F}$ will be called *a decreasing T-filter* if it admits a base $(D_m)_{m\in\mathbb{N}}$ with $D_m = d(a_m, d_m) \cap D \setminus (\cap_{m\in\mathbb{N}} d(a_m, r_m))$ together with a sequence of natural integers $(q_m)_{m\in\mathbb{N}}$ such that, putting $\Sigma_m = C(a_m, d_m)$, the sequences $(a_m)_{m\in\mathbb{N}}, (d_m)_{m\in\mathbb{N}}, (q_m)_{m\in\mathbb{N}}$ satisfy

$$\lim_{m\to\infty} \gamma_D(a, d_m, q_m)\prod_{j=1}^{m-1}\left(\frac{d_m}{d_j}\right)^{q_j} = 0. \tag{18.2}$$

In particular, this definition holds when $\mathcal{F}$ is a decreasing filter of center a and diameter S, and then we can take $a_m = a$ for every $m \in \mathbb{N}$.

Another way to define T-filters consists of introducing T-sequences. We will call *a weighted sequence* a sequence $(T_{m,i}, q_{m,i})_{1\le i\le s(m), m\in\mathbb{N}}$ with $(T_{m,i})_{1\le i\le s(m), m\in\mathbb{N}}$ a monotonous distances holes sequence and $(q_{m,i})_{1\le i\le s(m), m\in\mathbb{N}}$ a sequence of integers.

A weighted sequence $(T_{m,i}, q_{m,i})_{1\le i\le s(m), m\in\mathbb{N}}$ will be said to be *idempotent* if $q_{m,i} = 0$ or 1 for all (m, i), $(1 \le i \le s(m),\ m \in \mathbb{N})$.

All definitions given about monotonous distances holes sequences will apply in the same way to weighted sequences: a weighted sequence $(T_{m,i}, q_{m,i})_{1\le i\le s(m), m\in\mathbb{N}}$ will be said *to be something* (or *to satisfy a certain property*) if *so is* (or *so does*) the monotonous distances holes sequence $(T_{m,i})_{1\le i\le s(m), m\in\mathbb{N}}$. In particular, the diameter, the monotony, the piercing of the monotonous distances holes sequence $(T_{m,i})_{1\le i\le s(m), m\in\mathbb{N}}$ will be called, respectively, *the diameter, the monotony, the piercing* of the weighted sequence $(T_{m,i}, q_{m,i})_{1\leq i\leq s(m), m\in\mathbb{N}}$.

Now let $(T_{m,i}, q_{m,i})_{1\leq i\leq s(m), m\in\mathbb{N}}$ be an increasing (resp. decreasing) weighted sequence of monotony $(d_m)_{m\in\mathbb{N}}$, associated to $\mathcal{F}$ and for every (i, m), $(1 \leq i \leq s(m), m \in \mathbb{N})$, let $T_{m,i} = d(a_{m,i}, \rho^-_{m,i})$ and let $q_m = \sum_{i=1}^{s(m)} q_{m,i}$.

The weighted sequence will be said to be a *T-sequence* if it satisfies

$$\lim_{m\to\infty}\left(\sup_{1\leq j\leq s(m)}\left[\left(\frac{d_m}{\rho_{m,j}}\right)^{q_{m,j}}\prod_{\substack{i\neq j\\1\leq i\leq s(m)}}\left(\frac{d_m}{|a_{m,i}-a_{m,j}|}\right)^{q_{m,i}}\right]\prod_{n=1}^{m-1}\left(\frac{d_n}{d_m}\right)^{q_n}\right) = 0 \tag{18.3}$$

(resp.

$$\lim_{m\to\infty}\left(\sup_{1\leq j\leq s(m)}\left[\left(\frac{d_m}{\rho_{m,j}}\right)^{q_{m,j}}\prod_{\substack{i\neq j\\1\leq i\leq s(m)}}\left(\frac{d_m}{|a_{m,i}-a_{m,j}|}\right)^{q_{m,i}}\right]\prod_{n=1}^{m-1}\left(\frac{d_m}{d_n}\right)^{q_n}\right) = 0 \tag{18.4}$$

).

Given a T-sequence $(T_{m,i}, q_{m,i})_{1\leq i\leq s(m), m\in\mathbb{N}}$, a monotonous filter $\mathcal{F}$ will be said *to admit the T-sequence* $(T_{m,i}, q_{m,i})_{1\leq i\leq s(m), m\in\mathbb{N}}$ if it is associated to the monotonous distances holes sequence $(T_{m,i})_{1\leq i\leq s(m), m\in\mathbb{N}}$.

Remark: Given a T-sequence $(T_{m,i}, q_{m,i})_{1\leq i\leq s(m), m\in\mathbb{N}}$, for every $t \in \mathbb{N}$, the weighted sequence $(T_{m,i}, q_{m,i})_{1\leq i\leq s(m), m\geq t}$ is a T-sequence again.

The proofs of statements 18.3 to 18.8 are given in Chapter 36 of [30].

Lemma 18.3 *Let $(T_{m,i}, q_{m,i})_{1\leq i\leq s(m), m\in\mathbb{N}}$ be an increasing (resp. decreasing) weighted sequence, let $q_m = \sum_{i=1}^{s(m)} q_{m,i}$ and let $(C(a_m, d_m))_{m\in\mathbb{N}}$ be a sequence of circles that runs the weighted sequence. The weighted sequence is a T-sequence if and only if there exists a sequence of monic polynomials $(Q_m)_{m\in\mathbb{N}}$ such that for each $(m, i)_{1\leq i\leq s(m)}$, Q_m admits exactly* *a zero of order $q_{m,i}$ in $T_{m,i}$ and has no other zero in K, satisfying further*

$$\lim_{m\to\infty}\left(\left(\varphi_{a_m,d_m}(Q_m)\|\frac{1}{Q_m}\|_{C(a_m,d_m)\cap D}\right)\prod_{n=1}^{m-1}\left(\frac{d_n}{d_m}\right)^{q_n}\right)=0$$

$$\left(\text{resp. } \lim_{m\to\infty}\left(\left(\varphi_{a_m,d_m}(Q_m)\|\frac{1}{Q_m}\|_{C(a_m,d_m)\cap D}\right)\prod_{n=1}^{m-1}\left(\frac{d_m}{d_n}\right)^{q_n}\right)=0\right).$$

Lemma 18.4: *Let $\mathcal{F}$ be a monotonous filter on D. Then $\mathcal{F}$ is a T-filter if and only if there exists a T-sequence associated to $\mathcal{F}$.*

Proposition 18.5: *Let $\mathcal{F}$ be an increasing (resp. decreasing) T-filter of center α and diameter r. Let $h(x)=\frac{1}{x-a}$, let $D'=h(D)$ and $\mathcal{F}'=h(\mathcal{F})$. If $|a-\alpha|<r$, $\mathcal{F}'$ is a decreasing (resp increasing) T-filter of center α and diameter $\frac{1}{r}$. If $|a-\alpha|\geq r$, $\mathcal{F}'$ is an increasing (resp. decreasing) T-filter of center $h(\alpha)$ and diameter $\frac{r}{|a-\alpha|^2}$.*

Proposition 18.6: *Let D admit a T-sequence $(T_{m,i},q_{m,i})_{1\leq i\leq k(m),\ m\in\mathbb{N}}$, and let $(V_{m,j})_{1\leq j\leq l(m),\ m\in\mathbb{N}}$ be a monotonous distance holes sequence of an infraconnected set $E\subset D$ such that for every (m,i), $1\leq i\leq k(m)$, $m\in\mathbb{N}$, $T_{m,i}$ is included in certain hole $V_{m,j}$ of E. For every (m,j) $(1\leq j\leq l(m))$, we denote by $\mathcal{T}_{m,j}$ the set of the (m,i) such that $T_{m,i}$ is included in $V_{m,j}$ and we put $s_{m,j}=\sum_{(m,i)\in\mathcal{T}_{m,j}}q_{m,i}$ when $\mathcal{T}_{m,j}\neq\emptyset$, and $s_{m,j}=0$ when $\mathcal{T}_{m,j}=\emptyset$. Then the weighted sequence $(V_{m,j},s_{m,j})_{1\leq j\leq l(m),\ m\in\mathbb{N}}$ is a T-sequence of E.*

Lemma 18.7 is elementary:

Lemma 18.7: *Let $(d_m)_{m\in\mathbb{N}}$ be a strictly monotonous sequence of limit r. Then $\lim_{m\to\infty}\sum_{j=0}^{m-1}|\log d_m-\log d_j|_\infty=+\infty$ if and only if $\sum_{j=0}^{\infty}|\log r-\log d_j|_\infty=+\infty$.*

Theorem 18.8 is an immediate application of Lemma 18.8

Theorem 18.8: *Let $(T_m)_{m\in\mathbb{N}}$ be a well pierced monotonous distances holes sequence of monotony $(d_m)_{m\in\mathbb{N}}$ of diameter r. There exists an idempotent T-sequence $(T_m,u_n)_{m\in\mathbb{N}}$ if and only if $\sum_{j=0}^{\infty}|\log r-\log d_j|_\infty=+\infty$.*

Lemma 18.9 is elementary:

Lemma 18.9: *Let E be a set which is not countable and let f be a function from D into $\mathbb{R}_+$. There exists a sequence $(x_m)_{m\in\mathbb{N}}$ in E and $\lambda>0$ such that $f(x_m)\geq\lambda$ for all $m\in\mathbb{N}$.*

As an improvement of Theorem 36.10 in [30], we can state:

Theorem 18.10: *Let K be strongly valued. Let $a \in \widetilde{D}$ be such that $\delta(a, D) < diam(D)$. We assume that for every $r \in]\delta(a, D), diam(D)[\cap|K|$, each class of $C(a, r)$, but maybe finitely many ones, contains at least one hole of D.*

Then for each $r \in]\delta(a, D), diam(D)]$ there exists on D an increasing idempotent T-sequence and a decreasing idempotent T-sequence of center a and diameter r.

Proof: The proof is the same as that of Theorem 36.10 in [30]. Let $r \in]\delta(a, D), diam(D)]$. Let $J =]\delta(a, D), diam(D)[\cap|K|$, and let $(r_n)_{n\in\mathbb{N}}$, $(r'_n)_{n\in\mathbb{N}}$ be sequences of J satisfying $r_n < r'_n < r_{n+1}$, $\lim_{n\to\infty} r_n = r$. By hypothesis, either the residue class field $\mathcal{K}$ of K is not countable, or $|K|$ is not.

First we suppose that $|K|$ is not countable. Let n be fixed. By Lemma 18.9 there exists $\rho_n > 0$ together with an infinite family of circles $C(a, r)$, with $r_n < r < r'_n$, every one of them contains at least one hole of diameter bigger than ρ_n. Let $q_n \in \mathbb{N}$ satisfy

$$\left(\frac{r_n}{r'_n}\right)^{q_n} \frac{r'_{n+1}}{\rho_{n+1}} < \frac{1}{n+1}. \tag{18.5}$$

Thus, we have defined the sequences $(\rho_n)_{n\in\mathbb{N}}$, $(q_n)_{n\in\mathbb{N}}$. Now, for each $n \in \mathbb{N}$, we put $t_n = \sum_{h=1}^{n} q_h$. From above, we can clearly find q_n circles $C(a, d_m)_{t_n \le m < t_{n+1}}$, with $r_n < d_m < d_{m+1} < r'_n$, which all contain at least one hole T_m of diameter bigger than ρ_n.

Then by definition T_m is included in $C(a, d_m)$ while $T_h \cap C(a, d_m) = \emptyset$ whenever $h \neq m$. Moreover when $t_n \le m < t_{n+1}$, by hypothesis we have $diam(T_m) \ge \rho_n$. Thus by (18.5) the weighted sequence $(T_m, 1)$ is satisfies

$$\frac{d_m}{diam(T_m)} \prod_{j=1}^{m} \frac{d_j}{d_m} \le \frac{1}{n} \quad \text{whenever } t_n \le m < t_{n+1}.$$

This shows that

$$\lim_{m\to\infty} \frac{d_m}{diam(T_m)} \prod \frac{d_j}{d_m} = 0$$

and therefore the weighted sequence $(T_m, 1)$ is an idempotent T-sequence.

Now we suppose that the residue class field $\mathcal{K}$ is not countable. By Lemma 18.9, for each $n \in \mathbb{N}$, there exist $\rho_n > 0$ together with an infinite family of classes of $C(a, r_n)$, such that each contains at least one hole of diameter bigger than ρ_n. In the same way as the previous case, for each $n \in \mathbb{N}$, we take $q_n \in \mathbb{N}$ satisfying (18.5) again and then, we have defined sequences $(\rho_n)_{n\in\mathbb{N}}$, $(q_n)_{n\in\mathbb{N}}$. For each $n \in \mathbb{N}$, we put $t_n = \sum_{h=1}^{n} q_h$. Then in $C(a, r_n)$, we can clearly find q_n different classes such that each contains at least one hole of diameter bigger than ρ_n. Let $(T_m)_{t_n \leq m < t_{n+1}}$ be these holes. Thus by definition, we have $\delta(T_j, T_h) = r_n$ whenever h, j such that

$$t_n \leq h < t_{n+1}, t_n \leq j < t_{n+1}, h \neq j. \tag{18.6}$$

By (18.6) it is easily checked that $\gamma_D(a, r_n, q_n) \leq \frac{r_n}{\rho_n}$. Therefore the weighted sequence $(T_m, 1)_{m\in\mathbb{N}}$ is an idempotent T-sequence because by (18.5) we have

$$\gamma_D(a, r_n, q_n) \prod_{j=1}^{n} \left(\frac{r_j}{r_n}\right)^{q_j} \leq \frac{1}{n}.$$

Symmetrically, we can prove the existence of a decreasing idempotent T-sequence of center a and diameter r: in the proof, we just have to replace (18.5) by

$$\left(\frac{r'_n}{r_n}\right)^{q_n} \frac{r_{n+1}}{\rho_{n+1}} < \frac{1}{n+1}.$$

□

T-filters are used to describe the ability of certain analytic elements to vanish in a non trivial way. We have to recall this.

Definition: Let $\mathcal{F}$ be a monotonous filter of diameter S, on an infraconnected set $D \subset K$, and let $f \in H(D)$. First, f is said to be *vanishing along* $\mathcal{F}$ if $\lim_{\mathcal{F}} f(x) = 0$.

When $\mathcal{F}$ is decreasing (resp. increasing) of center a, then f is said to be *strictly vanishing* along $\mathcal{F}$ if it is vanishing along $\mathcal{F}$ and if there exists $S' > S$ (resp. $S' < S$) such that for every $r \in]S, S']$ (resp. $r \in [S', S[$) we have $\varphi_{a,r}(f) \neq 0$.

When $\mathcal{F}$ is decreasing with no center in K, admitting a canonical base $(D_n)_{n\in\mathbb{N}}$ with $D_n = d(a_n, r_n) \cap D$, then f is said to be *strictly vanishing* along $\mathcal{F}$ if it is vanishing along $\mathcal{F}$ and if there exists $S' > S$ such that $\varphi_{a_{n+1},r}(f) \neq 0$ whenever $r \in [r_n, S']$, whenever $n \in \mathbb{N}$.

In Chapter 37 of [30] it is shown that given a monotonous filter $\mathcal{F}$ on an infraconnected set D, there exist elements $H(D)$ strictly vanishing along $\mathcal{F}$

if and only if $\mathcal{F}$ is a T-filter. Moreover, by Theorem 37.2 in [30], we have this theorem:

Theorem 18.11: *Let $\mathcal{F}$ be a T-filter on D, let $(T_{m,i},\ m_{q,i})_{1\leq i\leq s(m),m\in\mathbb{N}}$ be a T-sequence associated to $\mathcal{F}$, and $(\sigma_m)_{m\in\mathbb{N}}$ be the sequence of circles carrying this T-sequence. Let $D' = K\setminus(\cup_{1\leq i\leq s(m),m\in\mathbb{N}}T_{m,i})$, and for every (m,i), let $a_{m,i}\in T_{m,i}$. Let $\mathcal{F}'$ be the T-filter on D' associated to the T-sequence $(T_{m,i},q_{m,i})$. Let $\alpha\in\mathcal{C}(F')$. There exists $g\in H(D')$ satisfying these properties:*

- *(i) g is meromorphic in $T_{m,i}$, admits $a_{m,i}$ as a pole of order at most $q_{m,i}$ and has no other pole in $T_{m,i}$,*
- *(ii) g is strictly vanishing along $\mathcal{F}'$ and equal to zero in $\mathcal{P}_{D'}(\mathcal{F}')$,*
- *(iii) for every circular filter $\mathcal{G}$ different from $\mathcal{F}$ and secant with $\mathcal{C}(\mathcal{F}')$, ${}_{D'}\varphi_{\mathcal{G}}(g)$ is different from 0,*
- *(iv) $g(\alpha)\neq 0$.*

In this work, we will mainly use idempotent T-sequences. In particular, we will also need Proposition 45.3 in [30]:

Proposition 18.12: *Let D admit an increasing (resp. a decreasing) T-sequence $(T_{m,i},q_{m,i})_{1\leq i\leq k_m,m\in\mathbb{N}}$ of piercing $\rho>0$, of center α and diameter r. Let $D'=K\setminus(\cup_{1\leq i\leq k_m,m\in\mathbb{N}}T_{m,i})$. Let $\mathcal{F}$ be the T-filter on D' associated to this T-sequence. For each (m,i) $(1\leq i\leq k_m\quad m\in\mathbb{N})$, let $b_{m,i}\in T_{m,i}$. Let $\epsilon\in]0,+\infty[$. There exists $\phi_\epsilon\in\mathcal{I}_0(\mathcal{F})$, meromorphic on each hole $T_{m,i}$, satisfying further*

- *(i) $|\phi_\epsilon(x)-1|\leq\epsilon$ whenever $x\in D\cap d(\alpha,r(1-\epsilon))$ (resp. $x\in D\setminus d(\alpha,(\frac{r}{1-\epsilon})^-)$),*
- *(ii) $\|\phi_\epsilon\|_{D'}\leq(\frac{r}{\rho})^3+r^3\epsilon$,*
- *(iii) $b_{m,i}$ is a pole of ϕ_ϵ of order $u_{m,i}\leq q_{m,i}$ and ϕ_ϵ has no pole different from $b_{m,i}$ in $T_{m,i}$ whenever $i=1,\ldots k_m,\ m\in\mathbb{N}$.*

Chapter 19

Applications of T-Filters and T-Sequences

T-filters help us characterize quasi-invertible analytic elements and solve various other problems (Theorem 38.4 in [30]). In particular, they provide a characterization of quasi-invertible elements and overall they let certain analytic elements be different from zero in an area, and identically zero in another area.

Notation: Throughout this chapter D denotes an infraconnected set of diameter $S \in \overline{\mathbb{R}}_+$.

Theorem 19.1: *Let D be open closed and bounded and let $f \in H(D)$. Then f is quasi-invertible if and only if it is not vanishing along a T-filter.*

From Theorem 19.1 here, and from Theorems 16.2, 38.9, and 42.5 in [30] we must recall the characterization of principal K-algebras $H(D)$.

Theorem 19.2: *Let D be infraconnected, open closed and bounded. Then the following statements are equivalent:*

(*i*) *D has no T-filter;*
(*ii*) *$H(D)$ is noetherian;*
(*iii*) *$H(D)$ is a principal ideal ring;*
(*iv*) *every ideal of $H(D)$ is generated by a polynomial whose zeros lie in D;*
(*v*) *Every element of $H(D)$ is quasi-invertible.*

Moreover, when the conditions above are satisfied, every maximal ideal $\mathcal{M}$ of $H(D)$ has codimension 1, and is associated to a point $a \in D$ such that $\mathcal{M} = \{f \in H(D) \mid f(a) = 0\}$ and for every $f \in H(D)$, we have $sp(f) = f(D)$.

Proof: According to Theorem 19.1 here and to Theorems 16.2, 38.9, and 42.5 in [30], the only new statement here is that for every $f \in H(D)$, we have $sp(f) = f(D)$, and this comes from the fact that every maximal ideal $\mathcal{M}$ of $H(D)$ has codimension 1, and is associated to a point $a \in D$ such that $\mathcal{M} = \{f \in H(D) \mid f(a) = 0\}$. $\square$

Notation: Let $\mathcal{F}$ be a circular filter or a monotonous filter on D. Henceforth, we will denote by $\mathcal{Y}(\mathcal{F})$ the set of the $f \in H(D)$ such that $\lim_{\mathcal{F}} f(x) = 0$. Hence $\mathcal{Y}(\mathcal{F})$ is equal to $Ker(\varphi_{\mathcal{F}})$, and therefore is a closed prime ideal of $H(D)$.

If $\mathcal{F}$ is a monotonous filter on D, we will denote by $\mathcal{Y}_0(\mathcal{F})$ the closed ideal of the $f \in H(D)$ such that $f(x) = 0$ whenever $x \in \mathcal{P}(\mathcal{F})$.

Lemma 19.3: *Let D be a closed bounded infraconnected set of diameter s admitting a partition of the form $(\mathcal{C}(\mathcal{F}_i))_{i \in I}$ where each $\mathcal{F}_i$ is an increasing T-filter, and assume that D has no other T-filter. Let $\mathcal{G}$ be the D-peripheral circular filter. Then $\mathcal{Y}(\mathcal{F}_i) = \mathcal{Y}(\mathcal{G})$, $\forall i \in I$.*

Proof: Let $i \in I$ be fixed, let a_i be a center of $\mathcal{F}_i$, and let $r_i = diam(\mathcal{F}_i)$. Let $f \in \mathcal{Y}(\mathcal{F}_i)$, so we have $\varphi_{a_i,r_i}(f) = 0$. Suppose there exists $r > r_i$ such that $\varphi_{a_i,r}(f) > 0$. Let $u = \inf\{r \mid \varphi_{a_i,r}(f) > 0\}$. Then f is strictly vanishing along the decreasing filter of center a_i and diameter u, so this filter is a T-filter, a contradiction to the hypothesis. Consequently, we have $\varphi_{a_i,s}(f) = 0$, and therefore $\mathcal{Y}(\mathcal{F}_i) \subset \mathcal{Y}(\mathcal{G})$.

Now, let $f \in \mathcal{Y}(\mathcal{G})$, so we have $\varphi_{a_i,r}(f) = 0$. Suppose that $\varphi_{a_i,r_i}(f) > 0$. Since $\varphi_{a_i,r}(f) = 0$, we can consider $u = \inf\{r \mid \varphi_{a_i,r}(f) = 0\}$. Then f is strictly vanishing along the increasing filter of center a_i and diameter u, so this filter is a T-filter, a contradiction to the hypothesis. Consequently, we have $\varphi_{a_i,r_i}(f) = 0$, and therefore $\mathcal{Y}(\mathcal{F}_i) = \mathcal{Y}(\mathcal{G})$. $\square$

Corollary 19.4: *Let D be a closed bounded infraconnected, let $a \in D$ and $r > 0$ such that $D \cap d(a, r^-)$ admit a partition of the form $(\mathcal{C}(\mathcal{F}_i))_{i \in I}$ where each $\mathcal{F}_i$ is an increasing T-filter, and assume that every other T-filter on $D \cap d(a, r^-)$ surrounds all the $\mathcal{F}_i$. Then, $\mathcal{Y}(\mathcal{F}_i) = \mathcal{Y}(\mathcal{F}_j)$, $\forall i, j \in I$.*

Definition: A partition $(A_i)_{i \in I}$ of an infraconnected set E is said to be *T-optimal* if for each $i \in I$, there exists an increasing T-filter $\mathcal{F}_i$ of D which is not surrounded by any increasing T-filter different from $\mathcal{F}_i$, such that $\mathcal{C}(\mathcal{F}_i) = A_i$.

If D admits a T-optimal partition $(\mathcal{C}(\mathcal{F}_i))_{i\in I}$, this partition will be said to be *T-specific* if for every decreasing T-filter $\mathcal{G}$ on D there exists $h \in I$ such that $\mathcal{G}$ is secant with $\mathcal{C}(\mathcal{F}_h)$.

Recall Lemma 41.3 in [30]:

Lemma 19.5: *If D admits a T-optimal partition, it is unique.*

Definition: A circular filter $\mathcal{F}$ on D is said to be *distinguished* if $D\setminus Q(\mathcal{F})$ either is empty, or is of the form $\mathcal{C}(\mathcal{G})$, with $\mathcal{G}$ a decreasing T-filter, and if $Q(\mathcal{F})$ either is empty or admits a T-specific partition.

A distinguished circular filter is said to be *regular* if either $Q(\mathcal{F}) = \emptyset$ or all elements of the T-specific partition of $Q(\mathcal{F})$ have a diameter equal to $diam(\mathcal{F})$. Else, it will be said to be *irregular*.

In the same way, a T-specific partition $(A_i)_{I\in I}$ of an infraconnected set E will be said to be *regular* if $diam(A_i) = diam(E)\ \forall i \in I$. Else, it will be said to be *irregular*.

According to the proof of Theorem 41.7 in [30] we can state Proposition 19.6:

Proposition 19.6: *If K is strongly valued, all T-specific partition of infraconnected subsets of K are regular. If K is weakly valued, there exists a bounded open closed infraconnected set E such that $\widetilde{E} = d(0,1)$, admitting $d(0,(\frac{1}{2})^-)$ as a hole, with an irregular T-specific partition whose elements are the sets $C(0,r) \cap d(a,r^-)$, whenever $|a| = r$, $\frac{1}{2} < r < 1$.*

By Proposition 42.4, Corollaries 42.7, 42.8 and Theorems 42.9, 42.10 in [30], we have the following statements:

Proposition 19.7: *Let $\mathcal{F}$ be a distinguished circular filter such that $\mathcal{Q}(\mathcal{F}, D) \neq \emptyset$. For every increasing T-filter $\mathcal{F}_i$ of the T-specific partition of $\mathcal{Q}(\mathcal{F}, D)$ we have $\mathcal{Y}(\mathcal{F}_i) = \mathcal{Y}(\mathcal{F})$.*

Corollary 19.8: *The restriction of the mapping $\mathcal{Y}$ to the set of distinguished circular filters is a bijection from this set onto $Max_\infty(H(D))$.*

By Theorem 16.20, we can state Corollary 19.9:

Corollary 19.9: *Let D be a closed bounded infraconnected set with no T-filter. Then the mapping ϕ from D into the set of ideals of $H(D)$ defined by $\phi(a) = \mathcal{Y}(a)$ is a bijection from D onto $Max(H(D))$.*

Remark: In particular, we notice that if $D = d(0,1^-)$, then all maximal ideal are of the form $\mathcal{Y}(a)$, $a \in D$, hence every $\psi \in Mult_m(H(D), \|\ .\ \|_D)$

a kernel of the form $\mathcal{Y}(a)$, hence $\psi(x) = |a| < \|x\|_D$. This shows that there exists no $\psi \in Mult_m(H(D), \| \,.\, \|_D)$ such that $\psi(x) = \|x\|_D$.

As a consequence of Proposition 19.6, we have Theorem 19.10 which is proven in [30] as Theorem 42.10.

Theorem 19.10: *Let K be strongly valued and let D be closed and bounded. The mapping Θ from $Mult_m(H(D), \| \,.\, \|)$ into $Max(H(D))$ defined as $\Theta(\psi) = Ker(\psi)$ is a bijection.*

Theorem 19.11: *If K is weakly valued there exists a closed bounded infraconnected set E in K such that $H(E)$ admits multiplicative semi-norms $\psi_1, \psi_2 \in Mult_m(H(E), \| \,.\, \|_E)$, with $\psi_1 \neq \psi_2$ such that $Ker(\psi_1) = Ker(\psi_2)$.*

The following Theorem 19.12 is a consequence of Theorem 38.10 in [30]:

Theorem 19.12: *Let D be closed and bounded and have finitely many infraconnected components $D_1, \dots, D_q$. Then $H(D)$ is isomorphic to the direct product $H(D_1) \times \dots \times H(D_q)$.*

In [7] (and [30]) we characterized continuous multiplicative semi-norms in a K-algebra $H(D)$ which are absolute values:

Theorem 19.13: *Let $\mathcal{F}$ be a large circular filter on D. Then $\varphi_{\mathcal{F}}$ is not an absolute value if and only if it satisfies one of these conditions:*

(*i*) *There exists a T-filter $\mathcal{G}$ on D such that $\mathcal{F}$ is secant with $\mathcal{P}(\mathcal{G})$,*
(*ii*) *$\mathcal{F}$ is T-filter.*

Corollary 19.14: *All non punctual continuous multiplicative seminorms of $H(D)$ are absolute values if and only if D has no T-filter.*

Definitions and notations: Let D be closed and bounded. Let $incT(D)$ (resp. $decT(D)$) be the set of the increasing (resp. decreasing) T-filters on D. We will denote by $<$ the relation defined on $incT(D)$ (resp. $decT(D)$) by $\mathcal{F}_1 < \mathcal{F}_2$ if $\mathcal{C}(\mathcal{F}_2) \subset \mathcal{C}(\mathcal{F}_1)$ and $\mathcal{C}(\mathcal{F}_2) \neq \mathcal{C}(\mathcal{F}_1)$. In other words, denoting by $\widetilde{\mathcal{F}}$ the circular filter less thin than a monotonous filter $\mathcal{F}$, if $\mathcal{F}_1$ and $\mathcal{F}_2$ are decreasing T-filters, then $\mathcal{F}_1 < \mathcal{F}_2$ is equivalent to say that $\widetilde{\mathcal{F}_1} \prec \widetilde{\mathcal{F}_2}$, and if $\mathcal{F}_1$ and $\mathcal{F}_2$ are increasing T-filters, then $\mathcal{F}_1 < \mathcal{F}_2$ is equivalent to say that $\widetilde{\mathcal{F}_2} \prec \widetilde{\mathcal{F}_1}$. An increasing (resp. a decreasing) T-filter $\mathcal{F}$ will be said to be *maximal* if it is maximal in $incT(D)$ (resp. in $decT(D)$) with respect to this relation.

We will call *an ascending chain of increasing (resp. decreasing) T-filters* a sequence of increasing (resp. decreasing) T-filters $(\mathcal{F}_n)_{n\in\mathbb{N}}$ such that $\mathcal{F}_n < \mathcal{F}_{n+1}$ whenever $n \in \mathbb{N}$.

Let $(\mathcal{F}_n)_{n\in\mathbb{N}}$ be an ascending chain of increasing T-filters. For each $n \in \mathbb{N}$ let $r_n = diam(\mathcal{F}_n)$. Since the sequence $(r_n)_{n\in\mathbb{N}}$ is decreasing, we put $r = \lim_{n\to\infty} r_n$, and then r will be named *the diameter of the chain.*

We put $A = \cap_{n\in\mathbb{N}}\mathcal{C}(\mathcal{F}_n)$ and for each $n \in \mathbb{N}, D_n = \mathcal{C}(\mathcal{F}_n) \setminus A$. The sequence $(D_n)_{n\in\mathbb{N}}$ is then a base of a filter $\mathcal{F}$ on D of diameter r.

If $r = 0$, since D is closed, A is a point a of D, hence $\mathcal{F}$ is the filter of the neighbourhoods of a in D.

If $r > 0$, $\mathcal{F}$ is a decreasing filter on D of diameter r.

In both cases $\mathcal{F}$ will be called *the returning filter of the ascending chain* $(\mathcal{F}_n)_{n\in\mathbb{N}}$.

Now let $(\mathcal{F}_n)_{n\in\mathbb{N}}$ be an ascending chain of decreasing T-filters and let $a \in \mathcal{P}(\mathcal{F}_n)$ for some $n \in \mathbb{N}$. The sequence $(r_n)_{n\in\mathbb{N}}$ is an increasing sequence of limit $r \in]0, +\infty]$, and r will be named *the diameter of the chain.* Since D bounded, we notice that $r < +\infty$, and then we will call *the returning filter of the ascending chain* $(\mathcal{F}_n)_{n\in\mathbb{N}}$ the increasing filter $\mathcal{F}$ of center a and diameter r (it is seen that $\mathcal{F}$ is not depending on the point $a \in \mathcal{P}(\mathcal{F}_n)$, whenever $n \in \mathbb{N}$).

Now, we can recall the characterisation of K-algebras $H(D)$ without non zero divisors of zero which admit no continuous absolute value [7].

Theorem 19.15: *Let $H(D)$ be a K-Banach algebra without non zero divisors of zèro. Then $Mult(H(D), \| \, . \, \|_D)$ contains no norm if and only if D admits an ascending chain of T-filters $(\mathcal{F}_n)_{n\in\mathbb{N}}$ whose returning filter is either a T-filter or a Cauchy filter.*

Remark: By Theorem 19.15 we notice that if a K-Banach algebra $H(D)$ has no non zero divisors of zero and has no continuous absolute value, then $Max(H(D))$ admits a partition of the form $\{F, G\}$, where F is a non empty closed subset for Jacobson's topology, and G is of the form $\cup_{n=1}^{\infty} F_n$, where each F_n is a non empty closed subset for Jacobson's topology, such that $F_n \subset F_{n+1}$, $F_n \neq F_{n+1}$. Indeed, each set $\mathcal{P}(\mathcal{F}_n)$ is a closed subset of $Max(H(D))$ which is identified with D.

Chapter 20

Analytic Elements on Classic Partitions

Classic partitions let us generalize the notion of holes for a subset D of K, relatively to a disk containing D. So, we can generalize algebras of analytic elements, a notion introduced in [23].

Definitions: Given a set closed set $S \subset K$, we call *a classic partition of* S a partition of S of the form $\left(d(b_j, r_j^-)\right)_{j \in J}$. The disks $d(b_j, r_j^-)$ are called *the holes* of the partition.

Let $\mathcal{O}$ be a classic partition of $d(a,r)$. A closed infraconnected set E included in $d(a,r)$, will be said to be a *sub-$\mathcal{O}$-set* if every hole of E is a hole of $\mathcal{O}$. Moreover, a sub-$\mathcal{O}$-set E will be called a $\mathcal{O}$-set if $\widetilde{E} = d(a,r)$.

A weighted sequence $(T_{m,i}, q_{m,i})_{1 \leq i \leq k_m,\ m \in \mathbb{N}}$ of any $\mathcal{O}$-set will be called *a weighted sequence of* $\mathcal{O}$. In particular, a T-sequence of any $\mathcal{O}$-set will be called *a T-sequence of* $\mathcal{O}$, and an idempotent weighted sequence of any $\mathcal{O}$-set will be called *an idempotent weighted sequence of* $\mathcal{O}$.

Let $a \in K$ and $r > 0$, let S be a closed subset of $d(a,r)$, and let $\mathcal{O} = \left(d(b_j, r_j^-)\right)_{j \in I}$ be a classic partition of $d(a,r) \setminus S$. An annulus $\Gamma(b, r', r'')$ included in $d(a,r) \setminus S$ will be said to be *$\mathcal{O}$-minorated* if there exists $\lambda > 0$ such that $r_j \geq \lambda$ for every $j \in I$ such that $d(b_j, r_j^-) \subset \Gamma(b, r', r'')$.

Remark: Given a set $D \subset K$, the set $\widetilde{D} \setminus \overline{D}$ admits a unique classic partition $\left(d(b_j, r_j^-)\right)_{j \in J}$ such that $r_j = \delta(b_j, D)\ \forall j \in J$, and then the disks $d(b_j, r_j^-)$ $(j \in J)$ are just the holes of D. This partition will be called the *natural partition of* $\widetilde{D} \setminus D$.

Theorem 20.1 (N. Mainetti [29,37]): *Let $\mathcal{O} = (d(\alpha_j, r_j^-))_{j\in J}$ be a classic partition of the annulus $\Gamma(a, r', r'')$. There exists $h \in J$ and a $\mathcal{O}$-minorated annulus of the form $\Gamma(\alpha_h, r_h, \rho)$ included in $\Gamma(a, r', r'')$.*

Proof: Suppose the claim is false. Without loss of generality we can assume $a = 0$. For every $i \in J$, we put $T_i = d(\alpha_i, r_i^-)$. Since our claim is false, for every $\epsilon > 0$, we can find an index $i \in J$ such that $r_i < \epsilon$. Let us fix $\epsilon > 0$, and let $h \in J$ be such that $r_h < \min(\epsilon, r')$ and Let $s_h \in]r_h, r_h + \epsilon[$. Then $\Gamma(\alpha_h, r_h, s_h)$ is included in $\Gamma(a, r', r'')$ and is not included in any hole T_i. Consequently, $\Gamma(\alpha_h, r_h, s_h)$ admits a classic partition of the form $\mathcal{O}_h = (T_i)_{i\in J_h}$, with $J_h \subset J$. Since our claim is false, we can find another index $k \in J_h$ such that $r_k < \frac{\epsilon}{2}$. In this way, by induction we can construct a sequence $(i_n)_{n\mathbb{N}}$ of J such that $T_{i_{n+1}} \subset \Gamma(a_{i_n}, r_{i_n}, r_{i_n} + \frac{\epsilon}{2^n})$ and $r_{i_n} < \frac{\epsilon}{2^n}$ $\forall n \in \mathbb{N}$. The sequence (α_n) is a Cauchy sequence whose limit α obviously lies in $\Gamma(0, r', r'')$. Hence α belongs to a certain hole T_l. Therefore, when n is big enough then α_{i_n} lies in T_l, a contradiction to the definition of $\mathcal{O}$. □

Proposition 20.2: *Let $S \subset d(a, r)$ and let $\mathcal{O}$ be a classic partition of $d(a, r) \setminus S$ admitting a $\mathcal{O}$-minorated annulus $\Gamma(b, r'r'')$. For all $\rho \in]r', r''[$, there exists an increasing idempotent T-sequence of $\mathcal{O}$ of center b and diameter ρ, together with a decreasing idempotent T-sequence of $\mathcal{O}$ of same center and diameter.*

Theorem 18.10 applied to classic partitions enables us to state Proposition 20.3.

Proposition 20.3: *Let K be strongly valued. Let S be a bounded set included in a disk $d(a, r)$, and let $\mathcal{O}$ be a classic partition of $d(a, r) \setminus S$. We assume that for every $r \in]\delta(a, S),\ diam(S)[\cap|K|$, each class of $C(a, r)$, except maybe finitely many ones, contains at least one hole of $\mathcal{O}$. Then for each $r \in]\delta(a, S), diam(S)]$ there exists in $\mathcal{O}$ an increasing idempotent T-sequence and a decreasing idempotent T-sequence of center a and diameter r.*

Corollary 20.4: *Let K be strongly valued. Let $\mathcal{O} = (d(b_j, r_j^-))_{j\in J}$ be a classic partition of the annulus $\Gamma(a, r', r'')$ and let $s \in]r', r''[$. Then $\mathcal{O}$ admits an increasing idempotent T-sequence and a decreasing idempotent T-sequence of center a and diameter s.*

Corollary 20.5: *Let K be strongly valued. Let S be a countable set, included in a disk $d(a, r)$ and let $\mathcal{O}$ be a classic partition of $d(a, r) \setminus S$.*

For every $l \in]0, r[$, $\mathcal{O}$ admits an increasing and a decreasing idempotent T-sequence of center a and diameter l.

Notation: Let $\mathcal{D}$ be a compact subset of $Mult(K[x])$ and let $D = \{a \in K \mid \varphi_a \in \mathcal{D}\}$. The K-algebra $R(D)$ is provided with the semi-multiplicative norm $\| \, . \, \|_{\mathcal{D}}$ defined as $\|h\|_{\mathcal{D}} = \sup\{\psi(h) \mid \psi \in \mathcal{D}\}$. Then we check that $\|h\|_{\mathcal{D}} \geq \|h\|_D \; \forall h \in R(D)$.

Let E be a closed subset of the disk $d(a,r)$ and let D be a closed subset of E. Let $\mathcal{O} = (d(b_i, r_i^-)_{i \in J})$ be a classic partition of $d(a,r) \setminus E$. We will denote by $\Phi(D, \mathcal{O})$ the set of circular filters $\mathcal{F}$ on K such that all elements $B \in \mathcal{F}$ contain points of D or holes of $\mathcal{O}$. In particular, if $D = \emptyset$, we just put $\Phi(\mathcal{O}) = \Phi(\emptyset, \mathcal{O})$.

Lemma 20.6 is an obvious corollary of Theorem 13.10:

Lemma 20.6: *$\Psi(\Phi(D, \mathcal{O}))$ is the closure of $Mult(R(D), \| \, . \, \|_D) \cap \{\varphi_T \mid T \in \mathcal{O}\}$ in $Mult(K[x])$.*

Notation: Let E be a closed subset of the disk $d(a,r)$ and let D be a closed subset of E. Let $\mathcal{O} = (d(b_i, r_i^-)_{i \in J})$ be a classic partition of $d(a,r) \setminus E$. Let $\mathcal{D} = \Psi(\Phi(D, \mathcal{O}))$. Then on $R(D)$ we put $\| \, . \, \|_{D,\mathcal{O}} = \| \, . \, \|_{\mathcal{D}}$.

We will denote by $H(D, \mathcal{O})$ the completion of $K(x)$ for the norm $\| \, . \, \|_{D,\mathcal{O}}$. In particular, if $D = \emptyset$, we denote by $H(\mathcal{O})$ the completion of $K(x)$ for the norm $\| \, . \, \|_{\mathcal{O}}$.

If $D = \emptyset$, we just put $\|h\|_{\mathcal{O}} = \sup\{\phi(h) \mid \phi \in \Phi(\mathcal{O})\}$. If $\mathcal{O}$ is the partition of holes of an infraconnected bounded set D, then by definition we have $\|f\|_{D,\mathcal{O}} = \|f\|_D \; \forall f \in R(D)$. And if $E = D$, then $\Sigma_0(\mathcal{O}) \subset \Sigma(\mathcal{O}) \subset \Phi(D, \mathcal{O})$

Each $\psi \in Mult(R(D), \| \, . \, \|_{D,\mathcal{O}})$ has continuation to $H(D, \mathcal{O})$ and will be denoted by ψ again.

By Lemma 5.4 we have Lemma 20.7:

Lemma 20.7: *Let $V = d(a,r)$, let E be a closed subset of V, let D be a closed subset of E and let $\mathcal{O}$ be a classic partition of $V \setminus E$. Then $\| \, . \, \|_{D,\mathcal{O}}$ is a semi-multiplicative norm of K-algebra on $R(D)$.*

Remark: Any K-algebra $H(D, \mathcal{O})$ is a uniform Banach K-algebra.

Lemma 20.8: *Let D be a bounded closed infraconnected subset of K and let $(T_i)_{i \in J}$ be the family of holes of D. Let $\mathcal{O} = (T_i)_{i \in J}$. Then $\| \, . \, \|_{D,\mathcal{O}} = \| \, . \, \|_D$.*

Proof: Let $T_i = d(b_i, r_i^-)$, $(i \in J)$ be a hole of $\mathcal{O}$, hence a hole of D. Since D is infraconnected, by Corollary 10.6 the circular filter $\mathcal{F}_i$ of center

b_i and diameter r_i is secant with D, therefore $\varphi_{\mathcal{F}_i}(f) \leq \|f\|_D \ \forall f \in R(D)$, so the claim is obvious. □

Lemma 20.9: *Let $\mathcal{O}$ be a classic partition of a disk $d(a,r)$, and let D be a sub-$\mathcal{O}$-set. Then we have $\|h\|_D \leq \|h\|_{\mathcal{O}}$ for every $h \in R(D)$. Moreover, if D is a $\mathcal{O}$-set, then $\|h\|_D = \|h\|_{\mathcal{O}}$ for every $h \in R(D)$.*

Proof: Let $h \in R(D)$. By Theorem 16.14 there exists a D-bordering filter $\mathcal{F}$ such that $\|h\|_D = \varphi_{\mathcal{F}}$. And $\varphi_{\mathcal{F}}$ is either $\varphi_{a,r}(h)$, or some $\varphi_{b,\rho}(h)$, where $d(b,\rho^-)$ is a hole of D. But since D is sub-$\mathcal{P}$-set, in all cases, $\varphi_{\mathcal{F}}$ belongs to $Mult(R(\mathcal{O}), \| \, . \, \|_{\mathcal{O}})$. Consequently, we have $\|h\|_D \leq \|h\|_{\mathcal{O}}$.

Now, suppose that D is a $\mathcal{O}$-set. Given a hole $d(b,\rho^-)$ of $\mathcal{O}$, this hole is a hole of D because D is a $\mathcal{O}$-set. And since D is infraconnected, this hole defines a circular filter secant with D. Therefore, we have $\varphi_{b,\rho}(h) \leq \|h\|_D$, and consequently $\|h\|_{\mathcal{O}} \leq \|h\|_D$. □

Corollary 20.10: *Let $\mathcal{O}$ be a classic partition of a disk $d(a,r)$, and let E be a $\mathcal{O}$-set. Then $H(E)$ is isometrically isomorphic to a K-subalgebra of $H(\mathcal{O})$.*

Henceforth, given a classic partition $\mathcal{O}$ of a disk $d(a,r)$, and a sub-$\mathcal{O}$-set E, we will consider $H(E)$ as a K-subalgebra of $H(\mathcal{O})$.

Chapter 21

Holomorphic Properties on Partitions

Throughout this chapter D is a closed bounded subset of the disk $d(a,r)$. Given a classic partition, we have a Mittag-Leffler Theorem which generalizes the well known Theorem 16.12 for analytic elements on an infraconnected set [29,37]. Next, properties of T-filters also apply to analytic elements on partitions.

Theorem 21.1: *Let $\mathcal{O}$ be a classic partition of $d(a,r) \setminus D$. Let $f \in H(D,\mathcal{O})$. There exists a unique sequence of holes $(T_n)_{n\in\mathbb{N}^*}$ of $\mathcal{O}$ and a unique sequence $(f_n)_{n\in\mathbb{N}}$ in $H(D,\mathcal{O})$ such that $f_0 \in H(d(a,r))$, $f_n \in H_0(K\setminus T_n)$ $(n>0)$, $\lim_{n\to\infty} f_n = 0$ satisfying*

(*i*) $f = \sum_{n=0}^{\infty} f_n$ *and* $\|f\|_{D,\mathcal{O}} = \sup_{n\in\mathbb{N}} \|f_n\|_{D,\mathcal{O}}$.

Moreover for every hole $T_n = d(a_n, r_n^-)$, we have

(*ii*) $\|f_n\|_{D,\mathcal{O}} = \|f_n\|_{K\setminus T_n} = \varphi_{a_n,r_n}(f_n) \le \varphi_{a_n,r_n}(f) \le \|f\|_{D,\mathcal{O}}$.

(*iii*) $\|f_0\|_{D,\mathcal{O}} = \|f_0\|_{\widetilde{D}} = \varphi_{a,r}(f_0) \le \varphi_{a,r}(f) \le \|f\|_{D,\mathcal{O}}$.

Let $D' = d(a,r)\setminus(\cup_{n=1}^{\infty} T_n)$. Then $f \in H(D')$ and its decomposition in $H(D')$ is given again by (i) and satisfies $\|f\|_{D'} = \|f\|_{D,\mathcal{O}}$.

Proof: Suppose first that f lies in $R(D)$. Without loss of generality, we may denote by $T_1,\dots,T_q$ the holes of $\mathcal{O}$ that contain poles of f. Let $E = d(a,r)\setminus \cup_{j=1}^{q} T_j$. Then E is a $\mathcal{O}$-set. Since E is infraconnected, we can apply Theorem 16.12 that gives all statements above. Now, consider $f \in H(D,\mathcal{O})$ and let $(h_m)_{m\in\mathbb{N}}$ be a sequence in $R(D)$ converging to f in $H(D,\mathcal{O})$. The set of holes of $\mathcal{O}$ containing at least one pole of one term h_m of this sequence is obviously countable and may be defined as a sequence

$(T_n)_{n\in\mathbb{N}^*}$. Thus, each h_m has a unique decomposition $h = \sum_{n=0}^{\infty} h_{m,n}$, with $h_{m,0} \in H(d(0,r))$, and $h_{m,n} \in H_0(K \setminus T_n)$ $\forall n \in \mathbb{N}^*$. So, similarly to the classical proof of the Mittag-Leffler Theorem for infraconnected sets, for any m, $q \in \mathbb{N}$, we have $\|h_m - h_q\|_{D,\mathcal{O}} = \sup_{n\in\mathbb{N}}(\|h_{m,n} - h_{q,n}\|_{D,\mathcal{O}})$ and then, for each fixed $n \in \mathbb{N}^*$, (resp. $n = 0$) the sequence $(h_{m,n})_{m\in\mathbb{N}}$ converges in $H_0(K \setminus T_n)$ (resp. in $H(d(a,r))$) to a limit f_n. We check that the series $\sum_{n=0}^{\infty} h_n$ converges to f in $H(D,\mathcal{O})$, and satisfies all statements above. □

Corollary 21.2: *Let $(T_i)_{i\in I}$ be the family of holes of D. Let J be a subset of I and let $L = I \setminus J$. Let $E = D \cup (\cup_{i\in J} T_i)$ and let $F = K \setminus \cup_{i\in L} T_i$. Then we have $H(D) = H(E) \oplus H_0(F)$, and for each $g \in H_0(E)$, $h \in H(F)$, we have $\|g + h\|_D = \max(\|g\|_E, \|h\|_F)$.*

Definitions and notation: Let $f \in H(D,\mathcal{O})$. We consider the series $\sum_{n=0}^{\infty} f_n$ obtained in Theorem 21.1, whose sum is equal to f in $H(D,\mathcal{O})$, with $f_0 \in H(\widetilde{D}), f_n \in H(K \setminus T_n) \setminus \{0\}$ and with T_n holes of $\mathcal{O}$. Each T_n will be called a *f-hole* and f_n will be called the *Mittag-Leffler term of f associated to T_n*, whereas f_0 will be called the *principal term* of f.

For each f-hole T of $\mathcal{O}$, the Mittag-Leffler term of f associated to T will be denoted by $\overline{f_T}$ whereas the principal term of f will be denoted by $\overline{f_0}$. The series $\sum_{n=0}^{\infty} f_n$ (with $f_n = \overline{f_{T_n}}$ $\forall n \in \mathbb{N}^*$) will be called *the Mittag-Leffler series of f on $(D,\mathcal{O})$.*

Similarly to what was done with sets $Mult(H(D), \|\cdot\|_D)$, here we have a Garandel-Guennebaud's Theorem for algebras $H(D,\mathcal{O})$.

Theorem 21.3: *Let E be a closed set such that $D \subset E \subset d(a,r)$, and let $\mathcal{O}$ be a classic partition of $d(a,r) \setminus E$. For every $\mathcal{F} \in \Phi(D,\mathcal{O})$, the multiplicative semi-norm $\varphi_{\mathcal{F}}$ defined on $R(D)$ extends by continuity to an element ${}_{D,\mathcal{O}}\varphi_{\mathcal{F}}$ of $Mult(H(D,\mathcal{O}), \|\cdot\|_{D,\mathcal{O}})$ such that ${}_{D,\mathcal{O}}\varphi_{\mathcal{F}}(f) = \lim_{\mathcal{F}} |f(x)|$ whenever $f \in R(D)$. Moreover, the mapping Ψ from $\Phi(D,\mathcal{O})$ into $Mult(H(D,\mathcal{O}), \|\cdot\|_{D,\mathcal{O}})$, defined as $\Psi(\mathcal{F}) = {}_{D,\mathcal{O}}\varphi_{\mathcal{F}}$, is a bijection.*

Notation: Let E be a closed set such that $D \subset E \subset d(a,r)$, and let $\mathcal{O}$ be classic partition of $d(a,r) \setminus E$. As we did in $Mult(H(D), \|\cdot\|_D)$, in order to avoiding a too heavy notation, the extension ${}_{D,\mathcal{O}}\varphi_{\mathcal{F}}$ of $\varphi_{\mathcal{F}}$ to $Mult(H(D,\mathcal{O}), \|\cdot\|_{D,\mathcal{O}})$ will just be denoted by $\varphi_{\mathcal{F}}$. Similarly, if $\mathcal{F}$ is the circular filter of center a and diameter ρ, then $\varphi_{a,\rho}$ will still denote the extension of $\varphi_{a,\rho}$ to $(H(D,\mathcal{O}), \|\cdot\|_{D,\mathcal{O}})$.

Let $\mathcal{O} = (d(b_i, r_i^-))_{i\in J}$ be a classic partition of $d(a,r)\setminus D$. We will denote by $R'(D,\mathcal{O},(b_i)_{i\in J})$ (resp. $R''(D,\mathcal{O},(b_i)_{i\in J})$) the K-subvector space of

$R(D)$ consisting of the $h \in R(D)$ of the form $\sum_{i\in I} \frac{\lambda_i}{x-b_i}$, with I a finite subset of J (resp. $\sum_{i\in L} \frac{\lambda_i}{x-b_i} + \frac{\mu_i}{(x-b_i)^2}$). The completion of $R'(D, \mathcal{O}, (b_i)_{i\in J})$ (resp. $R''(D, \mathcal{O}, (b_i)_{i\in J})$) is a Banach K-subvector space of $H(D, \mathcal{O})$ that we will denote by $H'(D, \mathcal{O}, (b_i)_{i\in J})$ (resp. $H''(D, \mathcal{O}, (b_i)_{i\in J})$).

By Theorem 21.1, Proposition 21.4 is immediate:

Proposition 21.4: *Let $\mathcal{O} = (d(b_i, r_i^-)_{i\in J})$ be a classic partition of $d(a,r) \setminus D$. Then $H'(D, \mathcal{O}, (b_i)_{i\in J})$ (resp. $H''(D, \mathcal{O}, (b_i)_{i\in J})$) is equal to the set of $f \in H(D, \mathcal{O})$ of the form $\sum_{i\in I} \frac{\lambda_i}{x-b_i}$, (resp. $\sum_{i\in I} \frac{\lambda_i}{x-b_i} + \frac{\mu_i}{(x-b_i)^2}$) with I a countable subset of J, and $\lim_{\mathcal{H}} \frac{|\lambda_i|}{r_i} = 0$, whereas $\mathcal{H}$ is the filter of complements of finite subsets of I.*

Proposition 21.5 will be useful in the following chapters.

Proposition 21.5: *Let $\mathcal{O} = (d(b_i, r_i^-)_{i\in J})$ be a classic partition of $d(a,r) \setminus D$, and suppose that $\mathcal{O}$ admits an increasing (resp. a decreasing) idempotent T-sequence (T_n) of center b and diameter l. Let $s \in]0, l[$ (resp. $s \in]l, r[$) and let $\epsilon \in]0,1[$. There exists $f \in H'(D, \mathcal{O}, (b_i)_{i\in J})$ satisfying*
(i) $|\psi(f-1)|_\infty < \epsilon \ \forall \psi \in Mult(H(D,\mathcal{O}), \|\ .\ \|_{D,\mathcal{O}})$ such that $\psi(x-b) \le s$ and $\psi(f) = 0 \ \forall \psi \in Mult(H(D,\mathcal{O}), \|\ .\ \|_{D,\mathcal{O}})$ such that $\psi(x-b) \ge l$,
(ii) $\psi(f) \ne 0 \ \forall \psi \in Mult(H(D,\mathcal{O}), \|\ .\ \|_{D,\mathcal{O}}) \setminus Mult_a(H(D,\mathcal{O}), \|\ .\ \|_{D,\mathcal{O}})$ such that $s < \psi(f) < l$.
(resp.
(i) $\psi(f-1) \le \epsilon \ \forall \psi \in Mult(H(D,\mathcal{O}), \|\ .\ \|_{D,\mathcal{O}})$ such that $\psi(x-b) \ge s$ and $\psi(f) = 0 \ \forall \psi \in Mult(H(D,\mathcal{O}), \|\ .\ \|_{D,\mathcal{O}})$ such that $\psi(x-b) \le l$
(ii) $\psi(f) \ne 0 \ \forall \psi \in Mult(H(D,\mathcal{O}), \|\ .\ \|_{D,\mathcal{O}}) \setminus Mult_a(H(D,\mathcal{O}), \|\ .\ \|_{D,\mathcal{O}})$ such that $l < \psi(f) < s$.)

Proof: We set $E = K \setminus (\cup_{n=0}^{\infty} T_n)$. Since the sequence $(T_n, 1)_{n\in\mathbb{N}}$ is an idempotent T-sequence of E, of center b and diameter l, by Theorem 18.11, there exists $f \in H(E)$, strictly vanishing along the increasing (resp. decreasing) T-filter of center b and diameter l, meromorphic on each hole T_n and admitting each b_n as a pole of order at most one, and having no other pole, satisfying further
$|f(x)| = 1$ for all $x \in D \cap d(b,s)$,
$f(x) = 0$ for all $x \in D \setminus d(b, l^-)$ (resp. $|f(x)-1| \le \epsilon$ for all $x \in D \setminus d(b, s^-)$, $f(x) = 0$ for all $x \in D \cap d(b,l)$), and for every circular filter $\mathcal{G}$ of diameter $u < l$, secant with $d(b, l^-)$, $\varphi_{\mathcal{G}}(g) \ne 0$.

Moreover, by Theorem 21.1 for each hole T_n, $\overline{f}_{T_n}$ is of the form $\frac{\lambda_n}{x-b_n}$, so all claims are shown. □

Corollary 21.6: *Let $\mathcal{O}$ be a classic partition of $d(a,r) \setminus D$, and suppose that $\mathcal{O}$ admits a $\mathcal{O}$-minorated annulus $\Gamma(b,r',r'')$. Let l, $\rho', \rho'' \in]r', r''[$ satisfy $\rho' < l < \rho''$. There exist f, $g \in H(D,\mathcal{O})$ satisfying*

(i) $\psi(f) = 1\ \forall \psi \in Mult(H(D,\mathcal{O}), \| \,.\, \|_{D,\mathcal{O}})$ such that $\psi(x-b) \leq \rho'$ and $\psi(f) = 0\ \forall \psi \in Mult(H(D,\mathcal{O}), \| \,.\, \|_{D,\mathcal{O}})$ such that $\psi(x-b) \geq l$,

(ii) $\psi(g) = 1\ \forall \psi \in Mult(H(D,\mathcal{O}), \| \,.\, \|_{D,\mathcal{O}})$ such that $\psi(x-b) \geq \rho''$ and $\psi(g) = 0\ \forall \psi \in Mult(H(D,\mathcal{O}), \| \,.\, \|_{D,\mathcal{O}})$ such that $\psi(x-b) \leq l$.

Lemma 21.7: *Let K be strongly valued, let $\mathcal{O}$ be a classic partition of $d(a,r)$ and let $\mathcal{F}$, $\mathcal{G} \in \Phi(\mathcal{O})$, $\mathcal{F} \neq \mathcal{G}$. There exists $f \in H(\mathcal{O})$ such that $\varphi_{\mathcal{F}}(f) = 0$, $\varphi_{\mathcal{G}}(f) \neq 0$.*

Proof: Suppose first that $\mathcal{F}$ and $\mathcal{G}$ are not comparable with respect to $\preceq$. Then, by Proposition 11.9 we can find disks $d(b,l) \in \mathcal{F}$ and $d(c,m) \in \mathcal{G}$ such that $d(b,l) \cap d(c,m) = \emptyset$. Let $t = |b-c|$. Then by Proposition 11.9 again, we have $t > \max(l,m)$, hence, denoting by $\mathcal{O}'$ the set of holes of $\mathcal{O}$ included in $\Gamma(b,l,t)$, this annulus admits $\mathcal{O}'$ as a classic partition.

Suppose now that $\mathcal{F}$ and $\mathcal{G}$ are comparable with respect to $\preceq$. So, we can suppose that $\mathcal{G}$ surrounds $\mathcal{F}$. Let $\rho = diam(\mathcal{F})$, $t = diam(\mathcal{G})$ and let $l \in]\rho, t[$. By Lemma 10.11, there exists a unique disk $d(b,l)$ which belongs to $\mathcal{F}$, and then denoting by $\mathcal{O}'$ the set of holes of $\mathcal{O}$ included in $\Gamma(b,l,t)$, this annulus admits $\mathcal{O}'$ as a classic partition.

Now, in both situations, we take $s \in]l,t[$. By Corollary 20.4. $\mathcal{O}$ admits an increasing idempotent T-sequence and a decreasing idempotent T-sequence, both of center b and diameter s. Therefore, the conclusion comes from Proposition 21.5. □

Theorem 21.8: *Let K be strongly valued and let $\mathcal{O}$ be a classic partition of $d(a,r)$. Then $Mult_m(H(\mathcal{O}), \| \,.\, \|_{\mathcal{O}}) = Mult(H(\mathcal{O}), \| \,.\, \|_{\mathcal{O}})$.*

Proof: Let $\psi \in Mult(H(\mathcal{O}, \| \,.\, \|_{\mathcal{O}})$ and let $\mathcal{M}$ be a maximal ideal of $H(\mathcal{O})$ containing $Ker(\psi)$. Then ψ is of the form $\varphi_{\mathcal{F}}$, with $\mathcal{F} \in \Phi(\mathcal{O})$. On the other hand, by Theorem 6.15 there exists $\phi \in Mult_m(H(\mathcal{O}), \| \,.\, \|_{\mathcal{O}})$ such that $Ker(\phi) = \mathcal{M}$, and ϕ is of the form $\varphi_{\mathcal{G}}$, with $\mathcal{G} \in \Phi(\mathcal{O})$. Suppose $\mathcal{G} \neq \mathcal{F}$. Then by Lemma 21.7 there exists $g \in H(\mathcal{O})$ such that $\varphi_{\mathcal{F}}(g) = 0$, and $\varphi_{\mathcal{G}}(g) \neq 0$, a contradiction to the hypothesis $Ker(\psi) \subset \mathcal{M}$. Consequently, $\mathcal{M} = Ker(\psi)$. □

Chapter 22

Shilov Boundary for Algebras $H(D, \mathcal{O})$

In Chapter 6 we proved that, given a commutative ultrametric Banach K-algebra with unity A, a Shilov boundary does exist for $(A, \|\ .\ \|_{si})$. However, this existence is abstract and doesn't let us determine its nature. When we consider a Banach K-algebra $H(D)$, we are able to describe very precisely its Shilov boundary by using the characterization of multiplicative semi-norms by circular filters.

Definitions and notation: In all this chapter D is a closed and bounded subset of the disk $d(a, r)$ and $\mathcal{O} = (d(b_i, r_i^-)_{i \in J})$ is a classic partition of $d(a, r) \setminus D$.

A circular filter $\mathcal{F}$ will be said to be *strictly $\mathcal{O}$-bordering* if it is the peripheral of a hole of $\mathcal{O}$ or of $d(a, r)$. A circular filter $\mathcal{O}$ will be said to be *$\mathcal{O}$-bordering* if either it is strictly $\mathcal{O}$-bordering or all elements $B \in \mathcal{F}$ contain holes of $\mathcal{O}$.

We will denote by $\Sigma(\mathcal{O})$ the set of $\varphi_{\mathcal{F}} \in Mult(K[x])$ such that $\mathcal{F}$ is $\mathcal{O}$-bordering and by $\Sigma_0(\mathcal{O})$ the set of $\varphi_{\mathcal{F}} \in Mult(K[x])$ such that $\mathcal{F}$ is strictly $\mathcal{O}$-bordering.

Lemma 22.1: *Let $\mathcal{O}$ be a classic partition of $d(a, r) \setminus D$. A circular filter $\mathcal{F}$ on K which is $\mathcal{O}$-bordering but not strictly $\mathcal{O}$-bordering is secant with $d(a, r) \setminus D$ but not secant with any hole of $\mathcal{O}$. Moreover, $\Sigma_0(\mathcal{O})$ is dense in $\Sigma(\mathcal{O})$.*

Proof: Let $\mathcal{O} = (d(b_i, r_i^-)_{i \in J})$ and let $\varphi_{\mathcal{F}} \in \Sigma(\mathcal{O}) \setminus \Sigma_0(\mathcal{O})$. Then every element of $\mathcal{F}$ contains holes of $\mathcal{O}$, hence is secant with $d(a, r) \setminus D$. Suppose that $\mathcal{F}$ is secant with a hole $T = d(b, \rho^-)$ of $\mathcal{O}$. Since $\mathcal{F}$ is not the peripheral of T it is secant with a disk $d(c,s) \subset d(b,\rho^-)$ with $s < \rho$, and of course it has elements B of diameter $l \in]s,\rho[$ that contain no hole of $\mathcal{O}$, a contradiction. So, $\mathcal{F}$ is not secant with any hole of $\mathcal{O}$. Now, let $h_1,\dots,h_n \in R(D,\mathcal{O})$, let $\epsilon > 0$. Since $\mathcal{F}$ admits a base of affinoids sets, we can find an affinoid set $E \in \mathcal{F}$ such that

$$|\,|h_j(x)| - \varphi_{\mathcal{F}}(h_j)|_\infty < \epsilon \ \forall x \in E, \forall j = 1,\dots,n.$$

Then there exists a hole T of $\mathcal{O}$ included in E. Consequently,

$$|\varphi_T(h_j) - \varphi_{\mathcal{F}}(h_j)|_\infty < \epsilon \ \forall j = 1,\dots,n.$$

This shows that $\Sigma_0(\mathcal{O})$ is dense in $\Sigma(\mathcal{O})$. □

Theorem 22.2: *Let $\mathcal{O}$ be a classic partition of $d(a,r) \setminus D$. Then $\Sigma(\mathcal{O})$ is equal to the boundary of $Mult(H(D,\|\cdot\|_{D,\mathcal{O}})$ inside $Mult(K[x])$.*

Proof: The proof roughly will follow the same way as this of Theorem 14.1. Let $\mathcal{O} = (d(b_i,r_i^-)_{i\in J})$. Since $\Sigma_0(\mathcal{O})$ is dense inside $\Sigma(\mathcal{O})$, and since the boundary of $Mult(H(D,\|\cdot\|_{D,})$ inside $Mult(K[x])$ is obviously closed, it is sufficient to show that $\Sigma_0()$ is included in this boundary. Let $\varphi_{\mathcal{F}} \in \Sigma_0()$. If $\mathcal{F}$ is the peripheral of $d(a,r)$, it is obvious that $\varphi_{\mathcal{F}}$ belongs to the closure of the subset of $Mult(K[x]) \setminus Mult(H(D,\mathcal{O}),\|\cdot\|_{D,\mathcal{O}})$ consisting of all $\varphi_{\mathcal{G}}$ such that $\mathcal{G}$ is secant with $K \setminus d(a,s)$ for some $s > r$. Consequently, we are led to assume that $\mathcal{F}$ is the peripheral of certain hole $d(b_j,r_j^-)$ of $\mathcal{O}$, and then symmetrically, it is also obvious that $\varphi_{\mathcal{F}}$ belongs to the closure of $Mult(K[x]) \setminus Mult(H(D,\mathcal{O}),\|\cdot\|_{D,\mathcal{O}})$ inside $Mult(K[x])$.

Conversely, let $\varphi_{\mathcal{F}} \in Mult(H(D,\mathcal{O}),\|.\|_{D,\mathcal{O}})$ belong to the closure of $Mult(K[x]) \setminus Mult(H(D,\mathcal{O}),\|.\|_{D,\mathcal{O}})$ in $Mult(K[x])$. If $\varphi_{\mathcal{F}}$ is the peripheral of $d(a,r)$, it obviously belongs to $\Sigma(\mathcal{O})$. So, we can suppose it is not the peripheral of $d(a,r)$. We can find an affinoid set $E \in \mathcal{F}$ such that $E \subset d(a,r)$. Then $Mult(H(E),\|\cdot\|_E)$ is a neighborhood of $\varphi_{\mathcal{F}}$, hence it contains an open neighborhood W of $\varphi_{\mathcal{F}}$. Therefore

$$W \cap (Mult(K[x]) \setminus Mult(H(D,\mathcal{O}),\|.\|_{D,\mathcal{O}})) \neq \emptyset.$$

Moreover, $W \cap (Mult(K[x]) \setminus Mult(H(D,\mathcal{O}),\|.\|_{D,\mathcal{O}}))$ is open in $Mult(K[x])$ because so are W and $Mult(K[x]) \setminus Mult(H(D,\mathcal{O}),\|.\|_{D,\mathcal{O}})$. Since $Mult_a(K[x])$ is dense inside $Mult(K[x])$, there exists

$$\varphi_a \in (Mult_a(K[x]) \setminus Mult(H(D,\mathcal{O}),\|.\|_{D,\mathcal{O}})) \cap Mult(H(E),\|\cdot\|_E).$$

Consequently, a lies in $E \setminus D$. But since $E \subset d(a,r)$, a must belong to Y, which proves that $\mathcal{F}$ is secant with Y. Therefore the boundary of

$Mult(H(D,\mathcal{O}),\|.\|_{D,\mathcal{O}})$ inside $Mult(K[x])$ is included in $\Sigma(\mathcal{O})$, and finally these two sets are equal. □

Theorem 22.3: *Let $\mathcal{O} = (d(b_i, r_i^-)_{i\in J})$ be a classic partition of $d(a,r)\setminus D$. Then $\Sigma(\mathcal{O})$ is the Shilov boundary for $(H(D,\mathcal{O}),\|\cdot\|_{D,\mathcal{O}})$.*

Proof: First, we notice that $\Sigma(\mathcal{O})$ is closed inside $Mult(K[x])$ because it is the boundary of $Mult(H(D,\mathcal{O}),\|\cdot\|_{D,\mathcal{O}})$. By Theorem 21.1 it is easily seen that $\Sigma_0(\mathcal{O})$ is a Shilov set for $H(D,\mathcal{O})$.

Now, we will show that $\Sigma(\mathcal{O})$ is the smallest closed Shilov set. Suppose it is not the smallest. Then, there exists another closed Shilov set S which does not contain $\Sigma(\mathcal{O})$. Since S is closed, and since $\Sigma_0(\mathcal{O})$ is dense in $\Sigma(\mathcal{O})$ there exists $\psi \in \Sigma_0(\mathcal{O}) \setminus S$. Let $\mathcal{G}$ be the strictly $\mathcal{O}$-bordering filter such that $\psi = \varphi_{\mathcal{G}}$. Thus ψ is of the form $\varphi_{b,s}$.

Since S is closed there exists an affinoid set E of the form $d(b,s'') \setminus \cup_{j=1}^{q} d(c_j, s')$, with $s' < s < s''$ such that $Mult(H(E),\|\cdot\|_E)E \cap S = \emptyset$. This means that for every $\varphi_{\mathcal{F}} \in S$, $\mathcal{F}$ is not secant with E. Suppose first that $s = r$. Consider $g(x) = \prod_{j=1}^{q}(x - c_j)$. Then $\varphi_{\mathcal{G}}(h) = r^q$, and for every $\varphi_{\mathcal{F}} \in S$ we can check that $\varphi_{\mathcal{F}}(g) \le s'^q$, a contradiction to the hypothesis "S is a Shilov set".

Consequently we have $s < r$, and therefore, since $\varphi_{\mathcal{G}} \in \Sigma_0(D)$, the disk $d(b, s^-)$ is a hole of $\mathcal{O}$. Let

$$h(x) = \frac{\prod_{j=1}^{q}(x - c_j)}{(x-b)^{q+1}}.$$

Then $\varphi_{\mathcal{G}}(h) = \frac{1}{s}$ and for every $\varphi_{\mathcal{F}} \in S$ we can check that $\varphi_{\mathcal{F}}(h) = \max(\frac{s'}{s^2}, \frac{1}{s''}) < \frac{1}{s}$, a contradiction to the hypothesis S *is a Shilov set.* Thus, we have proven that any closed Shilov set must contain $\Sigma(\mathcal{O})$, and this finishes proving that $\Sigma(\mathcal{O})$ is the Shilov boundary for $H(D,\mathcal{O})$. □

Theorem 22.4 was given in [8]:

Theorem 22.4 (K. Boussaf): *$\Sigma(D)$ is the Shilov boundary for $(H(D),\|\cdot\|_D)$.*

Proof: By Theorem 14.1 $\Sigma(D)$ is the boundary of $Mult(H(D),\|\cdot\|_D)$ inside $Mult(K[x])$, hence it is closed inside $Mult(K[x])$.

Suppose first that D is infraconnected. By considering the natural partition $\mathcal{O}$ of $\widetilde{D} \setminus D$, we have $\Sigma(\mathcal{O}) = \Sigma(D)$, so we can apply Theorem 22.3 showing that $\Sigma(D)$ is the Shilov boundary of $H(D)$. Now, we shall consider the general case.

First, we check that $\Sigma(D)$ is a Shilov set. Let $h \in R(D)$. If there exists an infraconnected component E of D such that $\|h\|_D = \|h\|_E$, then as we have seen, there exists $\varphi_{\mathcal{F}} \in \Sigma(E)$ such that $\varphi_{\mathcal{F}}(h) = \|h\|_D$, and since of course $\varphi_{\mathcal{F}} \in \Sigma(D)$, hence $\|h\|_D$ is reached in $\Sigma(D)$. Now, suppose that there exists no infraconnected component E of D such that $\|h\|_D = \|h\|_E$. Then there exists a sequence $(D_n)_{n\in\mathbb{N}}$ of infraconnected components of D such that $\|h\|_{D_n} > \|h\|_D - \frac{1}{n}$. Consequently, for each $n \in \mathbb{N}$, we can take a point $a_n \in D_n$ such that $|h(a_n)| > \|h\|_D - \frac{1}{n}$. By Theorem 9.1, from the sequence (a_n) we can extract a subsequence which is either a monotonous distances sequence, or an equal distances sequence, or a Cauchy sequence. Here, as D is closed, if we could extract a Cauchy sequence, there would exist $a \in D$ such that $|h(a)| = \|h\|_D$, which is excluded. Hence, we can extract a subsequence which is either a monotonous distances sequence, or an equal distances sequence. So, without loss of generality, we can assume that the sequence (a_n) is either a monotonous distances sequence, or an equal distances sequence, and therefore by Proposition 10.7 there exists a large circular filter $\mathcal{G}$ less thin than the sequence (a_n). Then $\mathcal{G}$ is obviously secant with D and satisfies $\varphi_{\mathcal{G}}(h) = \|h\|_D$. But by hypothesis, $\mathcal{G}$ is not secant with any infraconnected component because for each component E of D we have $\|h\|_D > \|h\|_E$. Consequently, $\mathcal{G}$ is secant with $K \setminus D$, and therefore belongs to $\Sigma(D)$. This shows that $\Sigma(D)$ is a Shilov set.

Let $\Sigma_1(D)$ be the set of $\varphi_{\mathcal{F}} \in \Sigma(D)$ which are B-bordering for one of the infraconnected components B of D. We will show that $\Sigma_1(D)$ is dense in $\Sigma(D)$. Let $\varphi_{\mathcal{F}} \in \Sigma(D)$. If $\mathcal{F}$ is secant with one of the infraconnected components B of D, by definition $\mathcal{F}$ is D-bordering, hence $\varphi_{\mathcal{F}} \in \Sigma(D)$. Now, suppose that $\mathcal{F}$ is not secant with any infraconnected component of D. Consider an affinoid set $E \in \mathcal{F}$. Then $E \cap D \neq \emptyset$, hence there exists an infraconnected component B of D such that $E \cap B \neq \emptyset$. If $B \subset \widetilde{E}$ then the peripheral of B is clearly secant with E, hence $\varphi_{\mathcal{F}} \in \Sigma(D)$. Now, suppose that B is not included in $\widetilde{E}$. Then, $\widetilde{E} \subset \widetilde{B}$, but since $E \cap K \setminus \neq \emptyset$, there exists a hole $T = d(b, r^-)$ of B included in $\widetilde{E}$ and then, $\varphi_{b,r}$ belongs to $\Sigma_1(D)$, and is secant to E. This shows that $Mult(H(E), \|\cdot\|_E) \cap \Sigma_1(D) \neq \emptyset$, and consequently $\Sigma_1(D)$ is dense inside $\Sigma(D)$.

Suppose now that $\Sigma(D)$ is not the smallest Shilov set. So, there exists a Shilov set S which does not contain $\Sigma(D)$. More precisely, since S is closed there exists $\varphi_{\mathcal{H}} \in \Sigma_1(D) \setminus S$. Then $\mathcal{H}$ has center $b \in B$ and diameter s. Since S is closed there exists an affinoid set E of the form $d(b, s'') \setminus \cup_{j=1}^{q} d(c_j, s')$, with $s' < s < s''$, $|c_i - c_j| = |c_i - b| = s\ \forall i \neq j$ and such that for every $\varphi_{\mathcal{F}} \in S$, $\mathcal{F}$ is not secant with E.

Suppose first that $\mathcal{H}$ is the peripheral of an infraconnected component B of D. There exists an annulus $\Gamma(b, z', z'')$ included in $K \setminus D$ such that $s < z' < z'' < \sqrt[q]{s^q s''}$. Let $c \in \Gamma(b, z', z'')$, let $l = |c|$ and

$$h(x) = \frac{\prod_{j=1}^{q}(x - c_j)}{(x - c)^{q+1}}.$$

Then we have $\varphi_{\mathcal{H}}(h) = \frac{s^q}{l^{q+1}}$. Now, let $\varphi_{\mathcal{F}} \in S$ be such that $\mathcal{F}$ is secant with $d(b, s)$. Since $\mathcal{F}$ is not secant with E, we check that

$$\varphi_{\mathcal{F}}(h) \leq \frac{s'^q}{l^{q+1}} < \varphi_{\mathcal{H}}(h).$$

And now, let $\varphi_{\mathcal{F}} \in S$ be such that $\mathcal{F}$ is not secant with $d(b, s)$. Since $\mathcal{F}$ is not secant with E, it can't be secant with $d(b, s'')$. Consequently, we have

$$\varphi_{\mathcal{F}}(h) \leq \frac{1}{s''} < \frac{s^q}{l^{q+1}} = \varphi_{\mathcal{H}}(h).$$

Thus if $\mathcal{H}$ is the peripheral of an infraconnected component B of D, we are led to a contradiction, and this shows that $\mathcal{H}$ is just the peripheral of a hole $T = d(b, s^-)$ of an infraconnected component B of D. Since B is an infraconnected component of D, there exists an annulus $\Gamma(c, z', z'')$ included in $K \setminus D$ such that $s' < z' < z'' < s$ with $c \in d(b, s^-)$. So, $T = d(c, s^-)$. Let $a \in \Gamma(c, z', z'')$, let $l = |a - c|$ and let

$$h(x) = \frac{\prod_{j=1}^{q}(x - c_j)}{(x - a)^{q+1}}.$$

Then we have $\varphi_{\mathcal{H}}(h) = \frac{1}{s}$. Now, let $\varphi_{\mathcal{F}} \in S$. If $\mathcal{F}$ is secant with $C(c, s)$, since $\mathcal{F}$ is not secant with E, we check that

$$\varphi_{\mathcal{F}}(h) \leq \frac{s'}{s^2} < \varphi_{\mathcal{H}}(h).$$

And if $\mathcal{F}$ is secant with $K \setminus d(c, s)$, it must be secant with $K \setminus d(c, s''^-)$, and then we check that

$$\varphi_{\mathcal{F}}(h) \leq \frac{1}{s''} < \frac{1}{s} = \varphi_{\mathcal{H}}(h).$$

Finally, suppose that $\mathcal{F}$ is secant with $d(c, s^-)$. Then one of the c_j lies in $d(c, s)$ and we can assume $c = c_1$. Therefore $\mathcal{F}$ must be secant with

$d(c_1, s'^-)$, and we check that

$$\varphi_{\mathcal{F}}(h) \le \frac{s' s^{q-1}}{l^{q+1}} < \frac{1}{s} = \varphi_{\mathcal{H}}(h).$$

This finishes showing that $\varphi_{\mathcal{F}}(h) < \varphi_{\mathcal{H}}(h)$ for every $\varphi_{\mathcal{F}} \in S$, and therefore we can conclude that any closed Shilov set must contain $\Sigma(D)$. Consequently, $\Sigma(D)$ is a Shilov boundary for $H(D)$. □

Remark: A boundary is not necessarily closed. Let $(a_n)_{n\in\mathbb{N}^*}$ be sequence of K such that $|a_{n+1} < |a_n| < 1$, $\lim_{n\to\infty} |a_n| = r > 0$, and let $D = \{x \in K \mid |x| < 1,\ |x - a_n| \ge |a_n|\ \forall n \in \mathbb{N}^*\}$. For every $n \in \mathbb{N}^*$, let $\mathcal{F}_n$ be the circular filter of center a_n and diameter $|a_n|$, let $\phi_n = \varphi_{\mathcal{F}_n}$, let $\mathcal{G}$ be the circular filter of center 0 and diameter 1, and let $\mathcal{G}'$ be the circular filter of center 0 and diameter r, let $\psi = \varphi_{\mathcal{G}}$, and let $\psi' = \varphi_{\mathcal{G}'}$. Clearly, $\Sigma_0(D) = \{\phi_n \mid n \in \mathbb{N}^*\} \cup \{\psi\}$, and then by Theorem 16.14 this is a boundary for $(H(D), \|\ .\ \|_D)$. On the other hand by Theorem 22.4 the Shilov boundary for $(H(D), \|\ .\ \|_D)$ is $\Sigma(D) = \{\phi_n \mid n \in \mathbb{N}^*\} \cup \{\psi,\ \psi'\}$. But $\Sigma_0(D)$ is not closed because ψ' is the limit of the sequence (ϕ_n) inside $Mult(H(D), \|\ .\ \|_D)$, with respect to the topology of simple convergence. One can ask whether $\Sigma_0(D)$ is equal to $Min(H(D), \|\ .\ \|_D)$.

By Theorems 14.1, 14.2 and 22.4, we can state Corollary 22.5:

Corollary 22.5: *Let D be closed bounded and have finitely many infraconnected components. The 4 following sets are equal:*

(*i*) $\Sigma(D)$,
(*ii*) *The boundary of $Mult(H(D), \|\ .\ \|_D)$ inside $Mult(K[x])$ with respect to the topology of simple convergence,*
(*iii*) *The Shilov boundary for $(H(D), \|\ .\ \|_D)$.*

Proof: Theorem 22.6 is just a corollary of Corollary 22.5 when D is infraconnected. Now, suppose that D has finitely many infraconnected components $D_1, \dots, D_q$. Since by Theorem 19.12 $H(D) = H(D_1) \times \dots \times H(D_q)$, and the theorem holds for each algebra $H(D_j)$, so we can easily see that the Shilov boundary of $H(D)$ is equal to the union of Shilov boundaries of all algebras $H(D_j)$, $1 \le j \le q$, hence the conclusion follows. □

More precisely, we can state Corollary 22.6:

Corollary 22.6: *Let D be closed bounded having finitely many D-bordering filters. The following four sets are equal:*

(*i*) $\Sigma(D)$,
(*ii*) *The boundary of $Mult(H(D), \| \cdot \|_D)$ inside $Mult(K[x])$ with respect to the topology of simple convergence,*
(*iii*) *The boundary of $Mult(H(D), \| \cdot \|_D)$ inside $Mult(K[x])$ with respect to the metric topology,*
(*iv*) *The Shilov boundary for $(H(D), \| \cdot \|_D)$.*

Proof: If D has finitely many D-bordering filters, then it has finitely many infraconnected components. □

Remark: In particular, Theorem 22.6 applies to $H(D)$ when D is an affinoid set.

Chapter 23

Holomorphic Functional Calculus

Holomorphic functional calculus is well known and helpful in complex Banach algebra. We can also define it, from the spectrum of an element x, by considering first the rational functions without poles in $sp(x)$. Now, considering an ultrametric Banach K-algebra A, thanks to specific ultrametric properties given in Lemma 23.1 bellow, this calculus has continuation from a Banach algebra of the form $H(sp(x), \mathcal{O})$ to A, where $\mathcal{O}$ is a natural partition defined by the norm of A. This calculus will let us solve several problems in spectral theory and about idempotents. The holomorphic functional calculus was first defined in [23].

Notation: In this chapter, A is a commutative Banach K-algebra with unity whose norm $\| \ . \ \|$ is ultrametric.

Recall that given $\phi \in Mult(K[x])$, the unique circular filter $\mathcal{G}$ on K such that $\varphi_{\mathcal{G}} = \phi$ is $\Psi^{-1}(\phi)$.

Lemma 23.1: *Let $x \in A$ be invertible and let $b \in K$ be such that $|b| < \frac{1}{\|x^{-1}\|}$. Then $x - b$ is invertible and satisfies $\|(x-b)^{-1}\| = \|x^{-1}\|$.*

Proof: By hypothesis, we have $|b| < \frac{1}{\|x^{-1}\|_{si}}$, hence

$$\lim_{n \to \infty} \|(bx^{-1})^n\| = 0.$$

Consequently, the series $\sum_{n=0}^{\infty} (\frac{b}{x})^n$ is converging in A. Then we check that $(x-b)\sum_{n=0}^{\infty}(b^n x^{-n-1}) = 1$. Now since $\|bx^{-1}\| < 1$, we have

$$\|x^{-1}\| > \|b^n x^{-n-1}\| \ \forall n \in \mathbb{N}^*.$$

But since $\| \ . \ \|$ is ultrametric, we have $\| \sum_{n=0}^{\infty} b^n x^{-n-1} \| = \|x^{-1}\|$, and therefore $\|(x-b)^{-1}\| = \|x^{-1}\|$. $\square$

Corollary 23.2: *Let $x \in A$ and for each $b \in K \setminus sp(x)$, let $\rho(b) = \frac{1}{\|(x-b)^{-1}\|}$. Then $\rho(a) = \rho(b)$ whenever a, $b \in K \setminus sp(x)$ such that $|a-b| < \frac{1}{\|(x-b)^{-1}\|}$. Let $r \geq \|x\|_{si}$. Then $d(0,r) \setminus sp(x)$ admits a classic partition of the form $(d(b,\rho(b)^-))_{b\in d(0,r)\setminus sp(x)}$, with $d(b,\rho(b)^-) = d(a,\rho(a)^-)$ whenever a, $b \in K \setminus sp(x)$ such that $|a-b| < \frac{1}{\|(x-b)^{-1}\|}$.*

Definitions and notation: Let $x \in A$, let $a \in K$ and let $r \in [\|x-a\|_{si}, \|x\|]$, with $r > \|x-a\|_{si}$ if A is not uniform. We will call *x-normal partition of center a and diameter r* a classic partition $\mathcal{O} = (d(b_i, r_i^-)_{i\in J})$ of $d(a,r) \setminus sp(x)$ satisfying $\frac{1}{\|(x-b_i)^{-1}\|} = r_i \ \forall i \in J$.

In the same way, we will call *x-spectral partition of center a and diameter r* a classic partition $\mathcal{O} = (d(b_i, r_i^-)_{i\in J})$ of $d(a,r) \setminus sp(x)$ satisfying $\frac{1}{\|(x-b_i)^{-1}\|_{si}} = r_i \ \forall i \in J$.

A x-normal (resp. x-spectral) partition will be said to be *centered* if it has center $a \in \widetilde{sp(x)}$.

A x-spectral partition of center a will be said to be *wide* if it has a diameter $r > \|x-a\|_{si}$. We will call *strict x-spectral partition* the unique centered x-spectral partition of diameter $\|x\|_{si}$. Then, if A is uniform, the strict x-spectral partition is a x-normal partition too.

Let $r = \|x-a\|$ and let $\mathcal{O} = (d(b_i, r_i^-)_{i\in J})$ be a centered x-normal partition of $d(a,r) \setminus sp(x)$. Then $x \in A$ will be said to have *bounded normal ratio* if the family $\{\frac{\|(x-b_j)^{-1}\|}{\|(x-b_j)^{-1}\|_{si}} \mid i \in J\}$ is bounded. The K-algebra A will be said *to have bounded normal ratio* if every element of A has bounded normal ratio.

Theorem 23.3: *Let $t \in A$. There exists a unique homomorphism Θ_t from $R(sp(t))$ into A such that $\Theta_t(P) = P(t)$ for all $P \in K[X]$. Moreover, Θ_t is injective if and only if t has a null minimal polynomial. For every $h \in R(sp(t))$, we have $sp(h(t)) = h(sp(t))$ and $sa(h(t) = h(sa(t))$.*

Proof: Let $D = sp(t)$. We may obviously define Θ_t from $K[x]$ to A as $\Theta_t(P) = P(t)$. Now let $Q \in K[X]$ have its zeros in $K \setminus D$. Then $Q(t)$ is invertible in A, so we may extend to $R(D)$ the definition of Θ_t, as $\Theta_t(\frac{P}{Q}) = P(t)\,Q(t)^{-1}$, for all rational function $\frac{P}{Q} \in R(D)$ (with $(P,Q) = 1$). The uniqueness to Θ_t is then obvious. Next, $Ker(\Theta_t)$ is an ideal of $R(D)$ which is obviously generated by a polynomial G. Then $G = 0$ if and only if t has a null minimal polynomial.

Now, let $h = \frac{P}{Q} \in R(sp(t))$ (with $(P,Q) = 1$). Let $\lambda \in sp(t))$, and let $\chi \in \mathcal{X}(A)$ be such that $\chi(h(t)) = \lambda$. Then $\chi(h(t)) = h(\lambda)$, hence $h(sp(t)) \subset sp(h(t))$. Conversely, let $\mu \in sp(h(t))$, let $\chi \in \mathcal{X}(A)$ be such that $\chi(h(t)) = \mu$, and let $\alpha = \chi(t)$. Then, we have $\chi(P(t)) - \mu\chi(Q(t)) = 0$, hence α is a zero of the polynomial $P(X) - \mu Q(X)$. Therefore α lies in K because K is algebraically closed. But then, as $t - \alpha$ belongs to the kernel of χ, α lies in $sp(t)$. Hence we have $sp(h(t)) = h(sp(t))$. Finally, $sa(h(t)) = \{\chi(h(t)) \mid \chi \in \mathcal{X}(A, K)\} = \{h(\chi(t)) \mid \chi \in \mathcal{X}(A, K)\} = h(sa(t))$. □

Remark: When the homomorphism Θ_t defined in Theorem 23.3 is injective, the K-subalgebra $B = \Theta_t(R(sp(t)))$ is isomorphic to $R(sp(t))$, and in fact is the full subalgebra generated by t in A. So, in such a case, we will consider $R(sp(t))$ as a K-subalgebra of A.

Definition and notation: Let $t \in A$. In this chapter and in the sequel we will denote by Θ_t the canonical homomorphism from $R(sp(t))$ into A defined in Theorem 23.3, and we will call it *the canonical homomorphism associated to t*. Meanwhile, when it is injective, we will currently confound a rational function $h \in R(sp(t))$ with its image $\Theta_t(h)$.

Given $\psi \in Mult(A, \| \, . \, \|)$ we will put $\psi_t = \psi \circ \Theta_t$. Thus, we have $\psi_t(P) = \psi(P(t)) \ \forall P \in K[x]$.

We will put $\sigma_A(t) = \{\phi_t \mid \phi \in Mult(A, \| \, . \, \|)$, and when there is no risk of confusion on the algebra A, we will put $\sigma(t)$ instead of $\sigma_A(t)$.

Lemma 23.4: *Let $t \in A$. Then $\sigma_A(t)$ is a compact in $Mult(K[x])$. Let $a \in sp(t)$, let $s = \|t - a\|_{si}$, let $b \in d(a, s) \setminus sp(t)$ and let $r = \frac{1}{\left\| \frac{1}{t-b} \right\|_{si}}$. Let $\psi \in Mult(A, \| \, . \, \|)$ and let $\mathcal{F}$ be the circular filter such that $\psi_t = \varphi_{\mathcal{F}}$. Then $\mathcal{F}$ is secant with $d(a, s) \setminus d(b, r^-)$.*

Proof: The mapping which associates to each $\phi \in Mult(A, \| \, . \, \|)$ its image $\phi_t \in Mult(K[x])$ is clearly continuous, hence $\sigma_A(t)$ is compact. By definition of ψ_t, it is obvious that $\mathcal{F}$ is secant with $d(a, s)$. Next, it is clear that $\psi_t(t - b) \geq r$, hence $\mathcal{F}$ is secant with $d(a, s) \setminus d(b, r^-)$. □

Theorem 23.5: *Let $t \in A$ and let $\mathcal{O}$ be a t-spectral partition. Θ_t satisfies $\|\Theta_t(f)\|_{si} = \|\Theta_t(f)\|_{\sigma(t)} \leq \|f\|_{sp(t),\mathcal{O}}$ for all $f \in R(sp(t), \mathcal{O})$.*

Proof: The equality $\|\Theta_t(f)\|_{si} = \|\Theta_t(f)\|_{\sigma(t)}$ is obvious and just comes from the definition of $\sigma(t)$. Let $D = sp(t)$, let $a \in sp(t)$, let $r_0 = \|f - a\|_{si}$. Let $\psi \in Mult(A, \| \, . \, \|_{si})$ and let $\mathcal{F}$ be the circular filter such that $\varphi_{\mathcal{F}} = \psi_t$. Let $(d(b_j, r_j))_{1 \leq j \leq n}$ be the family of holes of $\mathcal{O}$ containing at least one pole of f and for each $j = 1, \ldots, n$, let $\mathcal{F}_j$ be the circular filter of center b_j and diameter r_j. Finally, let $\mathcal{F}_0$ be the circular filter of center a and diameter r_0. According to Theorem 21.1 f is of the form $\sum_{j=0}^{n} f_j$, with $f_0 \in K[x]$ and $f_j \in R_0(K \setminus d(b_j, r_j^-))$, and we have $\|h\|_{sp(t),\mathcal{O}} = \max_{0 \leq j \leq n} \|h_j\|_{D,\mathcal{O}}$.

For each $j = 1, \dots, n$, since by Lemma 23.4 $\mathcal{F}$ is secant with $K \setminus d(b_j, r_j^-)$, we have $\varphi_{\mathcal{F}}(h_j) \leq \varphi_{b_j,r_j}(h_j)$. Similarly, as $\mathcal{F}$ is secant with $d(a, r_0)$, we have $\varphi_{\mathcal{F}}(h_0) \leq \varphi_{a,r_0}(h_0)$. Consequently, we obtain $\psi(h(t)) \leq \|h\|_{D,\mathcal{O}}$. This is true for all $\psi \in Mult(A, \| \, . \, \|)$, and therefore by Theorem 6.19 we obtain $\|f(t)\|_{si} \leq \|f\|_{D,\mathcal{O}}$. □

Remark: Particularly, Theorem 23.5 is true when $\mathcal{O}$ is the strict t-spectral partition.

Theorem 23.6: *Let $t \in A$ and let $\mathcal{O}$ be a t-normal partition. Θ_t is continuous and has continuation to $H(sp(t), \mathcal{O})$. Moreover, if A is uniform, then Θ_t is continuous as mapping from $(R(sp(t)), \| \, . \, \|_{\sigma(t)})$ to A and has continuation to $H(\sigma(t))$ and to $H(sp(t), \mathcal{O})$. So, it satisfies $\|\Theta_t(f)\| = \|f\|_{\sigma(t)} \leq \|f\|_{sp(t),\mathcal{O}}$ for all $f \in H(sp(t), \mathcal{O})$.*

Proof: The claim about $H(\sigma(t))$ is immediate and comes from Theorem 23.5. Now, let $\mathcal{O} = (d(b_i, r_i^-)_{i \in J})$ be a t-normal partition of diameter r. Let $D = sp(t)$. Let $T = d(a, \rho^-)$ be a hole of $\mathcal{O}$. It is easily checked that

$$\Big\|\Theta_t\Big(\frac{1}{(t-a)^k}\Big)\Big\| \leq \Big\|\Theta_t\big(\frac{1}{t-a}\big)\Big\|^k$$
$$= \Big(\frac{1}{\rho}\Big)^k = \Big(\Big\|\frac{1}{t-a}\Big\|_{D,\mathcal{O}}\Big)^k = \Big\|\frac{1}{(t-a)^k}\Big\|_{D,\mathcal{O}}.$$

Now, let $h \in R(D)$, let $T_1, \dots, T_q$ be the h-holes of $\mathcal{O}$ and let $\sum_{n=0}^{q} g_n$ be the Mittag-Leffler series of h on $(D, \mathcal{O})$, with $g_n = \overline{h_{T_n}}$ $\forall j = 1, \dots, q$. Then we have $\|\Theta_t(g_j)\| \leq \|h\|_{D,\mathcal{O}}$ for each $j = 1, \dots q$.

Finally, we check that there exists a constant $L \geq 1$ such that $\|t^n\| \leq Lr^n$. Indeed, if A is uniform, we just take $L = 1$. And if A is not uniform, then we have $r > \|t\|_{si}$, hence $\lim_{n\to\infty} \frac{\|t^n\|}{r^n} = 0$. Thus we have shown the existence of L in all cases. Consequently, in $H(D, \mathcal{O})$, we check that $\|\Theta_t(g_j)\| \leq L\|g_j\|_{D,\mathcal{O}}$ for each $j = 0, \dots, q$. But by Theorem 21.1, we have $\|h\|_{D,\mathcal{O}} = \max_{0 \leq j \leq q} \|g_j\|_{D,\mathcal{O}}$. Therefore, we conclude $\|\Theta_t(h)\| \leq L\|h\|_{D,\mathcal{O}}$, which finishes proving that Θ_t is continuous. Consequently, it has continuation to $H(D, \mathcal{O})$. If A is uniform, we can take $L = 1$, hence $\|\Theta_t(h)\| \leq \|h\|_{D,\mathcal{O}}$. □

Corollary 23.7: *Let $t \in A$, let $\mathcal{O}$ be a t-normal partition, and let $\psi \in Mult(A, \| \, . \, \|)$. Then ψ_t belongs to $Mult(R(sp(t), \mathcal{O}), \| \, . \, \|_{D,\mathcal{O}})$.*

Definitions and notation: Let $x \in A$, let $a \in sp(x)$, let $s = \|x - a\|_{si}$. An annulus $\Gamma(a, r', r'') \subset d(b, s)$ will be said to be *x-clear* if there exist

ψ', $\psi'' \in Mult(A, \| \,.\, \|)$ such that $\psi'(x-a) = r'$, $\psi''(x-a) = r''$, and $\psi(x-a) \notin]r', r''[\ \forall \psi \in Mult(A, \| \,.\, \|)$.

Remark: If $\Gamma(a, r', r'')$ is a x-clear annulus, in particular we have $\Gamma(a, r', r'') \cap sp(x) = \emptyset$. But the converse is not true.

Theorem 23.8: *Let $t \in A$. If there exists a t-clear annulus, then both $Mult(A, \| \,.\, \|)$ and $\sigma(t)$ are not connected. Moreover, there exists a t-clear annulus if and only if $\sigma(t)$ is not connected.*

Proof: Suppose first that there exists a t-clear annulus $\Gamma(a, r, s)$. Let $F = \{\phi \in Mult(A, \| \,.\, \|) \mid \phi(x-a) \leq r\}$ and let $G = \{\phi \in Mult(A, \| \,.\, \|) \mid \phi(x-a) \geq s\}$. Then, F and g are two closed subsets making a partition of $Mult(A, \| \,.\, \|)$, hence $Mult(A, \| \,.\, \|)$ is not connected.

Moreover, the sets $F_t = \{\phi_t \mid \phi \in F\}$ and $G_t = \{\phi_t \mid \phi \in G\}$ are two closed subsets of $\sigma(t)$ making a partition of $\sigma(t)$, hence $\sigma(t)$ is not connected. Conversely, suppose now that $\sigma(t)$ is not connected. By Theorem 14.6 there exists an annulus $\Xi = \Gamma(a, r_0, s_0)$ together with filters $\mathcal{F}$, $\mathcal{G} \in T$ such that $\mathcal{F} \in \Psi^{-1}(\sigma(t))$ secant with $\mathcal{I}(\Xi)$, $\mathcal{G}\Psi^{-1}(\sigma(t))$ secant with $\mathcal{E}(\Xi)$, such that none of circular filters $\mathcal{H} \in \Psi^{-1}(\sigma(t))$ is secant with Ξ.

Now, let $r = \sup\{\phi(t-a), \mid \phi \in Mult(A, \| \,.\, \|) \ \phi(t-a) \leq r_0\}$ and $s = \inf\{\phi(t-a), \ \phi \in Mult(A, \| \,.\, \|) \ \phi(t-a) \geq s_0\}$. Then, $\Gamma(a, r, s)$ is a t-clear annulus. □

Lemma 23.9 is obvious:

Lemma 23.9: *Let $x \in A$, let $a \in sp(x)$, let $\Gamma(a, r', r'')$ be a x-clear annulus. For all $s', s'' \in]r', r''[$ such that $s' < s''$, for every $\psi \in Mult(A, \| \,.\, \|)$, $\Psi^{-1}(\psi_x)$ is not secant with $\Gamma(a, s', s'')$.*

Theorem 23.10: *Let $x \in A$, let $a \in sp(x)$, let $s = \|x-a\|_{si}$, let $D = sp(x)$ and let $\mathcal{O}$ be the strict x-spectral partition. The equality $\|h(x)\|_{si} = \|h\|_{D,\mathcal{O}}$ holds for every $h \in R(sp(x))$ if and only if there exists no x-clear annulus.*

Proof: Suppose first that there exists a x-clear annulus $\Gamma(a, r', r'')$ and let $\Gamma(a, s', s'') \subset \Gamma(a, r', r'')$ be such that $\Psi^{-1}(\phi_x)$ is not secant with $\Gamma(a, s', s'')$, whenever $\phi \in Mult(A, \| \,.\, \|)$, while there exists $\phi', \phi'' \in Mult(A, \| \,.\, \|)$ such that $\Psi^{-1}(\phi'_x)$ is secant with $K \setminus d(a, r''^-)$, and $\Psi^{-1}(\phi''_x)$ is secant with $d(a, r')$. Let $r \in]s', s''[\cap |K|$, let $b \in C(a, r)$ and $f(x) = \frac{x-a}{(x-b)^2}$. We can easily check that

$$\|f\|_{si} \leq \max\left(\frac{s'}{r^2}, \frac{1}{s''}\right) < \frac{1}{r}.$$

Now, by definition of $\mathcal{O}$, it is clear that $d(b,r^-)$ belongs to $\mathcal{O}$, therefore $\|f\|_{D,\mathcal{O}} \geq \varphi_{b,r}(f) = \frac{1}{r}$: this shows that the two norms are not equal when there exists a x-clear annulus.

Conversely, we now suppose that there exists no x-clear annulus. Let $h \in R(D)$. By Theorem 23.5 we have

$$\|h(x)\|_{si} \leq \|h\|_{D,\mathcal{O}}. \tag{23.1}$$

Let $T = d(b,r^-)$ be a hole of $\mathcal{O}$ containing a pole of h and let $\mathcal{F}$ be the circular filter of center b and diameter r. We will show that there exists $\psi \in Mult(A, \| \,.\, \|)$ such that $\psi_x = \varphi_{\mathcal{F}}$. Indeed, suppose that it is not true. By definition of $\mathcal{O}$, we have $r = \inf\{\xi(x-b) \mid \xi \in Mult(A, \| \,.\, \|)\}$. Consequently, either there exists $\zeta \in Mult(A, \| \,.\, \|)$ such that $\zeta(x-b) = r$, or there exists a sequence $(\zeta_n)_{n\in\mathbb{N}}$ such that $\zeta_n(x-b) > r$ and $\lim_{n\to\infty} \zeta_n(x-b) = r$. If there exists a sequence $(\zeta_n)_{n\in\mathbb{N}}$ such that $\zeta_n(x-b) > r$ and $\lim_{n\to\infty} \zeta_n(x-b) = r$, then the sequence $(\zeta_n)_{n\in\mathbb{N}}$ admits a point of adherence ψ in $Mult(A, \| \,.\, \|)$ such that $\psi(x-b) = r$ and then, since $\zeta_n(x-b) > r \ \forall n \in \mathbb{N}$, we can see that $\psi_x = \varphi_{\mathcal{F}}$.

Now, we just suppose that there exists $\zeta \in Mult(A, \| \,.\, \|)$ such that $\zeta(x-b) = r$. Let $\mathcal{G}$ be the circular filter such that $\zeta_x = \varphi_{\mathcal{G}}$. Since we suppose that $\mathcal{G} \neq \mathcal{F}$, then $\mathcal{G}$ is secant with a class $d(c,r^-)$ of $C(b,r)$ and has a diameter $\rho < r$. Then by Lemma 10.11 we can find a disk $d(c,l)$ which belongs to $\mathcal{G}$, with $\rho < l < r$. Since there exists no x-clear annulus, for every $n \in \mathbb{N}^*$ there exists $\zeta_n \in Mult(A, \| \,.\, \|)$ such that $r - \frac{1}{n} < \zeta_n(x-c) < r$. Therefore the sequence $(\zeta_n)_{n\in\mathbb{N}}$ admits a point of adherence ψ in $Mult(A, \| \,.\, \|)$ such ψ_x is the limit of the sequence $((\zeta_n)_x)_{n\in\mathbb{N}}$, and thereby we have again $\psi_x = \varphi_{\mathcal{F}}$.

So, we have proven in all cases that there exists $\psi \in Mult(A, \| \,.\, \|)$ such that $\psi_x = \varphi_{\mathcal{F}}$. Let $\mathcal{F}_0$ be the circular filter of center a and diameter s. Let $(d(b_j,r_j))_{1\leq j\leq n}$ be the family of holes of $\mathcal{O}$ containing at least one hole of h and for each $j = 1,\ldots,n$, let $\mathcal{F}_j$ be the circular filter of center b_j and diameter r_j. As we just saw, for every $j = 1,\ldots,n$, there exists $\psi_j \in Mult(A, \| \,.\, \|)$ such that $(\psi_j)_x = \varphi_{\mathcal{F}_j}$.

In the same way, we can easily show that there exists $\psi_0 \in Mult(A, \| \,.\, \|)$ such that $(\psi_0)_x = \varphi_{\mathcal{F}_0}$. Consequently, we have $\|h\|_{si} \geq \max_{0\leq j\leq n} \varphi_{\mathcal{F}_j}(h)$. On the other hand by Theorem 21.1 we know that $\max_{0\leq j\leq n} \varphi_{\mathcal{F}_j}(h) = \|h\|_{D,\mathcal{O}}$, hence by (23.1) we obtain $\|h(x)\|_{si} = \|h\|_{D,\mathcal{O}}$. □

Theorem 23.11: *Let $t \in A$ have bounded normal ratio and let $\mathcal{O}$ be a wide t-spectral partition of diameter r. Let $(d(b_i,r_i^-))_{i\in J}$ be the family of* *holes of $\mathcal{O}$. The restriction of Θ_t to $(R'(sp(t),(b_i)_{i\in J}),\|\ .\ \|_{sp(t),\mathcal{O}})$ is a continuous K-linear mapping and has continuation to $H'(s(t),(b_i)_{i\in J}),\mathcal{O})$.*

Proof: Let $D = sp(t)$. Let $f \in (R'(sp(t),\mathcal{O},(b_i)_{i\in J})$. Let $\overline{f}_0 + \sum_{n=1}^{\infty} \overline{f}_{T_n}$ be the Mittag-Leffler series of f on $\mathcal{O}$. For each $n \in \mathbb{N}^*$, let $b_n \in T_n$. Then $\overline{f}_{T_n}$ is of the form $\frac{\lambda_n}{x-b_n}$. Since t has bounded normal ratio, there exists a constant M such that $\|\overline{f}_{T_n}\|_{D,\mathcal{O}} \leq M\|\Theta_t(\overline{f}_{T_n})\|\ \forall n \in \mathbb{N}^*$. Finally, let $\overline{f}_0 = \sum_{m=0}^{q} a_m x^m$. By hypothesis, we have $\lim_{m\to\infty} |a_m| r^m = 0$. If $\|t\|_{si} = \|t\|$, then $r = \|x\|$, hence $\|\Theta_t(\overline{f}_0)\| \leq \|\overline{f}_0\|_{D,\mathcal{O}}$.

Now, if $\|t\|_{si} < \|t\|$, then $r > \|t\|_{si}$, hence there exists a constant N such that $\|t^m\| < Nr^m\ \forall m \in \mathbb{N}$, and therefore, $\|\Theta_t(\overline{f}_0)\| \leq N\|\overline{f}_0\|_{D,\mathcal{O}}$. Thus, putting $Q = \max(M,N)$, we have $\|\Theta_t(f)\| \leq Q\|f\|_{D,\mathcal{O}}\ \forall f \in (R'(sp(x),(b_i)_{i\in J})$. □

By Lemma 10.11, we have Lemma 23.12.

Lemma 23.12: *Let $t \in A$ have a null minimal polynomial. Let $a \in K$, let $\phi \in Mult(A,\|\ .\ \|)$, let ϕ_t be the restriction of ϕ to $R(sp(t))$, and let $r = \phi(t-a)$. Then $\Psi^{-1}(\phi_t)$ is (a,r)-approaching.*

Proposition 23.13: *Let $t \in A$ be such that the mapping Θ_t is injective. Let $a \in K\setminus sp(t)$, and let $r = \|(t-a)^{-1}\|_{si}^{-1}$. There exists $\theta \in Mult(A,\|\ .\ \|)$ whose restriction to $R(sp(t))$ has a circular filter (a,r)-approaching.*

Proof: We consider $R(sp(t))$ as a K-subalgebra of A. Let $\phi \in Mult(A,\|\ .\ \|)$. If $\Psi^{-1}(\phi_t)$ is secant with a disk $d(a,\rho)$ for some $\rho \in]0,r[$, then clearly we have $\phi(t-a) \leq \rho$ hence $\phi((t-a)^{-1}) > \frac{1}{r}$ and therefore $\|(t-a)^{-1}\|_{si} > \frac{1}{r}$ which contradicts the hypothesis. So $\Psi^{-1}(\phi_t)$ is secant with $K \setminus d(a,r^-)$.

Suppose that there exists $\rho > r$ such that, for every $\phi \in Mult(A,\|\ .\ \|)$, $\Psi^{-1}(\phi)$ is not secant with $d(a,\rho)$. Clearly we have $\phi(t-a) \geq \rho$ for all $\phi \in Mult(A,\|\ .\ \|)$ and therefore $\|(t-a)^{-1}\|_{si} < \frac{1}{r}$, a contradiction. Consequently, for each $n \in \mathbb{N}^*$ we can find $\phi_n \in Mult(A,\|\ .\ \|)$ such that $\Psi^{-1}((\phi_n)_t)$ is secant with $d(a, r+\frac{1}{n})$. And since it is also secant with $K \setminus d(a,r^-)$, finally, it is secant with $\Gamma(a,r,r+\frac{1}{n})$. Since $Mult(A,\|\ .\ \|)$ is compact, we can take a point of adherence θ of the filter associated to the sequence $(\phi_m)_{m\in\mathbb{N}}$. So, for every $m \in \mathbb{N}$, there exists $q_m \in \mathbb{N}$ such that $|\phi_{q_m}(t) - \theta(t)|_\infty \leq \frac{1}{m}$. Since this is true for all $m \in \mathbb{N}$, and since $\lim_{n\to\infty} \phi_n(t-a) = r$, we have $\theta(t-a) = r$, thereby $\Psi^{-1}(\theta_t)$ is (a,r)-approaching. □

holes of $\mathbb{C}$. The restriction of Θ_t to $(R(\mathrm{sp}(t)), \Theta_t(\mathrm{se}_\tau), \ldots)$ [illegible] is a continuous K-linear mapping and has continuation to $H(\ldots)$ [illegible].

Proof. Let $D = \mathrm{sp}(t)$, and $f \in (R(\mathrm{sp}(t)), \mathcal{O}\ldots)$ [illegible]. Let $f_0 + \sum \ldots f_n$ be the Mittag-Leffler series of f on D. For each n [illegible] let b_n [illegible]. Then f_n is of the form $\frac{a_n}{\ldots}$ [illegible]. Since t has bounded normal [illegible] there exists a constant M such that [illegible] $\leq M$ [illegible]. Finally, let $f_0 = \sum \ldots$ [illegible]. By hypothesis, we have [illegible] $= 0$. If [illegible], then [illegible] $\leq$ [illegible].

Now if [illegible] $<$ [illegible], then [illegible], and so there exists a constant N such that [illegible] $\leq N$, and therefore [illegible] $\leq N$ [illegible]. Thus, [illegible], we have [illegible] $\leq$ [illegible] $(R(\mathrm{sp}(t)), \ldots)$. □

[illegible] Lemma [illegible].

Lemma [illegible] *Let* [illegible] *be a* [illegible] *of* [illegible] *let* [illegible] *be the* [illegible] *of* [illegible] $R(\mathrm{sp}(t))$ [illegible] *Then* [illegible] *(a) compact* [illegible].

Proposition [illegible] *Let* [illegible] *such that the mapping* [illegible] *is* [illegible] $\ldots \in K$ [illegible] *There exists* $\ldots \in M(\ldots)$ [illegible] *whose restriction to* $R(\mathrm{sp}(t))$ *has a* [illegible].

Proof. [illegible]

[illegible] and therefore [illegible] $(\ldots)$.

Suppose that [illegible] $M(\ldots)$ [illegible] is not [illegible] we have [illegible] for all $\ldots \in M(\ldots)$ [illegible] and therefore [illegible] a contradiction. Consequently [illegible] $\in \mathbb{N}$ [illegible] such that [illegible] with [illegible]. And [illegible] with [illegible]. Finally, [illegible] with [illegible]. Since [illegible] $M(\ldots)$ [illegible] we can [illegible] of the [illegible] associated to the [illegible] since [illegible] and since [illegible] we have [illegible] approaching. □

Chapter 24

Uniform *K*-Banach Algebras and Properties (*s*) and (*q*)

Notation: Throughout this chapter, A is a commutative ultrametric Banach K-algebra with unity, whose norm is $\| \, . \, \|$. Given $t \in A$ we will denote by Θ_t the canonical morphism associated to t.

The question whether semi-norms $\| \, . \, \|_{si}$ and $\| \, . \, \|_{sm}$ are equal is an old one. Actually, when K is weakly valued, there exist commutative ultrametric Banach K-algebras with unity where the semi-norm $\| \, . \, \|_{si}$ is strictly superior to the semi-norm $\| \, . \, \|_{sm}$ for certain elements. But if we assume that $\| \, . \, \|_{si}$ is a norm and that A is complete for this norm, then we can prove the equality [29, 37]. In Corollary 24.13 we find again a Theorem due to B. Guennebaud [34], stating that the completion of a field, with respect to a semi-multiplicative norm, admitting at least two continuous absolute values, has non zero divisors of zero. In particular, this shows why Corollary 4.3.4 in [2] is not true.

Lemma 24.1: *Let $t \in A$ have bounded normal ratio and let $\mathcal{O} = (d(a_i, r_i^-)_{i \in J})$ be a wide t-spectral partition. The restrictions of Θ_t to $R'(sp(t), \mathcal{O}, (a_i)_{i \in J})$ and $R''(sp(t), \mathcal{O}, (a_i)_{i \in J})$ are continuous.*

Proof: The restriction of Θ_t to $R'(sp(t), \mathcal{O}, (a_i)_{i \in J})$ is obviously continuous. Next, we check that the family of $\{\frac{\|(x-b_j)^{-2}\|}{\|(x-b_j)^{-2}\|_{si}} \mid i \in J\}$ is bounded because on one hand $\|(x-b_j)^{-2}\|_{si} = (\|(x-b_j)^{-1}\|_{si})^2$, and on the other hand $\|(x-b_j)^{-2}\| \leq (\|(x-b_j)^{-1}\|)^2$. □

Lemma 24.2: *Let $t \in A$. Assume that $Ker(\Theta_t) \neq \{0\}$. Then $\|t\|_{si} = \|t\|_{sm}$. Moreover, if $A \neq K$, then A has non zero divisors of zero.*

Proof: Let $D = sp(t)$ and let $B = \Theta_t(R(D))$. Then $Ker(\Theta_t)$ is an ideal of $R(D)$ generated by a monic polynomial $G(x) = \prod_{i=1}^{q}(x - a_i)$. Since $G(t) = 0$, for every $\psi \in Mult(A, \| \, . \, \|)$ we have $\psi(G(t)) = 0$, hence there exists an integer $l(\psi) \in \{1, \dots, q\}$ such that $\psi(t - a_{l(\psi)}) = 0$, hence $\psi(t) = |a_{l(\psi)}|$. Then $t - a_{l(\psi)}$ lies in $Ker(\psi)$ and therefore belongs to a maximal ideal $\mathcal{M}$ of A. But by Theorem 6.15, there exists $\theta_\psi \in Mult_m(A, \| \, . \, \|)$ such that $Ker(\theta_\psi) = \mathcal{M}$. Hence we have $\theta_\psi(t) = |a_{l(\psi)}| = \psi(t)$. Thus we have shown that $\psi(t) \leq \|t\|_{sm}$. But this is true for all $\psi \in Mult(A, \| \, . \, \|)$. Consequently, by Theorem 6.19 we have $\|t\|_{si} = \|t\|_{sm}$, which proves the first statement.

Next, we notice that $Ker(\Theta_t)$ admits a generator $G(x) \in K[x]$ whose zeros lie in D. If $\deg(G) = 1$, then t lies in K (considered as a K-subalgebra of A), and obviously we have $\psi(t) = |t| \; \forall \psi \in Mult(A, \| \, . \, \|)$, and therefore, $\sup\{|\lambda| \mid \lambda \in D\} = \|t\|_{si}$. And if $\deg(G) > 1$, then $Ker(\Theta_t)$ is not prime, hence A contains non zero divisors of zero, so the second statement is trivial. □

Theorem 24.3 ([29, 37]): *Let $t \in A$ have bounded normal ratio. Then $\|t\|_{si} = \|t\|_{sm}$.*

Proof: Let $D = sp(t)$. We put $r = \|t\|_{si}$, $r' = \|t\|_{sm}$ and we suppose $r' < r$. If $\|t\| = \|t\|_{si}$ we put $u = r$, and if $\|t\| > \|t\|_{si}$ we take $u \in]r, \|t\|[$. Let $B = \Theta_t(R(D))$. By Lemma 24.1 we can assume that $Ker(\Theta_t) = \{0\}$. Hence B is isomorphic to $R(D)$. Let $\mathcal{O}$ be a wide t-spectral partition of diameter u, and for convenience, we put $\| \, . \, \|_t = \| \, . \, \|_{D,\mathcal{O}}$. Let $(d(a_i, r_i))$, $(i \in J)$ be the family of holes of $\mathcal{O}$ and let $B = \Theta_t(R'(D, \mathcal{O}, (a_i)_{i\in J}))$.

By Theorem 23.6 the restriction Θ'_t of Θ_t to $R'(D, (a_i)_{i\in J}))$ is continuous once $R'(D, \mathcal{O}, (a_i)_{i\in J}))$ is provided with the norm $\| \, . \, \|_t$. Therefore, Θ'_t has continuation to a continuous K-vector space homomorphism from $H'(D, \mathcal{O}, (a_i)_{i\in J})$ into the closure $\overline{B}$ of B in A. We will still denote it by Θ'_t.

Let $s' \in]r', r[$. The annulus $\Gamma(0, s', r)$, admits a partition by a subfamily $\mathcal{S}$ of holes of $\mathcal{O}$. Hence by Theorem 20.1, $\Gamma(0, s', r)$, contains a $\mathcal{O}$-minorated annulus $\Gamma(b, \rho, v)$. Of course, we may choose v as close to ρ as we want. Particularly, if $|b| > \rho$, we take $v \in]\rho, |b|[$. Next, we take $\lambda \in]\rho, v[$. Clearly b does not lie in D. Let $r_b = \frac{1}{\|(x-b)^{-1}\|_{si}}$. By Lemma 23.12 there exists $\theta \in Mult(A, \| \, . \, \|)$ such that $\Psi^{-1}(\theta)$ is (b, r_b)-approaching. In fact, by definition, we have $r_b \leq \rho$, hence $C(b, r_b)$ is included in $d(b, \rho)$, hence $\Psi^{-1}(\theta)$ is secant with $d(b, \rho)$. On the other hand, there certainly exists $\phi \in Mult(A, \| \, . \, \|)$ such that $\Psi^{-1}(\phi)$ is (b, r)-approaching. Then by Propositions 21.5 there exist $f,\ g \in H'(D, \mathcal{O}, (a_i)_{i\in J})$ satisfying:

(i) $\psi(f) = 1$ for all $\psi \in Mult(H(D,\mathcal{O}), \| \,.\, \|_{D,\mathcal{O}})$ such that $\psi(t-b) > v$,
(ii) $\psi(f) = 0$ for all $\psi \in Mult(H(D,\mathcal{O}), \| \,.\, \|_{D,\mathcal{O}})$ such that $\psi(t-b) \leq \lambda$,
(iii) $\psi(g) = 1$ for all $\psi \in Mult(H(D,\mathcal{O}), \| \,.\, \|_{D,\mathcal{O}})$ such that $\psi(t-b) < \rho$,

and

(iv) $\psi(g) = 0$ for all $\psi \in Mult(H(D,\mathcal{O}), \| \,.\, \|_{D,\mathcal{O}})$ such that $\psi(t-b) \geq \lambda$.

We put $F = \Theta'_t(f)$, and $G = \Theta'_t(g)$. Though Θ_t is not supposed to be continuous on all $H(D, \mathcal{O}, (a_i)_{i\in J})$, it is continuous on $H'(D, \mathcal{O}, (a_i)_{i\in J})$. Given $\phi \in Mult(A, \| \,.\, \|)$, we will denote by ϕ_t the K-vector space semi-norm defined on $R(D) + H'(D,\mathcal{O}), \| \,.\, \|_{D,\mathcal{O}}$.

Suppose first $|b| \leq \rho$. It is seen that for every $\psi \in Mult_m(A, \| \,.\, \|)$, we have $\psi(t) \leq r'$, hence $\psi_t(x) \leq r'$. Consequently, $\psi_t(g) = 1$, and therefore $\psi(G) = 1$, hence G is invertible in A. On the other hand, $\Psi^{-1}((\phi)_t)$ is $(0, r)$-approaching, hence $(\phi)_t(g) = 0$, and therefore $(\phi)(G) = 0$, a contradiction to the conclusion G is invertible in A.

Finally, suppose $|b| > \rho$. in the same way we have $d(0, r') \subset K \setminus d(b, \lambda)$, and therefore, $\psi(F) = 1$ for all $\psi \in Mult_m(A, \| \,.\, \|)$, and $\psi(F) = 0$ for all ψ such that $\Psi^{-1}(\psi_t)$ is secant with $d(b, \lambda)$. So, F is invertible. But $\Psi^{-1}(\theta_t)$ is secant with $d(b, \lambda)$ and therefore, $\theta(F) = 0$, a contradiction to the property "F invertible". □

Corollary 24.4: *If A has bounded normal ratio, then A satisfies Property (s).*

Corollary 24.5: *If A is uniform, then A satisfies Property (s).*

Corollary 24.6: *If A has bounded normal ratio and has no maximal ideals of infinite codimension, then A satisfies Property (p).*

Corollary 24.7: *If A is uniform and has no maximal ideals of infinite codimension, then A satisfies Property (p).*

Corollary 24.8: *If A is uniform then its Jacobson radical is null.*

Proposition 24.9: *Let $t \in A$ have bounded normal ratio and assume that there exist $\psi',\ \psi'' \in Mult(A, \| \,.\, \|)$ and $r, r',\ r''$ such that $\psi'(t) \leq r' < r < r'' \leq \psi''(t)$ and $\Gamma(0, r', r'') \cap sp(t) = \emptyset$.*

Then A has non zero divisors of zero.

Proof: Let $D = sp(t)$, and let $B = \Theta_t(R(D))$. Let $\mathcal{O} = (d(a_i, r_i^-)_{i\in J})$ be a wide t-spectral partition. The annulus $\Gamma(0, r, r')$ admits a partition by a subfamily $\mathcal{S}$ of holes of $\mathcal{O}$. Hence by Theorem 20.1, $\Gamma(0, r, r')$ contains a $\mathcal{O}$-minorated annulus $\Gamma(b, \rho, \upsilon)$. Moreover, we notice that there exists at least one element ϕ of $Mult(A, \| \, . \, \|)$ such that $\Psi^{-1}(\phi_t)$ is secant with $d(b, \rho)$. Indeed, if $t - b$ is not invertible, this is obvious.

Now, suppose that $t - b$ is invertible. Since $\Gamma(b, \rho, \upsilon)$ admits a partition by holes of the t-spectral partition, according to Corollary 23.2 we have $\|(t-b)^{-1}\|^{-1} \leq \rho$ and by and Lemma 23.12 there exists $\phi \in Mult(A, \| \, . \, \|)$ such that $\Psi^{-1}(\phi_t)$ is secant with $d(b, \rho)$. Then by Corollary 21.6 there exist f, $g \in H'(D, \mathcal{O}, (a_i)_{i \in J})$ satisfying $fg = 0$ and $\psi'_t(f) = \phi_t(g) = 1$. By Theorem 23.6 we know that Θ_t is continuous on $R''(D, \mathcal{O}, (a_i)_{i \in J})$. And by Proposition 21.4 fg belongs to $H''(\mathcal{O}, (a_i)_{i \in J})$. Consequently, $\Theta_t(f)\Theta_t(g) = 0$. But $\psi'(\Theta_t(f)) = \phi(\Theta_t(f)) = 1$ hence $\Theta_t(f)$, $\Theta_t(g)$ are non zero divisors of zero in A. □

Theorem 24.10: *Let A be an integral domain and let $t \in A$ have bounded normal ratio and be such that $sp(t) \neq \emptyset$. Then $\tau(t) = \|t\|_{si}$.*

Proof: Suppose $\tau(t) < \|t\|_{si}$. Since $sp(t) \neq \emptyset$, by Corollary 6.3 there obviously exist $\psi_1 \in Mult(A, \| \, . \, \|)$ such that $\psi_1(t) = \tau(t)$ and by Corollary 6.10 there exists $\psi_2 \in Mult(A, \| \, . \, \|)$ such that $\psi_2(t) = \|t\|_{si}$. Putting $r_1 = \psi_1(t)$ and $r_2 = \psi_2(t)$, we have $\Gamma(0, r_1, r_2) \cap sp(t) = \emptyset$. Hence by Proposition 24.8 A has non zero divisors of zero, a contradiction. □

Corollary 24.11: *Let A be uniform and have no non zero divisors of zero, and be such that $sp(t) \neq \emptyset$ $\forall t \in A$. Then A satisfies Property (q).*

Corollary 24.12: *Let B be a commutative uniform ultrametric Banach K-algebra without non zero divisors of zero, with unity, such that $sp_B(t) \neq \emptyset$ $\forall t \in B$ and let ϕ be a K-algebra homomorphism from A to B. Let $\| \, . \, \|_{si}^A$ (resp. $\| \, . \, \|_{si}^B$) be the spectral norm of A (resp. B). Then ϕ satisfies*

$$\|\phi(x)\|_{si}^B \leq \|\phi(x)\|_{si}^A, \ \forall x \in A.$$

Thanks to Proposition 24.9, Guennebaud's Theorem [34] on the completion of a field which is a normed K-algebra having more than one continuous absolute value, appears as a simple corollary:

Theorem 24.13: *Let F be a field extension of K provided with a semi-multiplicative norm and admitting two different continuous absolute values. Then the completion of F has non zero divisors of zero.*

Proof: Let A be the completion of F, with respect to its norm. It is a uniform Banach K-algebra such that $Mult'(A, \| \, . \, \|)$ contains at least the expansions ψ_1, ψ_2 of two different continuous absolute values defined on F. Let $t \in F$ be such that $\psi_1(t) < \psi_2(t)$. As an element of F, the spectrum of t is empty, hence it is also empty as an element of A. Putting $r_1 = \psi_1(t)$, $r_2 = \psi_2(t)$, we can apply Proposition 24.9, therefore A has non zero divisors of zero. □

Remark: In the proof of Corollary 4.3.4 of [2], the completion of a field for a spectral norm (denoted by $\mathcal{H}(\Sigma)$) is assumed to be a field, which is not true.

However, Theorem 24.14 shows that, even in a uniform K-algebra Properties (s) and (q) are not equivalent.

Theorem 24.14: *There exists a uniform commutative K-Banach algebra with unity satisfying Property (s) but not Property (q).*

Proof: Let $l \in]0,1[$, let $D = d(0,l^-)$ and let $\mathcal{O}$ be the classic partition of $d(0,1)$ that consists of $d(0,l^-)$ and of the disks $d(a,|a|^-)$ for $a \in d(0,1)$ and $l \leq |a| \leq 1$. For every $h \in R(D)$, we have $\|h\|_{\mathcal{O}} = \sup\{\varphi_{0,r}(h), | \ l \leq r \leq 1\}$. Let A be the completion of $R(D)$ for the norm $\| \ . \ \|_{\mathcal{O}}$. By construction, A is complete for its norm $\| \ . \ \|_{si}$. Consequently, by Corollary 24.5 it satisfies Property (s). On the other hand for every $\lambda \in K \setminus d(0,l)$, $x - \lambda$ is invertible in $R(D)$, hence in A. Consequently, we have $\tau(x) < \|x\|_{si}$. So, A does not satisfy Property (q). □

Theorem 24.15: *Let A be a Banach K-algebra having a bounded normal ratio, such that $sp(x)$ is infraconnected for all $x \in A$. Then A satisfies Property (q).*

Proof: Since $sp(x)$ is infraconnected for all $x \in A$, we can see that $Ker(\Theta_x) = \{0\}$ $\forall x \notin K$. Suppose there exists $t \in A$ such that $\tau(t) < \|t\|$. Let $r' = \tau(t)$, let $r'' = \|t\|$ and let $\mathcal{O} = (d(a_i, r_i^-)_{i \in J})$ be a wide t-spectral partition. By Theorem 20.1 there exists a $\mathcal{O}$-minorated annulus $\Gamma(a, s', s'')$ included in $\Gamma(0, r', r'')$, and by Corollary 23.2 and Lemma 23.12 there exists $\phi \in Mult(A, \| \ . \ \|)$ such that $\Psi^{-1}(\phi_t)$ is secant with $d(a, s')$.

Let $r \in]s', s''[$. Suppose first $\Gamma(a, s', s'')$ admits 0 as a center. Since $\Gamma(0, s', s'')$ is a $\mathcal{O}$- minorated annulus, we can find an increasing distances holes sequence $(T'_n)_{n \in \mathbb{N}}$ of $\mathcal{O}$ of center 0 and diameter r such that the weighted sequence $(T_n, 1)_{n \in \mathbb{N}}$ makes a T-sequence. The T-filter of center 0 and diameter r will be denoted by $\mathcal{F}$.

Let $D = d(0, r'') \setminus \bigcup_{n \in \mathbb{N}} T_n$. Since D is clearly a $\mathcal{O}$-set, $H(D)$ is a K-subalgebra of $H(sp(t), \mathcal{O}, (a_i)_{i \in J})$. Now, we denote by D' the set

$D \cup (K \setminus d(0,r)$. Hence $H_b(D')$ is a Banach K-subalgebra of $H(D)$. Let $R'(D') = R_b(D') \cap R'((sp(t), \mathcal{O}, (a_i)_{i\in J})$ and $H'(D') = H_b(D') \cap H'((sp(t), \mathcal{O}, (a_i)_{i\in J})$. Particularly, by Theorem 23.11 the canonical homomorphism Θ_t is defined and continuous on $H'(D')$. Then, by Theorem 18.12, there exists $f \in H(D)$, meromorphic in each hole T_n, admitting in T_n a_n as unique pole, a_n having order at most 1, such that $|f(c)| = 1 \ \forall c \in d(0,s')$, $f(c) = 0 \ \forall c \in D \setminus d(0, r-)$.

Thus, we see that f belongs to $H'(D')$. Let $F = \Theta_t(f)$. Since $r'' > r$, there exists $\psi \in Mult(A, \| \ . \ \|)$ such that $\psi(t) > r$, hence $\psi(\Theta_t(f)) = 0$, hence F belongs to a maximal ideal containing $Ker(\psi)$, and therefore 0 belongs to $sp(F)$. On the other hand for every $c \in sp(x)$, we have $|f(c)| = 1$, hence $sp(F) \cap C(0,1) \neq \emptyset$. Actually, we will show that $sp(F) \subset C(0,1) \cup \{0\}$, i.e. $F - l$ is invertible for all $l \in K$ such that $0 < |l| < 1$.

Let $l \in K$ be such that $0 < |l| < 1$ and suppose that $F - l$ is not invertible in A. Let $\zeta \in \mathcal{X}(A)$ be such that $\zeta(F - l) = 0$. Let $\mathcal{M} = Ker(\zeta)$, and let $\Omega = \frac{A}{\mathcal{M}}$. Let $\phi \in Mult(A, \| \ . \ \|)$ be such that $Ker(\phi) = \mathcal{M}$. Then ϕ defines an element ϕ_t of $Mult(R(D))$ which of course is of the form $\varphi_{\mathcal{G}}$, with $\mathcal{G}$ a circular filter on K. Moreover, in $R_b(D')$, the norm $\| \ . \ \|_{D'}$ obviously satisfies $\|h\|_{D'} \geq \|h\|_{sp(t),\mathcal{O}}$, hence the restriction of $\varphi_{\mathcal{G}}$ to $R'(D')$ is clearly continuous with respect to the norm $\| \ . \ \|_{D'}$ and has extension to $H'(D')$. So, in particular, we have $\phi_t(f - l) = 0$, hence $\lim_{\mathcal{G}} f(x) - l = 0$, and therefore $\mathcal{G}$ is not secant with $d(0,s') \cup (K \setminus d(0,r))$. In particular, $\mathcal{G} \neq \mathcal{F}$. But since $\mathcal{F}$ is the only T-filter on D', this implies that $\mathcal{G}$ is a Cauchy filter on D', of limit a, hence we have $\zeta(t) = a$, a contradiction to the hypothesis $sp(t) \subset d(0,r')$. Consequently, $F - l$ is invertible in A for all $l \in K$ such that $0 < |l| < 1$, and therefore $sp(F)$ is not infraconnected.

We have a symmetric proof when 0 is not a center of $\Gamma(a, s', s'')$. Then $\Gamma(a, s', s'')$ is included in a class of $C(0, |a|)$, therefore $s'' < |a|$. Indeed, consider $u = \frac{1}{t-a}$. We can check that $sp(u) \subset C(0, \frac{1}{|a|})$, and since there exists $\psi \in Mult(A, \| \ . \ \|)$ such that $\Psi^{-1}(\psi_t)$ is secant with $d(a, r')$, we have $\psi(t - a) \leq s''$, hence $\psi(u) \geq \frac{1}{s''} > \frac{1}{|a|}$. Consequently, we check that $\tau(u) \leq \frac{1}{|a|}$, and $\|u\| > \frac{1}{s''}$, so we are led to the same situation with u instead of t. □

Chapter 25

Properties (o) and (q) in Uniform Banach K-Algebras

We shall show that in a uniform Banach K-algebra, Property (o) implies Property (q). Conversely, we shall see that Property (q) doesn't imply Property (o).

Notation: Throughout this chapter, A is a commutative ultrametric Banach K-algebra with unity. Several results come from [23].

Lemma 25.1: *Let $x \in A$ be such that $\tau(x) < \|x\|$. There exists $\phi \in Mult(A, \| \,.\, \|)$, a number $v > 0$ an affinoid set F which belongs to the filter $\mathcal{F} = \Psi^{-1}(\phi_x)$, such that $diam(\Psi^{-1}(\psi_x)) \geq v$, for all $\psi \in Mult(A, \| \,.\, \|)$ such that $\Psi^{-1}(\psi_x) \in \Phi(F)$.*

Proof: We assume Lemma is not true. Let $r = \tau(x)$ and let $s = \|x\|$, let $m \in]r, s[$, let $\mathcal{O}$ be the strict x-spectral partition, and let $\phi_0 \in Mult(A, \| \,.\, \|)$ be such that $\Psi^{-1}((\phi_0)_x)$ is secant with $\Gamma(0, m, s)$. Let $\mathcal{F}_0 = \Psi^{-1}((\phi_0)_x)$ and let $s_0 = diam(\mathcal{F}_0)$.

Suppose we have already constructed a finite sequence $(\phi_n)_{n \in \mathbb{N}}$ for $n = 0, \dots, q$, satisfying $diam(\Psi^{-1}((\phi_n)_x)) < \frac{s}{n+1}$ and

$$\delta(\Psi^{-1}((\phi_n)_x), \Psi^{-1}((\phi_{n+1})_x)) < \frac{2s}{n+1}.$$

Let $\mathcal{F}_n = \Psi^{-1}((\phi_n)_x)$. Since the Lemma is supposed to be false, there exists no affinoid $F \in \mathcal{F}_n$ such that $\mathcal{F}_n$ is the only circular filter of the form $\Psi^{-1}(\psi_x)$, $\psi \in Mult(A, \| \,.\, \|)$, which is secant with F.

Suppose that we can't find $\phi_{n+1} \in Mult(A, \| \,.\, \|)$ such that the filter $\mathcal{F}_{n+1} = \Psi^{-1}((\phi_{n+1})_x))$ satisfies $diam(\mathcal{F}_{n+1}) < \frac{s}{n+2}$, and

$\delta(\mathcal{F}_{n+1}, \mathcal{F}_n) < \frac{2s}{n+1}$. This means that for every $\psi \in Mult(A, \| \cdot \|)$ such that $\delta(\Psi^{-1}(\psi_x), \mathcal{F}_n) < \frac{2s}{n+1}$, we have $diam(\Psi^{-1}(\psi_x)) \geq \frac{1}{n+1}$. Let F be the unique disk of diameter $\frac{2s}{n+1}$ such that $\mathcal{F}_n \in \Phi(F)$. Then for all $\mathcal{G} \in \Phi(F)$ we have $\delta(\mathcal{F}, \mathcal{G}) \leq \frac{2s}{n+1}$, hence $diam(\Psi^{-1}(\psi_x)) \geq \frac{s}{n+1}$ for all $\psi \in Mult(A, \| \cdot \|)$ such that $\Psi^{-1}(\psi_x) \in \Phi(F)$, which proves the conclusion of the Lemma. Thus, since we have supposed that the conclusion is false, we can find $\phi_{n+1} \in Mult(A, \| \cdot \|)$ such that the filter $\mathcal{F}_{n+1} = \Psi^{-1}((\phi_{n+1})_x))$ satisfies

$$diam(\mathcal{F}_{n+1}) < \frac{s}{n+2}, \text{ and } \delta(\mathcal{F}_{n+1}, \mathcal{F}_n) < \frac{2s}{n+1}.$$

And this is true for all $n \in \mathbb{N}$, hence the sequence $((\phi_n)_x)_{n\in\mathbb{N}}$ is a Cauchy sequence with respect to the δ-topology. Consequently, its limit in $Mult(K[x])$ is of the form φ_α, with $\alpha \in \Gamma(0, r, s)$. On the other hand, since $Mult(A, \| \cdot \|)$ is compact for the topology of simple convergence, the sequence $(\phi_n)_{n\in\mathbb{N}}$ admits a point of adherence θ with respect to the topology of simple convergence. Consequently, θ_x is a point of adherence of the sequence $((\psi_n)_x)_{n\in\mathbb{N}}$. Next, since $Mult(K[x])$ is sequentially compact, we can extract from the sequence $(\psi_n)_x$ a subsequence converging to θ_x with respect to the topology of simple convergence. Thus, without loss of generality, we can assume that the sequence $(\psi_n)_x$ converges to θ_x with respect to the topology of simple convergence in $Mult(K[x])$. But since the sequence $(\psi_n)_x$ converges to φ_α for the δ-topology, so much the more does it converge to φ_α for the topology of simple convergence, hence $\varphi_\alpha = \theta_x$. Thus, θ_α is punctual, a contradiction to the fact that $\alpha \notin sp(x)$. Consequently, our hypothesis *the conclusion is false* is wrong. □

Theorem 25.2: *If A does not satisfy Property (q), then* $\mathtt{GM}_A$ *is not injective.*

Proof: Suppose that A does not satisfy Property (q) and let $x \in A$ be such that $\tau(x) < \|x\|$. Consider now the strict x-spectral partition $\mathcal{O}$ of $d(0, s)^{(}x)$. Let $r = \tau(x)$ and let $s = \|x\|$. By Lemma 25.1 there exists $\phi \in Mult(A, \| \cdot \|)$, a number $v > 0$ and an affinoid set F which belongs to the filter $\mathcal{F} = \Psi^{-1}(\phi_x)$, such that $diam(\Psi^{-1}(\psi_x)) \geq v$, for all $\psi \in Mult(A, \| \cdot \|)$ such that $\Psi^{-1}(\psi_x) \in \Phi(F)$. Whether or not $\mathcal{F}$ has a center, we are going to construct an element $G \in A$ such that, given $\psi \in Mult(A, \| \cdot \|)$, either $\phi(G) = 0$, or $diam(\Psi^{-1}(\psi_x)) \geq v$.

Suppose first that $\mathcal{F}$ has no center. By Lemma 10.11 there exists disks $d(a, t) \in \mathcal{F}$, $d(a, l) \in \mathcal{F}$ included in F, with $l < t$. By Proposition 21.5

there exists $g \in H(sp(x), \mathcal{O})$ such that $\varphi_{\mathcal{G}}(g) = 1$ $\forall \mathcal{G} \in \Phi(d(a,l))$ and $\varphi_{\mathcal{G}}(g) = 0$ $\forall \mathcal{G} \in \Phi(d(0,s) \setminus d(a,t^-))$.

Suppose now that $\mathcal{F}$ has center a and diameter m. Without loss of generality we can assume that F is of the form $d(a, m+\epsilon) \setminus \cup_{j=1}^{q} d(a_j, (m-\epsilon)^-)$, with $m + \epsilon < s$ when $m < s$. For each $j = 1, \ldots, q$, by Proposition 21.5 there exists $g_j \in H(sp(x), \mathcal{O})$ such that $\varphi_{\mathcal{G}}(g_j) = 1$ $\forall \mathcal{G}$ secant with $d(0,s) \setminus d(a,m)^-$ and $\varphi_{\mathcal{G}}(g_j) = 0$ $\forall \mathcal{G}$ secant with $d(a_j, m-\epsilon)$. Moreover, if $m = s$ we put $g_0 = 1$, and if $m < s$, by Proposition 21.5 there exists $g_0 \in H(sp(x), \mathcal{O})$ such that $\varphi_{\mathcal{G}}(g) = 1$ $\forall \mathcal{G}$ secant with $d(a,m)$ and $\varphi_{\mathcal{G}}(g) = 0$ $\forall \mathcal{G}$ secant with $d(0,s) \setminus d(a, (m+\epsilon)^-)$. Then we put $g = \prod_{j=0}^{q} g_j$.

In both cases, we now put $G = \Theta_x(g)$ and we can check that given $\psi \in Mult(A, \| \,.\, \|)$, either $\phi(G) = 0$, or $diam(\Psi^{-1}(\psi_x)) \geq v$. Let $\xi \in K$ satisfy $\|\xi G\| < v$ and let $y = x + \xi G$. Then $\|y - x\| < v$. Consider $\psi \in Mult(A, \| \,.\, \|)$, let $\varphi_{\mathcal{R}} = x^*(\psi)$, $\varphi_{\mathcal{T}} = y^*(\psi)$. If $\psi(G) = 0$, then $\psi(y-x) = 0$, hence by Theorem 15.14 we have $x^*(\psi) = y^*(\psi)$. And now, if $\psi(G) \neq 0$, then we can see that $\psi(y-x) < v \leq diam(\mathcal{R}) \leq diam(\sup(\mathcal{R}, \mathcal{T}))$, therefore $x^*(\psi) = y^*(\psi)$ again. Consequently, $x^*(\psi) = y^*(\psi)$ $\forall \psi \in Mult(A, \| \,.\, \|)$. However, $x \neq y$ because $\phi(G) \neq 0$. □

Corollary 25.3: *Let A be uniform, having at least one maximal ideal of codimension 1 and such that the intersection of all maximal ideals of codimension 1 is equal to $\{0\}$. Then A satisfies Property (q).*

Proof: In Theorem 15.13 we saw that $\mathtt{GM}_A$ is injective, hence by Theorem 25.2, A satisfies Property (q). □

Theorem 25.4: *Let A be uniform and satisfy Property (o) and $Max_a(A) \neq \emptyset$. Then A satisfies Property (p).*

Proof: Suppose A does not satisfy Property (p) hence it does not satisfy Property (q). Let $t \in A$ be such that $\tau(t) < \|t\|$. Since $Max_a(A) \neq \emptyset$ we have $sp(t) \neq \emptyset$. Taking $\alpha \in sp(t)$, we notice that $\tau(t - \alpha) = \tau(t)$, and $\|t - \alpha\| = \|t\|$. So, without loss of generality, we can assume $0 \in sp(t)$. Let $s = \|t\|$, let $r = \tau(t)$, and let $\mathcal{O}$ be the strict t-spectral partition. There exists a $\mathcal{O}$-minorated annulus $\Gamma(b, r', r'')$ included in $\Gamma(0, r, s)$.

Suppose 0 is (resp. is not) a center of $\Gamma(b, r', r'')$. We can find in $\Gamma(0, r', r'')$ (resp. in $\Gamma(b, r', r'')$) an increasing (resp. a decreasing) idempotent T-sequence $(T_n)_{n \in \mathbb{N}}$ of center 0 (resp. of center b) and diameter $\rho \in]r', r''[$ whose holes are holes of $\mathcal{O}$. Consider the $\mathcal{O}$-set D admitting each T_n as a hole, and no other holes. We notice that D admits a unique T-filter $\mathcal{F}$ run by the T-sequence $(T_n)_{n \in \mathbb{N}}$. By Theorem 18.11 there exists $f \in H(D)$ such that $\varphi_{0,\rho}(f) = 0$, $|f(a)| = 1\ \forall a \in d(0, r')$ (resp. $f(a) = 1\ \forall a \in d(0, s) \setminus d(0, r'')$), $f(a) = 0\ \forall a \in d(0, s) \setminus d(0, \rho^-)$, (resp. $f(a) = 0\ \forall a \in d(b, \rho)$), and such that f does not vanish along any large circular filter secant with $d(0, r)$ (resp. with $d(0, s) \setminus d(b, \rho)$). Then, since the T_n are the only holes of D, D admits no T-filter different from $\mathcal{F}$. Therefore, for every λinK^*, $f - \lambda$ is quasi-invertible.

On the other hand, since D is a $\mathcal{O}$-set, $H(D)$ is a K-subalgebra of $H(sp(t), \mathcal{O})$. And since A is uniform Θ_t is defined and continuous on $H(D)$. For every $\psi \in Mult(A, \| \,.\, \|)$ we put $\psi_t = \Theta_t \circ \psi$. Since $s > \rho$ we can find $\psi \in Mult(A, \| \,.\, \|)$ such that ψ_t has a filter secant with $d(0, s) \setminus d(0, \rho^-)$ (resp. since $\Gamma(b, r', r'')$ is not included in any hole of $\mathcal{O}$, we can find $\psi \in Mult(A, \| \,.\, \|)$ such that ψ_t has a filter secant with $d(0, r')$). Consequently, we have $\psi(\Theta_t(f)) = 0$. Let $F = \Theta_t(f)$: we see that F is not invertible in A.

Now, let $\chi \in \mathcal{X}(A, K)$. Since A is uniform, $\chi \circ \Theta_t$ defines a character $\chi_t \in \mathcal{X}(H(D), K)$ which by hypothesis satisfies $\chi_t(x) \leq r$, hence $sa(\Theta_t(f))$ is included in $C(0, 1)$. Let $\alpha \in K$ be such that $< |\alpha| < 1$. Since D has no T-filter except $\mathcal{F}$, we notice that $f - \alpha$ is not vanishing along any T-filter of D and therefore by Theorem 19.1 $f - \alpha$ is quasi-invertible, hence it is of the form $P(x)h$, with P a polynomial and h an invertible element of $H(D)$. Let $g = Ph$ and let $G = \Theta_t(g)$. Suppose that G is not invertible in A and let $\theta \in \mathcal{X}(A)$ be such that $\theta(G) = 0$. Then $\theta(P(t)) = 0$, hence $\theta(t)$ is one of the zeros of P, and therefore lies in $sp(t)$. Consequently, we have $\theta(t) \leq r$, hence $|\theta_t(f)| = 1$, hence $|\theta(F)| = 1$ and therefore we check that $|\theta(G)| = 1$ a contradiction. Thus, we see that G is invertible in A. Then, $sa(G) \subset C(0, 1)$, but α belongs to $sp(G)$ because F is not invertible in A. And since G is invertible in A, $sa(\frac{1}{G})$ is also included in $C(0, 1)$, but $\frac{1}{\alpha}$ belongs to $sp(\frac{1}{G})$. Thus, we have $\|\frac{1}{G}\|_{sa} = 1$, but $r(\frac{1}{G}) \geq |\frac{1}{\alpha}| > 1$, a contradiction to Property (o). □

On the other hand we will prove that there exist uniform Banach ultrametric commutative K-algebras with unity satisfying Property (q), but not property (o).

According to Corollary 15.5 and Theorem 25.4 we have Corollary 25.5:

Corollary 25.5: *Let B be a uniform normed commutative K-algebra with unity satisfying Property (o) and let ϕ be a K-algebra homomorphism from A to B. Let $\| \,.\, \|_{si}^B$ be the norm of B and let $\| \,.\, \|_{si}^A$ be the spectral semi-norm of A. Then ϕ is continuous and satisfies $\|\phi(x)\|_{si}^B \leq \|x\|_{si}^A\ \forall x \in A$.*

Definitions and notation: Let E be a commutative ultrametric normed K-algebra with unity. Let $n \in \mathbb{N}$ and let $\| \,.\, \|_0$ be the norm of E. The E-algebra of polynomials in n variables $E[X_1, \ldots, X_n]$ is provided with the *Gauss norm* $||| \,.\, |||$ defined as

$$\left|\left|\left| \sum_{i_1,,i_n} a_{i_1,\ldots,i_n} X_1^{i_1} \ldots X_n^{i_n} \right|\right|\right| = \sup_{i_1,\ldots,i_n} \|a_{i_1,\ldots,i_n}\|_0.$$

By Proposition 6.18 the norm is a K-algebra norm and if the norm of E is multiplicative, this norm $||| \,.\, |||$ is known to be a multiplicative norm of K-algebra.

Further, by Proposition 6.18, if E is complete, the completion of $E[X_1, \ldots, X_n]$ with respect to this norm, denoted by $E\{X_1, \ldots, X_n\}$, consists of the set of power series in n variables $\sum_{i_1,,i_n} a_{i_1,\ldots,i_n} X_1^{i_1} \ldots X_n^{i_n}$ such that $\lim_{i_1+\ldots+i_n \to \infty} a_{i_1,\ldots,i_n} = 0$. The elements of such an algebra are called *the restricted power series in n variables.*

Henceforth, in this chapter we will denote by M the disk $d(0, 1^-)$ and by A the K-algebra $H(M)\{Y\}$. We will denote by x the identical mapping on M, we fix $r \in]0, 1[$, we put $t = 1 - xY$ and by S the multiplicative subset generated in A by the $t - a$, $a \in d(0, r^-)$, and B will be the K-algebra $S^{-1}A$.

Finally, we will denote by $\mathcal{T}(r)$ the set of $\psi \in Mult(A, \| \,.\, \|)$ such that $\psi(X) \geq r$.

Lemma 25.6: *Let $P \in E[Y]$ and let $Q(Y) = P(1 - Y)$. Then $|||P||| = |||Q|||$.*

Proof: Let $P(Y) = \sum_{j=0}^{n} a_j Y^j$ and let $Q(Y) = \sum_{j=0}^{n} b_j Y^j$. Then each b_j is of the form $(-1)^j \sum_{k=j}^{n} {i \atop j} a_k$. Hence $|b_j| \leq |||P|||$ and thereby $|||Q||| \leq |||P|||$. But on the other hand $P(Y) = Q(1 - Y)$, hence $|||P||| \leq |||Q|||$, consequently the equality holds. □

Lemma 25.7: *Let $\psi \in \mathcal{T}(r)$ and let $h = \frac{f}{g} \in B$, with $f \in A$, $g \in S$. Then $0 \notin \psi(S)$. The mapping $\widetilde{\psi}$ defined in B by $\widetilde{\psi}\left(\frac{f}{g}\right) = \frac{\psi(f)}{\psi(g)}$ belongs to $Mult(B)$. Further, the mapping ϕ from B to $[0, +\infty]$ defined as $\phi(h) = \sup_{\psi \in \mathcal{T}(r)} \widetilde{\psi}(h)$ belongs to $SM(B)$.*

Proof: First, we will show that $\phi(h) < +\infty$. We can write g in the form $\prod_{j=1}^{n} (t - a_j)^{q_j}$, $a_j \in d(0, r^-)$, $q_j \in \mathbb{N}^*$. We set $q = \sum_{j=1}^{n} q_j$. Since $\psi(t) \geq r$, clearly $\psi(t - a_j) = \psi(t)$, and hence we have $\psi(g) \geq r^q$. Consequently,

$$\widetilde{\psi}\left(\frac{f}{g}\right) \leq \frac{\psi(f)}{r^q} \leq \frac{\|f\|}{r^q},$$

and therefore $\phi(h) \leq \frac{\|f\|}{r^q}$. Thus, $\phi \in SM(B)$. On the other hand, since the norm $\| \, . \, \|_M$ of $H(M)$ is an absolute value, we can find in $\mathcal{T}(r)$ a multiplicative norm which obviously lies in $\mathcal{T}(r)$. Consequently ϕ is a norm. □

Notation: Henceforth, $\| \, . \, \|_r$ will denote the norm defined on B in Lemma 25.7 and $\widehat{B}$ will denote the completion of B with respect to this norm. Then by definition, $\widehat{B}$ is a commutative uniform ultrametric Banach K-algebra with unity.

Given a normed K-algebra E, we will denote by $E\{\{Y\}\}_r$ the set of Laurent series with coefficients in E such that $\lim_{n\to+\infty} a_n = 0$, and $\lim_{n\to-\infty} \|a_n\| r^n = 0$.

In $K[Y]$ (resp. in $K\{Y\}$) we will denote by $K[Y]_0$ (resp. $K\{Y\}_0$) the ideal of polynomials (resp. of series) f such that $f(0) = 0$. And we put $W = H(M) + K\{Y\}_0$.

Lemmas 25.8 and 25.9 are immediate:

Lemma 25.8: *W is the direct sum of the K-vector spaces $H(M)$ and $K\{Y\}_0$, and is provided with the product norm $\| \, . \, \|$ defined as $\|g + h\| = \max(\|g\|_M, \|h\|)$, with $g \in H(M)$ and $h \in K\{Y\}_0$.*

Lemma 25.9: $B \subset H(M)\{t\}\{\frac{1}{t}\}_r \subset \widehat{B}$.

Lemma 25.10: *For every $a \in C(0,1)$, $(a - xY)A$ is a maximal ideal of infinite codimension of A. Moreover, we have $sa(xY) = M$, $sp(xY) = U$.*

Proof: Let $N = sa(xY)$ and let $D = sp(xY)$. Since $|||xY||| = 1$, $sp(xY)$ is obviously included in U. On the other hand D is clearly included in M because given $a \in M$, the homorphism $\chi_{a,1}$ from A to K defined as

$$\chi_{a,1}\left(\sum_{j=0}^{\infty} f_j Y^j\right) = \sum_{j=0}^{\infty} f_j(a) \text{ satisfies } \chi_{a,1}(xY) = a.$$

Consequently, we have

$$M \subset N \subset D \subset U. \tag{25.1}$$

Let $a \in U \setminus M$ and consider the ideal I generated by $a - xY$. Clearly $I \cap H(M) = \{0\}$. Consequently the canonical surjection θ from A onto the algebra $A' = \frac{A}{I}$ induces an injection from $H(M)$ to A'.

We will show that $\frac{A}{I}$ is just the completion of the field of fractions F of $H(M)$. Since the norm of A is multiplicative, I is obviously closed in A, hence A' is provided with the quotient norm, which makes it a Banach K-algebra. Let $a \in M$. In A', we have

$$\theta(x) = \left(\theta\left(\frac{Y}{a}\right)^{-1}\right), \text{ and } \left\|\theta\left(\frac{Y}{a}\right)\right\| \leq \left|\left|\left|\frac{Y}{a}\right|\right|\right| = 1,$$

and similarly $\|\theta(x)\| \leq 1$, hence $\|\theta(x)\| = \left\|\theta\left(\frac{Y}{a}\right)\right\| = 1$. Consequently, the series $\sum_{j=0}^{\infty} \frac{b^n}{(\theta(x))^{n+1}}$ converges in K for every $a \in M$, which proves that $\theta(b-x)$ is invertible in A' and admits for inverse $\sum_{j=0}^{\infty} \frac{b^n}{(\theta(x))^{n+1}}$. Thus, A' contains a field isomorphic to F and we will confound it with F, due to the fact that the restriction of θ to $H(M)$ may be considered as the identity.

In the field F, it is clearly seen that the relation $\|(\theta(x-b))^{-1}\| = \|(\theta(x-b))\|^{-1}$ holds for all $b \notin M$ because $x - b$ is invertible in $H(M)$, and that $|||\,.\,|||$ is an absolute value. Consider now $b \in M$. Indeed,

$$(\theta(b-x)^{-1}) = \sum_{j=0}^{\infty} \frac{b^n}{(\theta(x))^{n+1}} = \sum_{j=0}^{\infty} b^n \left(\theta\left(\frac{Y}{a}\right)\right)^{n+1},$$

which proves that $\|(\theta(b-x)^{-1})\| \leq 1$. So, actually, $\|(\theta(b-x)^{-1})\| = 1$, and of course $\|\theta(b-x)\| = 1$ because $|||b-x||| = 1$. Thus the norm of A' induces on F the continuation of the absolute value of $H(M)$. Since $\theta(H(M)[Y]) \subset F$, then F is dense in A', hence A' is the completion of F with respect to its absolute value. Therefore, I is a maximal ideal of infinite codimension. Thus, given $a \in U \setminus M$, then a belongs to $D \setminus N$, hence $U \setminus M = D \setminus N$, and consequently, by (25.1), $L = M$, and $D = U$. □

Theorem 25.11 ([37]): *$\widehat{B}$ is a commutative ultrametric uniform Banach K-algebra without non zero divisors of zero, satisfying Property (q) but not Property (o). Moreover, it is equal to the set of Laurent series $\sum_{-\infty}^{+\infty} a_n Y^n$, with $a_n \in W$, $\lim_{n\to+\infty} a_n = 0$, $\lim_{n\to-\infty} \|a_n\| r^n = 0$, and we have*

$$\left\|\sum_{-\infty}^{+\infty} a_n Y^n\right\| = \max\left(\sup_{n\geq 0} \|a_n\|, \sup_{n<0} \|a_n\| r^n\right).$$

Proof: First, we will show that $\widehat{B}$ does not satisfy Property o). Let $t = 1 - xY$. By Lemma 25.10 we have $sp_A(xY) = U$, $sa_A(xY) = M$. Consequently, we can see that $sa_A(t) = d(1, 1^-)$. Let $\chi \in \mathcal{X}(A, K)$. For all $g \in S$, we notice that $\chi(g) \neq 0$. Thus, χ has continuation to a K-algebra homomorphism $\widetilde{\chi}$ from B to K defined as $\widetilde{\chi}\left(\frac{f}{g}\right) = \frac{\chi(f)}{\chi(g)}$, $\forall f \in A$, $g \in S$. Since $|\chi|$ clearly lies in $\mathcal{T}(r)$, $\widetilde{\chi}$ is continuous with respect to the norm $\| \,.\, \|$ of B, therefore $\widetilde{\chi}$ has continuation to a K-algebra homomorphism $\widehat{\chi}$ from $\widehat{B}$ to K. Consequently, $sa_A(xY) \subset sa_{\widehat{B}}(xY)$. But of course, the inverse inequality is true because $A \subset \widehat{B}$, so we have $sa_{\widehat{B}}(xY) = M$, thereby

$$\left\| \frac{1}{t} \right\|_{sa}^{\widehat{B}} = 1. \tag{25.2}$$

Now, let $\chi \in \mathcal{X}(A)$ satisfy

$$|\chi(t)| \geq r \tag{25.3}$$

and let $\Lambda = \chi(A)$ be provided with an absolute value extending this of K. By (25.3) we have $\chi(g) \neq 0$ $\forall g \in S$. Thus, χ has continuation to a K-algebra homomorphism $\widetilde{\chi}$ from B to Λ such that $|\widetilde{\chi}| \in \mathcal{T}(r)$. Then $\widetilde{\chi}$ is continuous with respect to the norm $\| \,.\, \|$ of B, and therefore it has continuation to a K-algebra homomorphism from $\widehat{B}$ to Λ. Thus, we have proven that $sp_{\widehat{B}}(t) \supset \Delta(0, r, 1)$. But since B is a K-subalgebra of $\widehat{B}$, we have the inverse inclusion, and consequently, the equality holds. Hence $sp_{\widehat{B}}\left(\frac{1}{t}\right) = \Delta(0, 1, \frac{1}{r})$ and thereby

$$\sup\left\{0,\ |\lambda|\ \lambda \in sp_{\widehat{B}}\left(\frac{1}{t}\right)\right\} = \frac{1}{r}.$$

Thus, by (25.2) we can see that $\widehat{B}$ does not satisfy Property (o).

Now, we will show that $\widehat{B}$ has no non zero divisors of zero, and first we will prove that B is the K-vector space of the power series f of the form $\sum_{n=0}^{\infty} a_n t^n$, with $a_n \in W$, $\lim_{n\to+\infty} a_n = 0$, $\|f\| = \sup_{n\geq 0} \|a_n\|$.

Let $h \in H(M)$ and let $j \in \mathbb{N}^*$. Considering the power series of h at 0, by Theorem 16.8 we can write $h = \sum_{j=0}^{n-1} a_j x^j + x^n g$, with $g \in H(M)$, and then we have $\|h\|_M = (|a_0|, \dots |a_{n-1}|, \|g\|_M)$. Consequently,

$$h(x)Y^k = \sum_{j=0}^{k-1} a_j x^j Y^k + (xY)^k g = \sum_{j=0}^{k-1} (a_j)(xY)^j (Y^{k-j}) + (xY)^k g.$$

So, we can write

$$hY^k = \sum_{j=0}^{k-1} a_j(1-t)^j Y^{k-j} + (1-t)^k g. \tag{25.4}$$

Now consider the polynomial $P(Z) = \sum_{j=0}^{k-1} a_j Y^{k-j} Z^j \in K[Y]_0[Z]$ and let $Q(Z) = P(1-Z)$. By Lemma 25.6 we have $|||Q||| = |||P||| \sup_{0\le j\le k-1} |a_j|$. Putting $Q(Z) = \sum_{j=0}^{k-1} b_j(Y)Z^j$, we get

$$|||Q(Z)||| = \sup_{0\le j\le k-1} \|b_j(Y)\| = \sup_{0\le j\le k-1} |a_j|. \tag{25.5}$$

Consequently, we obtain

$$\left\|\sum_{j=0}^{k-1} a_j(1-t)^j Y^{k-j}\right\| = \left\|\sum_{j=0}^{k-1} b_j(Y)Z^j\right\| = |||Q||| = |||P||| = \sup_{0\le j\le k-1} |a_j|.$$

And finally,

$$\|Q(Y)\| = \max(|a_0|, \ldots, |a_{k-1}|). \tag{25.6}$$

On the other hand we have $|||g(1-Z)^j||| = |||gY^j||| = \|g\|$. But since $\|1-t\| = 1$, in A we obtain

$$\|g(1-t)^j\| = \|g\|. \tag{25.7}$$

By (25.4) we deduce $hY^j = Q(t) + (1-t)^j g$, $g \in H(M)$, $Q(Z) \in K[Y]_0[Z]$. And by (25.5), (25.6), (25.7) we obtain

$$\|h\| = \max(\|Q(t)\|, \|(1-t)^j g\|). \tag{25.8}$$

Now, consider $f = \sum_{j=0}^{+\infty} f_j Y^j \in A$. For each $j \in \mathbb{N}$, $f_j Y^j$ is of the form $Q_j(t) + g_j S_j(t)$, with $Q_j(Z) \in K[Y]_0[Z]$, $g_j \in H(M)$, $S_j(Z) \in K[Z]$, satisfying further

$$|||Q_j||| \le \|f_j\| \tag{25.9}$$

and

$$|||g_j||| \le \|f_j\|. \tag{25.10}$$

Consequently, we have

$$|||Q_j(Z) + g_j S_j||| \le \|f_j\|. \tag{25.11}$$

By (25.9), the series $\sum_{j=0}^{\infty} Q_j(Z)$ is clearly converging in $K\{Y\}_0\{Z\}$ to a limit $\sum_{j=0}^{\infty} \lambda_j(Y)Z^j$. In the same way, by (25.10) the series $\sum_{j=0}^{\infty} g_j S_j(t)$ converges in $H(M)\{Z\}$ to a limit $\sum_{j=0}^{\infty} \mu_j Z^j$. Moreover, by (25.9) and (25.10), we can see that $\sup_{j\in\mathbb{N}} \|\lambda_j\| \le \|f\|$ and $\sup_{j\in\mathbb{N}} \|\mu_j\| \le \|f\|$. On the other hand we have $\|f\| \le \max(\sup_{j\ge 0} \|\lambda_j\|, \sup_{j\ge 0} \|\mu_j\|)$, thereby

$$\|f\| = \max(\sup_{j\ge 0} \|\lambda_j\|, \sup_{j\ge 0} \|\mu_j\|).$$

Putting $a_j = \lambda_j + \mu_j$, we have $\|a_j\| = \max(\|\lambda_j\|, \|\mu_j\|)$. This shows A to be the set of series $G = \sum_0^{+\infty} a_n t^n$ with coefficients $a_j \in \Lambda$, satisfying $\lim_{n\to+\infty} a_n = 0$,

$$\|G\| = \sup_{n\ge 0} \|a_n\|.$$

Now, we will show that $\widehat{B}$ is the set Ω of series $G = \sum_{-\infty}^{+\infty} a_n t^n$, with coefficients $a_j \in W$, satisfying $\lim_{n\to+\infty} a_n = 0$, $\lim_{n\to-\infty} \|a_n\| r^n = 0$, and $\|G\| = \sup_{n\ge 0} \|a_n\|$. Since $\left\|\frac{1}{t}\right\| = \frac{1}{r}$, it is seen that $\Omega \subset \widehat{B}$. Given $a \in d(0, r^-)$, we notice that $\frac{1}{t-a} = \sum_{n=0}^{+\infty} \frac{a^n}{t^{n+1}}$ belongs to Ω. So, by Lemma 25.9 B is included in Ω, and therefore Ω is dense in $\widehat{B}$.

Since the norm of $H(M)$ is an absolute value, on $H(M)[Z]$ we can consider the two absolute values ψ_0, ψ_1 defined by

$$\psi_0\left(\sum_{j=0}^{q} a_j Z^j\right) = \sup_{0\le j\le q} \|a_j\|,\ \psi_1\left(\sum_{j=0}^{q} a_j Z^j\right) = \sup_{0\le j\le q} \|a_j\| r^j$$

$(a_j \in H(M), 0 \le j \le q)$. These two absolute values have continuation to $H(M)[Z, \frac{1}{Z}]$, so this apply to $H(M)[t, \frac{1}{t}]$. But $Y = \frac{1-t}{x}$ does belong to $H(M)[t, \frac{1}{t}]$. Therefore, we can consider the restrictions ψ_0', ψ_1' of ψ_0, ψ_1 to $H(M)[Y]$. Thus, by hypothesis we have

$$\psi_i'(Y) = \frac{\psi_i'(1-t)}{\psi' i(x)} = \psi'(1-t) = 1\ (i = 0, 1).$$

Then given $\sum_{k=0}^{q} b_k Y^k\ H(M)[Y]$, we have

$$\psi_i'\left(\sum_{k=0}^{q} b_k Y^k\right) \le \sup_{0\le k\le q} \|b_k\|\ (i = 0, 1).$$

Consequently, $\psi'(f) \le \|f\|\ \forall f \in H(M)[Y]$, hence each ψ_i' has continuation to an absolute value ψ_i'' defined on A, and actually belongs to

$Mult(A, \| \,.\, \|)$ $(i = 0, 1)$. On the other hand, we notice that each ψ_i'' belongs to $\mathcal{T}(r)$, and therefore has extension ψ_i''' to B and $\widehat{B}$ $(i = 0,\ 1)$. In particular we have

$$\psi_i'''(f) \leq \|G\| \ \forall G \in \widehat{B} \ (i = 0, 1). \tag{25.12}$$

Now, let $l = \sum_{-\infty}^{+\infty} a_n t^n \in \Omega$. By (25.12), we can check that

$$\|G\| \geq \psi_0(G) = \sup_{n\geq 0} \|a_n\| \text{ and } \|G\| \geq \psi_1(G) = \sup_{n>0} \|a_n\| r^n.$$

Thus we have $\|G\| = \max(\sup_{n\geq 0} \|a_n\|, \sup_{n<0} \|a_n\| r^n)$. Consequently, Ω is complete for the norm $\| \,.\, \|$ and therefore is equal to $\widehat{B}$.

Let F be the field of fractions of $H(M)$ and let $\widehat{F}$ be its completion with respect to the absolute value $\| \,.\, \|_M$ of $H(M)$ extended to F. Then, $\widehat{F}$ is provided with an absolute value that we still denote by $\| \,.\, \|_M$. We can consider $\widehat{F} + K\{Y\}_0$ as a K-vector space provided with the norm $\| \,.\, \|$ defined as it follows: given $g \in \widehat{F}$, $h \in K\{Y\}_0$, we set $\|g + h\| = \max(\|g\|_M, \|h\|)$. Let V be the K-vector space consisting of all series $\sum_{-\infty}^{+\infty} a_n t^n$ $a_n \in \widehat{F} + K\{Y\}_0)\{\{t\}\}_r$. It is provided with the norm defined by

$$\Big\| \sum_{-\infty}^{+\infty} a_n t^n \Big\| = \max(\sup_{k\geq 0} \|a_k\|, \sup_{k<0} \|a_k\| r^k).$$

Thus, $\widehat{B}$ is a K-subvector space of V which actually is included in $\widehat{F}\{\{t\}\}_r$ because so is $K\{Y\}$, due to the fact that $Y = \frac{1-t}{x}$ and $\left\|\frac{1}{x}\right\| = 1$. Since both the multiplication of $\widehat{B}$ and its norm are induced by those of $\widehat{F}\{\{t\}\}_r$, $\widehat{B}$ is a K-subalgebra of $\widehat{F}\{\{t\}\}_r$, and therefore $\widehat{B}$ is an integral domain. But the norm of $\widehat{B}$ is semi-multiplicative, and therefore $\widehat{B}$ is uniform, hence by Corollary 24.11 $\widehat{B}$ satisfies Property (q). $\square$

Chapter 26

Properties (o) and (q) and Strongly Valued Fields

In Chapter 25 we saw that in a commutative uniform Banach K-algebra, Property (o) implies Property (q). Here we examine whether this remains true when the algebra is not supposed to be uniform. Actually, this is depending on the ground field [23]: if it is strongly valued, then the property is always true. If it is weakly valued, counter-examples show that the property sometimes doesn't hold.

Notation: In the Chapter, A is a commutative ultrametric Banach K-algebra with unity.

Theorem 26.1: *K is supposed weakly valued. There exist a non semi-simple commutative ultrametric Banach K-algebra without non zero divisors of zero, with unity, all maximal ideals of which have codimension* 1, *such that $\| \,.\, \|_{si}$ is a norm.*

Proof: For convenience, we put $l = \frac{1}{2}$. For each $r \in [0, 1]$, we denote by $\mathcal{F}_r$ the circular filter of center 0 and diameter r. By Proposition 19.6 we can construct an open closed bounded infraconnected set $E \subset d(0, 1^-) \setminus d(0, l^-)$ such that $\widetilde{E} = d(0, 1)$, admitting an irregular T-specific partition whose elements are the sets $d(a, r^-) \cap E$, whenever $|a| = r,\ l < r < 1$.

Now, let $(r_n)_{n \in \mathbb{N}}$ be a strictly increasing sequence of $[l, 1[\cap|K|$ such that $\lim_{n\to\infty} r_n = 1$ and $|\sum_{n=0}^{\infty} \log r_n|_\infty = +\infty$, and for each $n \in \mathbb{N}$, let $a_n \in C(0, r_n)$ and let $T_n = d(a_n, r_n^-)$. Let $E' = E \setminus (\cup_{n\in\mathbb{N}} T_n)$ and $D = \Gamma(0, 1, 2) \cup E'$. Let $\mathcal{T} = \mathcal{Y}(\mathcal{F}_1)$. By Theorem 18.8 the weighted sequence $(T_n, 1)_{n\in\mathbb{N}}$ is an idempotent T-sequence of center 0 and diameter 1. And the subset E' of D admits a partition of the form $(\mathcal{C}_{E'}(\mathcal{G}_{i,r}),\ i \in I, r \in]l, 1[\cap|K|)$

where the $(\mathcal{G}_{i,r})_{i\in I}$ are T-filters of diameter r secant with $C(0,r)$, and where the only T-filter on E' which is not one of the $\mathcal{G}_{i,r}$, is the increasing filter of center 0 and diameter 1. Hence for every increasing filter $\mathcal{G}$ of D secant with any circle $C(0,r)$, by Corollary 19.4 we have $\mathcal{Y}(\mathcal{G}) = \mathcal{Y}(\mathcal{F}_r)$. In particular, since for all $r, r' \in]l, 1[$, $\mathcal{F}_r$ $\mathcal{F}'_r$ are secant with $\mathcal{P}_D(\mathcal{G})$, we check that $\mathcal{Y}(\mathcal{F}_r) = \mathcal{Y}(\mathcal{F}_{r'})$. So, we will just denote by $\mathcal{J}$ this closed prime ideal $\mathcal{Y}(\mathcal{F}_r)$ which actually does not depend on $r \in]l, 1[$.

Since D has no decreasing T-filter, as we just saw, each element f of $\mathcal{J}$ satisfies $f(a) = 0\ \forall a \in \Gamma(0,1,2)$, and therefore $\mathcal{J} \subset \mathcal{T}$. Moreover, since D has no decreasing T-filter, $\mathcal{F}_1 \cap D$ is not a distinguished circular filter, therefore D actually has no distinguished circular filter. Consequently every maximal ideal of $H(D)$ has codimension 1. For every $a \in D$, let $\mathcal{M}_a$ be the maximal ideal of $H(D)$ consisting of the $f \in H(D)$ such that $f(a) = 0$ and let $\chi_a \in \mathcal{X}(H(D), K)$ be defined as $\chi_a(f) = f(a)$. Then each maximal ideal of $H(D)$ is of the form $\mathcal{M}_a$. Let $A = \frac{H(D)}{\mathcal{J}}$ and let θ be the canonical surjection from $H(D)$ onto A. For each $a \in \Gamma(0,1,2)$, $\theta(\mathcal{M}_a)$ is a maximal ideal of codimension 1, because $\mathcal{M}_a \supset \mathcal{J}$. But for each $a \in d(0,1) \cap D$, $\mathcal{J}$ is not included in $\mathcal{M}_a$, hence $\theta(\mathcal{M}_a)$ is not a maximal ideal of A. Consequently, by Lemma 1.13 all maximal ideals of A have codimension 1 and are of the form $\theta(\mathcal{M}_a)$ with $a \in \Gamma(0,1,2)$. But since D admits an increasing T-sequence of center 0 and diameter 1, there exists in $H(D)$ an element $f \in \mathcal{T}$ satisfying $g(a) = 0\ \forall a \in \Gamma(0,1,2)$, and $\varphi_{0,r}(g) > 0\ \forall r \in]l, 1[$. Let $h = \theta(g)$. Then $h \neq 0$, but since $g(a) = 0\ \forall a \in \Gamma(0,1,2)$, we see that $\gamma(h) = 0\ \forall \gamma \in Mult_m(A, \|\ .\ \|)$. Therefore we have $\|h\|_{sm} = 0 < \|h\|_{si}$. By construction, A has no non zero divisors of zero. Now, suppose that an element y of A is such that $\|y\|_{si} = 0$. Let $f \in H(D)$ be such that $\theta(f) = y$. By definition, for each $r \in]l, 1[$, $\varphi_{0,r}$ is of the form $\psi_r \circ \theta$, with $\psi_r \in Mult(A, \|\ .\ \|)$, therefore $\varphi_{0,r}(f) = 0$, hence f belongs to $\mathcal{J}$, thereby $y = 0$, hence $\|\ .\ \|_{si}$ is a norm. □

Corollary 26.2: *If K is weakly valued, there exist commutative ultrametric Banach K-algebra with unity, without non zero divisors of zero, all maximal ideals of which have codimension* 1, *satisfying property* (o) *but neither Property* (q) *nor Property* (s).

Theorem 26.3: *Let K be strongly valued. Then A satisfies Property* (s).

Proof: Suppose Property (s) is not satisfied. So, there exists $t \in A$ such that $\|t\|_{si} > \|t\|_{sm}$, and therefore, there exists $\psi_0 \in Mult(A, \|\ .\ \|)$ such that $\psi_0(t) > \|t\|_{sm}$. Let $r = \|t\|_{sm}$ and $u = \psi_0(t)$. Thus, $sp(t)$ is included in $d(0,r)$. Let $\mathcal{O}$ be a t-normal partition. By Corollary 20.5 there exists an increasing idempotent T-sequence $(T_n)_{n\in\mathbb{N}}$ of center 0 and diameter u with all T_n included in the annulus $\Gamma(0, r, u)$. Therefore, by Proposition 21.5 there exists $f \in H(D, \mathcal{O})$ such that $\psi(f) = 1\ \forall\psi \in Mult(H(D,\mathcal{O}), \|\,.\,\|_{D,\mathcal{O}})$ such that $\psi(t) \leq r$, and $\psi(f) = 0\ \forall\psi \in Mult(H(D,\mathcal{O}), \|\,.\,\|_{D,\mathcal{O}})$ such that $\psi(t) \geq u$. Let $\mathcal{M}$ be a maximal ideal of A containing $Ker(\psi_0)$, and let $\phi \in Mult(A, \|\,.\,\|)$ such that $Ker(\phi) = \mathcal{M}$. Since $\mathcal{O}$ is t-normal partition, by Theorem 22.6 we have a canonical continuous morphism Θ_t from $H(D, \mathcal{O})$ into A. Let $F = \Theta_t(f)$. Then, $\psi_0(F) = 0$, and therefore $\phi(F) = 0$. But since $\phi \in Mult_m(A, \|\,.\,\|)$, we have $\phi(t) \leq r$. Let $\widetilde{\phi} = \phi \circ \Theta_t$. Thus, $\widetilde{\phi}$ is of the form $\varphi_{\mathcal{F}}$ and $\mathcal{F}$ is secant with $d(0, r)$, so we have $\widetilde{\phi}(f) = 1$, and therefore $\widetilde{\phi}(F) = 1$, a contradiction. □

Proposition 26.4: *Assume K is strongly valued. Suppose there exists $t \in A$, ψ_1, $\psi_2 \in Mult(A, \|\,.\,\|)$ such that $\psi_1(t) < \psi_2(t)$ and $\Gamma(0, \psi_1(t), \psi_2(t)) \cap sp(t) = \emptyset$. Then A admits non zero divisors of zero and A contains elements whose spectrum is not infraconnected. Moreover, if $\psi_1(t) = \tau(t)$, then A does not satisfy Property (o) and the intersection of all maximal ideals of codimension 1 is not null.*

Proof: Let $D = sp_A(t)$. We put $r_1 = \psi_1(t)$, $r_2 = \psi_2(t)$. Let $s \in]\|t\|_{si}, \|t\|[$. Let $\mathcal{O}$ be a t-normal partition of diameter s. Let $r \in]r_1, r_2[$. By Corollary 20.5 there exist an increasing T-sequence and a decreasing T-sequence of center 0 and diameter r, and therefore by Proposition 21.5 there exist $f, g \in H(D, \mathcal{O})$ such that:

$$\psi(f) = 1\ \forall\psi \in Mult(H(D,\mathcal{O}), \|\,.\,\|_{D,\mathcal{O}}) \text{ such that } \psi_t(t) \leq r_1, \text{ and}$$

$$\psi_t(f) = 0\ \forall\psi \in Mult(H(D,\mathcal{O}), \|\,.\,\|_{D,\mathcal{O}}) \text{ such that } \psi_t(t) \geq r, \tag{26.1}$$

$$\psi_t(f) \neq 0 \forall\psi \in Mult(A, \|\,.\,\|) \setminus Mult_a(A, \|\,.\,\|) \text{ such that } r_1 < \psi_t(f) < r, \tag{26.2}$$

$$\psi_t(g) = 1\ \forall\psi \in Mult(H(D,\mathcal{O}), \|\,.\,\|_{D,\mathcal{O}}) \text{ such that } \psi(t) \geq r_2, \text{ and}$$

$$\psi_t(g) = 0\ \forall\psi \in Mult(H(D,\mathcal{O}), \|\,.\,\|_{D,\mathcal{O}}) \text{ such that } \psi(t) \leq r. \tag{26.3}$$

Thus $fg = 0$. Since $\mathcal{O}$ is t-normal partition, by Theorem 23.6 we have the canonical continuous morphism Θ_t from $H(D, \mathcal{O})$ into A. Now, Θ_t is defined on $H(D, \mathcal{O})$. Let $F = \Theta_t(f)$, $G = \Theta_t(g)$. Let $\mathcal{F}_j$ be the circular filter of $(\psi_j)_t$ $(j = 1, 2)$. Then $\mathcal{F}_1$ is secant with $d(0, r_1)$, hence $(\psi_1)_t(f) = 1$, and therefore $\psi_1(F) = 1$. In the same way, $\mathcal{F}_2$ is secant with $d(0, s) \setminus d(0, r^-)$, hence $(\psi_2)_t(g) = 1$, and therefore $\psi_2(G) = 1$. Thus, both F, G are different from 0 though $FG = 0$ and therefore A contains non zero divisors of 0.

Now, we will show that $sp_A(F)$ is not infraconnected. Consider $l \in d(0, 1^-) \setminus \{0\}$, and suppose that $F - l$ is not invertible in A. Let $\zeta \in \mathcal{X}(A)$ be such that $\zeta(F-l) = 0$. Let $\mathcal{M} = Ker(\zeta)$. Let $\phi \in Mult(A, \| \, . \, \|)$ be such that $Ker(\phi) = \mathcal{M}$. Then ϕ defines an element ϕ_t of $Mult(R(D, \| \, . \, \|_D)$ which of course is of the form $\varphi_{\mathcal{G}}$, with $\mathcal{G}$ a circular filter on K. But since $0 < |l| < 1$, by (26.1) we can see that $\mathcal{G}$ is not secant with $d(0, s_1)$, and is not secant with $K \setminus d(0, r^-)$ either. Consequently, $\mathcal{G}$ approaches a circle $C(0, u)$, with $s_1 < u < r$. But by (26.2) we can see that $\varphi_{\mathcal{G}} \in Mult_a(A, \| \, . \, \|)$, hence there exists $\chi \in \mathcal{X}(A, K)$ such that $\varphi_{\mathcal{G}} = |\chi|$, therefore $|\chi(t)| \leq s_1$, a contradiction. Consequently, $F - l$ is invertible in A for all $l \in K$ such that $0 < |l| < 1$, and therefore $sp(F)$ is not infraconnected.

Now, suppose that $\tau(t) = \psi_1(t)$ and let $\chi \in \mathcal{X}(A, K)$. Then $|\chi(t)| \leq \tau(t) = s_1$, hence $\chi(G) = 0$, therefore G belongs to the intersection of all maximal ideals of codimension 1. So, this intersection is not null.

Since $(\psi_2)_t(f) = 0$, F is not invertible in A, hence $0 \in sp_A(F)$. Consider $l \in d(0, 1^-) \setminus \{0\}$, and let $y = \frac{1}{F-l}$. Thus, we have $sp_A(F - l) \subset C(0, 1) \cup \{-l\}$ and $sp_A(y) \subset C(0, 1) \cup \{\frac{1}{-l}\}$. More precisely, since F is not invertible in A, $-l$ does belong to $sp_A(F - l)$, and consequently, $\frac{1}{-l}$ does belong to $sp_A(y)$. Hence we have $\tau(y) = \frac{1}{|l|}$. However, for every $\chi \in \mathcal{X}(A, K)$, $\chi(t)$ lies in D, hence $\chi(x)$ belongs to $d(0, s_1)$, and therefore $|\chi(F)| = |f(\chi(t))| = 1$, thereby $|\chi(y)| = 1$. Consequently, we have $\|y\|_{sa} = 1$, which proves that Property (o) is not satisfied. □

Corollary 26.5: *Assume K is strongly valued. Suppose there exists $t \in A$ such that $sp(t)$ is not infraconnected. Then A admits non zero divisors of zero.*

Proof: Let $\Gamma(a, r_1, r_2)$ be an empty annulus of $sp_A(t)$. By Lemma 8.15 there exists ψ_1, $\psi_2 \in Mult(A, \| \, . \, \|)$ such that $r_1 = \psi_1(t), r_1 = \psi_2(t)$. So, we can apply Theorem 26.4. □

Theorem 26.6: *Assume K is strongly valued. There exists $t \in A$, ψ_1, $\psi_2 \in Mult(A, \| \, . \, \|)$ such that $\psi_1(t) < \psi_2(t)$ and $\Gamma(0, \psi_1(t), \psi_2(t)) \cap sp(t) = \emptyset$ if and only if A contains elements whose spectrum is not infraconnected.*

Proof: Suppose that there exists $t \in A$, ψ_1, $\psi_2 \in Mult(A, \| \, . \, \|)$ such that $\psi_1(t) < \psi_2(t)$ and $\Gamma(0, \psi_1(t), \psi_2(t)) \cap sp(t) = \emptyset$. Then by Proposition 26.4 A admits an element u such that $sp(u)$ is not infraconnected. Conversely, if A admits an element u such that $sp(u)$ is not infraconnected, $sp(u)$ admits an empty annulus $\Gamma(a, r_1, r_2)$ and then by Lemma 8.15 there exist ψ_1, $\psi_2 \in Mult(A, \| \; . \; \|)$ such that $\psi_1(t) = r_1$, $\psi_2(t) = r_2$. □

Theorem 24.7: *Assume K is strongly valued and let A be such that $Max_a(A) \neq \emptyset$. If A satisfies one of the following conditions bellow, then A satisfies Property (q):*

(i) Property (o),
(ii) A has no non zero divisors of zero,
(iii) $sp(x)$ is infraconnected for every $x \in A$,
(iv) the intersection of all maximal ideals of codimension 1 is null.

Proof: We assume that A does not satisfies Property q), and will show that A does not satisfy the three properties at the bottom of Theorem 26.7. Let $t \in A$ be such that $\tau(t) < \|t\|_{si}$. Let $s_1 = \tau(t)$, let $s_2 = \|t\|_{si}$. Both s_1, s_2 lie in the closure of $\{\phi(t) \mid \phi \in Mult(A, \| \; . \; \|)\ \}$, hence by Corollary 6.3 there exists $\psi_1 \in Mult(A, \| \; . \; \|)$ such that $\psi_1(t) = s_1$, and by Theorem 6.19 there exists $\psi_2 \in Mult(A, \| \; . \; \|)$ such that $\psi_2(t) = s_2$. By hypothesis we have $\Gamma(0, s_1, s_2) \cap sp(t) = \emptyset$ hence by Proposition 26.4, A has non zero divisors of zero, the intersection of all maximal ideals of codimension 1 is not null, $sp_A(t)$ is not infraconnected, and Property (o) is not satisfied. □

Chapter 27

Multbijective Banach K-Algebras

In Theorem 6.15 we saw that the mapping from $Mult_m(A, \|.\ \|)$ into $Max(A)$ associating to each $\psi \in Mult_m(A, \|.\ \|)$ its kernel, is surjective. The natural question coming next, is whether the mapping is also injective. The answer was given in [23].

Definition and notation: In all this chapter, A is a commutative ultrametric Banach K-algebra with unity.

A is said to be *multbijective* if for every maximal ideal $\mathcal{M}$ there exists only one $\psi \in Mult_m(A, \|.\ \|)$ such that $Ker(\psi) = \mathcal{M}$.

Remark: As recalled in Chapter 5, in $\mathbb{C}$ it is well known that every maximal ideal of a $\mathbb{C}$-Banach algebra is the kernel of one and only one multiplicative semi-norm, and that every multiplicative semi-norm has a kernel which is a maximal ideal.

A partial answer to the question above was first given in [21] and is recalled in Theorem 42.10 in [30]:

Theorem 27.1: *If K is weakly valued there exists a closed bounded infraconnected set E in K such that $H(E)$ admits multiplicative semi-norms $\psi_1, \psi_2 \in Mult_m(H(E), \|\ .\ \|_E)$, with $\psi_1 \neq \psi_2$ such that $Ker(\psi_1) = Ker(\psi_2)$.*

Proof: The proof is based upon the same construction as in Theorem 26.1. Here we just take again the set E constructed at the beginning of the proof of Theorem 26.1. Let $l = \frac{1}{2}$. For each $r \in [0, 1]$, we denote by $\mathcal{F}_r$ the circular filter of center 0 and diameter r. By Proposition 19.6 we can construct an open closed bounded infraconnected set $E \subset d(0,1^-) \setminus d(0,l^-)$ such that $\widetilde{E} = d(0,1)$, admitting an irregular T-specific partition $(\mathcal{C}(\mathcal{G}_{i,r}), i \in I, r \in]l,1[\cap|K|)$ and we denote by $\mathcal{F}$ the circular filter of center 0 and diameter 1. By Lemma 19.3 we have

$$\mathcal{Y}(\mathcal{G}_{i,r}) = \mathcal{Y}(\mathcal{G}_{j,s}) = \mathcal{Y}(\mathcal{F}),\ \forall i,j \in I, \forall r,s \in]l,1[\cap|K|.$$

Let $\mathcal{M}$ be this ideal $\mathcal{Y}(\mathcal{F})$ which actually does not depend on $i \in I$, and $r \in]l,1[\cap|K|$. By Corollary 19.8 $\mathcal{M}$ is a maximal ideal of $H(D)$ of infinite codimension. And for every $r \in]l,1[\cap|K|$, we have

$$Ker(\varphi_{0,r}) = \mathcal{Y}(\mathcal{G}_{i,r}) = \mathcal{M} \text{ (for any } i \in I).$$

Thus, all the $\varphi_{0,r}$ $(l < r < 1)$ have the same kernel. □

When K is strongly valued, we have a different answer:

Theorem 27.2: *Let K be strongly valued, and let F be a field extension of K which is an ultrametric normed K-algebra admitting two different continuous absolute values. Then the completion of F admits non zero divisors of zero.*

Proof: Let $\widehat{F}$ be the completion of F. Let ψ_1, ψ_2 be two different continuous absolute values have continuation to $\widehat{F}$. We can find $x \in F$ such that $\psi_1(x) < \psi_2(x)$. We put $r_1 = \psi_1(x)$, $r_2 = \psi_2(x)$. Since x does not belong to K, we have $sp(x) = \emptyset$. Consequently, by Theorem 26.4 $\widehat{F}$ has non zero divisors of zero. □

Remark: In Theorem 24.12 we saw that a commutative Banach K-algebra with unity which is the completion of a field extension F of K provided with a semi-multiplicative norm, has non zero divisors of zero. We then made no additional hypothesis on the ground field. Here we suppose K strongly valued, but we no longer assume that the norm is the spectral norm of F.

As a Corollary, we obtain Theorem 27.3:

Theorem 27.3: *If K is strongly valued, then A is multbijective.*

Proof: Suppose K is strongly valued. Let $\mathcal{M}$ be a maximal ideal of A, and suppose there exist ψ_1, $\psi_2 \in Mult_m(A, \|\ .\ \|)$ such that $Ker(\psi_1) = Ker(\psi_2) = \mathcal{M}$. Let $F = \frac{A}{\mathcal{M}}$. Thus, the field F admits two different absolute values $|\ .\ |_j$ quotients of ψ_j respectively, $(j = 1,2)$ which by construction, are continuous for the norm of F quotient of the norm of A. Since F is a

K-Banach algebra for its quotient norm, by Theorem 27.2 we are led to a contradiction. □

Corollary 27.4: *There exist non multbijective commutative ultrametric Banach K-algebras with unity if and only if K is weakly valued.*

Theorem 27.5: *Let K be strongly valued, let $\psi \in Mult(A, \| \, . \, \|)$ and suppose that the set of maximal ideals containing $Ker(\psi)$ is countable. Then ψ belongs to $Mult_m(A, \| \, . \, \|)$.*

Proof: Let $\mathcal{J} = Ker(\psi)$ and let $B = \frac{A}{\mathcal{I}}$ and let θ be the canonical surjection from A onto B. Then $Max(B) = \theta(Max(A))$ is countable. Suppose that $\mathcal{J} \notin Max(A)$. Let $\mathcal{M}$ be a maximal ideal of A containing $\mathcal{J}$ and let $\mathcal{M}' = \theta(\mathcal{M})$. Then B admits an absolute value ψ' such that $\psi = \psi' \circ \theta$. Moreover, there exists (a unique) $\phi \in Mult_m(A, \| \, . \, \|)$ such that $Ker(\phi) = \mathcal{M}$. Let $\phi' \in Mult_m(B, \| \, . \, \|')$ be such that $Ker(\phi') = \mathcal{M}'$. Since $\mathcal{M} \neq \mathcal{J}$, $\mathcal{M}'$ is not null. Let $t \in \mathcal{M}' \setminus \{0\}$: we have $\phi'(t) = 0$, $\psi'(t) \neq 0$. Let $D = sp(t)$, and let $l = \psi'(t)$ and let $\mathcal{S}$ be a t-normal partition. Since $Max(B)$ is countable, so is D. We notice that 0 lies in D and that none of the classes of the circles $C(0, r)$, with $0 < r < \|t\|$ meet D, except at most a countable subfamily. Consequently, by Theorem 18.10, $\mathcal{S}$ admits an increasing idempotent T-sequence of center 0 and diameter l. Therefore, by Proposition 21.5 there exists $f \in H(D, \mathcal{S})$ such that $f(0) = 1$ and $\varphi(f) = 0 \; \forall \varphi \in Mult(H(D, \mathcal{S}), \| \, . \, \|_{D,\mathcal{S}})$ such that $\phi(t) \geq l$. Since $\mathcal{O}$ is t-normal partition, by Theorem 23.6 we have a canonical continuous morphism Θ_t from $H(D, \mathcal{S})$ into A. Let $F = \Theta_t(f)$, let $L = \frac{B}{\mathcal{M}'}$ and let Ω be the canonical surjection from B onto L. Then Ω is continuous with respect to quotient topologies of B and L. And since $t \in \mathcal{M}'$, we have $\Omega(\Theta_t(x)) = 0$ and therefore, $\Omega(\Theta_t(f)) = f(0) = 1$. Consequently, $F \neq 0$. Next, as $\psi'(t) = l$, ψ' defines on $H(sp(t), \mathcal{S})$ an element ψ'_t such that $\psi'_t(x) = l$, hence $\psi'_t(f) = 0$ and therefore $\psi'(F) = 0$, a contradiction to the fact that ψ' is an absolute value on B. □

Remark: In Chapter 25 as in Chapter 24, several times, we have obtained results when the field K is strongly valued, (making no hypothesis on the norm of A), which look like results obtained in Chapters 22 and 23 when A is uniform, making no particular hypothesis on the field K. However, this comparison does not involve the multbijectivity.

On the other hand, counter-examples of non multbijective Banach K-algebras are very hard to construct, and seem to be a very strange case. It would be interesting to obtain sufficient conditions to prevent this kind of problem. We notice that our counter-example is an algebra of analytic elements $H(D)$. Hence in particular, $Mult_m(H(D), \| \cdot \|_D)$ is dense in $Mult(H(D), \| \cdot \|_D)$. Consequently, density does not imply multbijectivity. However, we are not able to construct a noetherian non multbijective ultrametric Banach K-algebra. So, we can ask whether noetherian ultrametric Banach K-algebras are multbijective, no mater what the field.

Chapter 28

Pseudo-Density of $Mult_m(A, \| \cdot \|)$

Thanks to Property (s), we can ask whether in a commutative uniform Banach K-algebra with unity A, $Mult_m(A, \| \cdot \|)$ is dense inside $Mult(A, \| \cdot \|)$. Actually, we can only show a weaker property: pseudo-density.

Notation: Throughout the chapter A is a commutative ultrametric Banach K-algebra with unity.

Definition: A subset S of $Mult(A, \| \cdot \|)$ will be said to be *pseudo-dense* in $Mult(A, \| \cdot \|)$ if for every $\psi \in Mult(A, \| \cdot \|)$ and if for every $\epsilon > 0$, there exists a neighborhood of the form $\mathcal{V}(\psi, f, \epsilon)$ $f \in A$, such that $\mathcal{V}(\psi, f, \epsilon) \cap S \neq \emptyset$ [29, 37].

Theorem 28.1 ([29]): *$Mult_m(A, \| \cdot \|)$ is pseudo-dense in $Mult(A, \| \cdot \|)$ if and only if A satisfies Property (s).*

Proof: Suppose first that $Mult_m(A, \| \cdot \|)$ is pseudo-dense in $Mult(A, \| \cdot \|)$. Let $x \in A$ and let $r = \|x\|_{si}$. Take $\epsilon > 0$. By Theorem 6.19 there exists $\psi \in Mult(A, \| \cdot \|)$ such that $\psi(x) = r$. But since $Mult_m(A, \| \cdot \|)$ is pseudo-dense in $Mult(A, \| \cdot \|)$, we can find $\phi \in Mult_m(A, \| \cdot \|)$ such that $\phi(x) > \psi(x) - \epsilon$, hence finally $\phi(x) > r - \epsilon$, so Property s) is clearly satisfied.

Now, suppose that Property (s) is satisfied, and that $Mult_m(A, \| \cdot \|)$ is not pseudo-dense in $Mult(A, \| \cdot \|)$. Then, there exist $x \in A$, $\psi \in Mult(A, \| \cdot \|)$ and $\lambda > 0$ such that $\mathcal{V}(\psi, x, \lambda) \cap Mult_m(A, \| \cdot \|) = \emptyset$. Let us recall that, as a consequence of Corollary 6.16 we have $|sp(x)| \subset \{\phi(x) \mid \phi \in Mult_m(A, \| \cdot \|)\}$.

Let $r = \psi(x)$. If $r = 0$, then $d(0, \lambda) \cap sp(x) = \emptyset$, hence x is invertible, a contradiction to the hypothesis $\psi(x) = 0$. So, we can assume $r > 0$. Then $sp(x) \cap \Gamma(0, r - \lambda, r + \lambda) = \emptyset$. Let $\nu \in]0, \frac{\lambda}{4}[$, $a, b \in K$ such that $r - \nu < |a| < r$, $\quad r + \lambda - \nu < |b| < r + \lambda$ and let $\theta \in Mult_m(A, \| \,.\, \|)$. Since $a \in \Gamma(0, r - \lambda, r + \lambda)$, $x - a$ is invertible in A. We put $u = \frac{x(x-b)}{(x-a)^2}$. Then we check that $\psi(x - a) = \psi(x)$, and $\psi(x - b) = |b|$. Thus, we have

$$\psi(u) = \frac{|b|}{r}. \tag{28.1}$$

Suppose first that $\theta(x) \geq r + \lambda$. Then $\theta(x - a) = \theta(x - b) = \theta(x)$, hence $\theta(u) = 1$. Consequently,

$$\psi(u) - \theta(u) = \frac{|b|}{r} - 1 \geq \frac{r + \lambda - \nu}{r} - 1 = \frac{\lambda - \nu}{r}.$$

But since $\nu < \frac{\lambda}{4}$, we obtain

$$\psi(u) - \theta(u) \geq \frac{\lambda}{2r}. \tag{28.2}$$

Suppose now that $\theta(x) \leq r - \lambda$. Then $\theta(x - a) = |a|$ and $\theta(x - b) = |b|$, hence $\theta(u) = \frac{\theta(x)|b|}{|a|^2}$. By (28.1) we obtain

$$\psi(u) - \theta(u) = |b|\Big(\frac{1}{r} - \frac{\theta(x)}{|a|^2}\Big) \geq |b|\Big(\frac{1}{r} - \frac{r - \lambda}{|a|^2}\Big) \geq (r + \lambda - \nu)\Big(\frac{1}{r} - \frac{r - \lambda}{(r - \nu)^2}\Big).$$

But since $\nu < \frac{\lambda}{4}$, we have

$$(r + \lambda - \nu)\Big(\frac{1}{r} - \frac{r - \lambda}{(r - \nu)^2}\Big) \geq \frac{(r + \lambda - \nu)(\lambda - 2\nu)}{r^2} \geq \frac{\lambda - 2\nu}{r} \geq \frac{\lambda}{2r}. \tag{28.3}$$

Thus, in both cases, by (28.1), (28.2) and (28.3) we have proven that $\psi(u) - \theta(u) \geq \frac{\lambda}{2r}$. Consequently, we see that

$$\theta(u) \leq \|u\|_{si} - \frac{\lambda}{2r}, \ \forall \theta \in Mult_m(A, \| \,.\, \|),$$

a contradiction to the hypothesis on Property (s). □

Corollary 28.2: *Let A have bounded normal ratio. Then $Mult_m(A, \| \,.\, \|)$ is pseudo-dense in $Mult(A, \| \,.\, \|)$.*

Proof: Indeed, since A has bounded normal ratio, by Theorem 24.3 A satisfies Property (s). □

Corollary 28.3: *Let K be strongly valued. Then $Mult_m(A, \|\ .\ \|)$ is pseudo-dense in $Mult(A, \|\ .\ \|)$.*

Proof: Indeed, since K is strongly valued, by Theorem 26.3 A satisfies Property (s). □

Lemma 28.4: *Let $\phi \in Mult(A, \|\ .\ \|)$, let $f_1, \ldots, f_q \in A$ let ϵ be > 0, and let $h_1, \ldots, h_q \in A$ be such that $\|f_j - h_j\| \leq \frac{\epsilon}{3}\ \forall j = 1, \ldots, q$. Then $\mathcal{V}(\phi, h_1, \ldots, h_q, \frac{\epsilon}{3})$ is included in $\mathcal{V}(\phi, f_1, \ldots, f_q, \epsilon)$.*

Proof: Let $\psi \in \mathcal{V}(\phi, h_1, \ldots, h_q, \frac{\epsilon}{3})$ and let $k \in \{1, \ldots, q\}$. We have

$$|\phi(f_k) - \psi(f_k)|_\infty \leq |\phi(f_k) - \phi(h_k)|_\infty + |\phi(h_k) - \psi(h_k)|_\infty + |\psi(h_k) - \psi(f_k)|_\infty.$$

But

$$|\phi(f_k) - \phi(h_k)|_\infty \leq \phi(f_k - h_k) \leq \|f_k - h_k\| \leq \frac{\epsilon}{3},$$

and in the same way,

$$|\psi(f_k) - \psi(h_k)|_\infty \leq \psi(f_k - h_k) \leq \|f_k - h_k\| \leq \frac{\epsilon}{3}.$$

Since by hypothesis $|\phi(h_k) - \psi(h_k)|_\infty \leq \frac{\epsilon}{3}$, the conclusion is clear. □

Theorem 28.5: *Let A have bounded normal ratio and contain a dense Luroth K-subalgebra B. Then $Mult_m(A, \|\ .\ \|)$ is dense in $Mult(A, \|\ .\ \|)$.*

Proof: By definition, B is of the form $K[h, x]$, with $h \in K(x)$. Let $\psi \in Mult(A, \|\ .\ \|)$, let $f_1, \ldots, f_q \in A$ and let ϵ be > 0. We can find $h_1, \ldots, h_q \in B$ such that $\|f_j - h_j\| \leq \frac{\epsilon}{3}\ \forall j = 1, \ldots, q$, and then by Lemma 28.3 we have

$$\mathcal{V}\left(\psi, h_1, \ldots, h_q, \frac{\epsilon}{3}\right) \subset \mathcal{V}(\psi, f_1, \ldots, f_q, \epsilon).$$

For every $\phi \in Mult(A, \|\ .\ \|)$, we denote by ϕ_x its restriction to B.

Let $\mathcal{F}$ be the circular filter of ψ_x. By Theorem 13.9 there exists $V \in \mathcal{F}$ such that

$$|\ |h_j(\alpha)| - \phi(h_j)|_\infty < \frac{\epsilon}{3}\ \forall \alpha \in V\ \forall j = 1, \ldots q.$$

Now, by Corollary 28.2 $Mult_m(A, \|\ .\ \|)$ is pseudo-dense in $Mult(A, \|\ .\ \|)$, hence there exists $\theta \in Mult_m(A, \|\ .\ \|)$ such that the circular filter $\mathcal{G}$ of θ_x

is secant with V. Then, since $\theta(h_j) = \lim_{\mathcal{F}_0} |h_j(\alpha)|$, it is clear that

$$|\theta(h_j) - \psi(h_j)|_\infty \leq \frac{\epsilon}{3}, \ \forall j = 1, \ldots, q.$$

Consequently, θ belongs to $\mathcal{V}(\psi, h_1, \ldots, h_q, \frac{\epsilon}{3})$ and therefore it belongs to $\mathcal{V}(\psi, f_1, \ldots, f_q, \epsilon)$. □

Chapter 29

Polnorm on Algebras and Algebraic Extensions

This chapter is designed to help us prove that a reduced affinoid algebra is complete for its spectral norm. In particular, this will be useful in characteristic $p \neq 0$. We will roughly follow the same way as in [3].

Definition and notation: A field E is said to be *perfect* if every algebraic extension of E is separable. Given a field E of characteristic $p \neq 0$, we denote by $E^{\frac{1}{p}}$ the extension of E containing all p^{th}-roots of elements of E. Let L be a subfield of K.

B is a commutative L-algebras with unity, without non zero divisors of zero and A is a L-subalgebra of B containing the unity of B, provided with a L-algebra semi-multiplicative semi-norm ψ.

Let $P(X) = \sum_{j=0}^{q} a_j X^j \in A[X]$ be monic. We put

$$\mathtt{S}(P, \psi) = \max_{0 \leq j < q} (\psi(a_j))^{\frac{1}{q-j}}.$$

When there is no risk of confusion on the L-algebra semi-multiplicative semi-norm ψ of A, we will only write $\mathtt{S}(P)$ instead of $\mathtt{S}(P, \psi)$. In particular, when the monic polynomial P belongs to $L[X]$, then $\mathtt{S}(P)$ will denote $\max_{0 \leq j < q}(|a_j|^{(\frac{1}{q-j})})$.

Now, suppose that B is integral over A. The function $\| \,.\, \|_{pol}^{A}$ defined on B as $\|y\|_{pol}^{A} = \mathtt{S}(irr(y, A), \psi)$ will be called *the A-polnorm of B.*

Theorem 29.1 is well known in fields theory:

Theorem 29.1: *Every field of characteristic* 0 *is perfect. A field E of characteristic $p \neq 0$ is perfect if and only if $E = E^{\frac{1}{p}}$.*

Lemma 29.2: *Let B be provided with a L-algebra semi-multiplicative semi-norm ψ. Let $P(X) \in A[X]$ and let $b \in B$ satisfy $P(b) = 0$. Then* $\mathrm{S}(P, \psi) \geq \psi(b)$.

Proof: Let $P(X) = \sum_{j=0}^{q} a_j X^j$ and suppose $\psi(b) > \mathrm{S}(P)$. Thus, we have $\psi(a_j) < \psi(b)^{q-j}$, hence $\psi(a_j b^j) \leq \psi(a_j)\psi(b)^j < \psi(b)^j \psi(b)^{q-j} = \psi(b)^q$ whenever $j = 0, \ldots, q-1$. Consequently, $\mathrm{S}(\sum_{j=0}^{q-1} a_j X^j) < \psi(b^q)$. But by Lemma 5.3 ψ is ultrametric. And since $P(b) = 0$ we have $\psi(\sum_{j=0}^{q-1} a_j b^j) = \psi(b^q)$, a contradiction. □

Lemma 29.3 is almost classical and comes from the behaviour of polynomials in an algebraically closed complete ultrametric field recalled in Lemmas 13.6 and 13.7.

Lemma 29.3: *Let $P(X) = \sum_{j=0}^{q} a_j X^j \in K[X]$ be monic, let $\alpha_1, \ldots, \alpha_q$ be its zeros in K. Then* $\max_{1 \leq j \leq q} |\alpha_j| = \mathrm{S}(P)$.

Proof: Let $\{|\alpha_1|, \ldots, |\alpha_q|\} = \{r_1, \ldots, r_h\}$, with $r_j < r_{j+1}$ $(1 \leq j < h)$. For each $j = 1, \ldots, h$, let m_j be the number of zeros of P inside $C(0, r_j)$, let $t_j = N^+(P, -\log(r_j))$, and $s_j = N^-(P, -\log(r_j))$. On one hand, by Corollary 17.3 we have $m_j = t_j - s_j$ $(1 \leq j < h)$ and $\log(r_j) = \frac{1}{t_j - s_j}(\log(|a_{s_j}| - \log(|a_{t_j})))$. In particular

$$\log(r_h) = \frac{1}{m_h} \log(|a_{s_h}|) = \max_{1 \leq j < h} \log(|\alpha_j|).$$

But since

$$N^-(P, -\log(r_h)) = s_h \text{ and } N^+(P, -\log(r_h)) = q,$$

by definition we have

$$r_h^q = |a_{s_h}|(r_h)^{s-h} \geq |a_j|(r_h)^j \ 1 \leq j < q,$$

hence

$$\frac{1}{q-j} \log(|a_j|) \leq \log(|a_{s_h}|) \leq r_h = \frac{1}{q - s_h} \log(|a_{s_h}|) \ \forall j = 0, \ldots, q-1.$$

Consequently

$$\max_{1 \leq j < q} \log(|\alpha_j|) = \frac{1}{q - s_h} \log(|a_{s_h}|) \geq \frac{1}{q-j} \log(|a_j|) \ \forall j = 0, \ldots, q-1,$$

therefore $\mathrm{S}(P) = \max_{0 \leq j \leq q} |\alpha_j|$. □

Proposition 29.4: *Let A be integrally closed such that B is finite over A. Let $\mathcal{T}$ be a set of L-algebra homomorphisms from A into K such that, for each $t \in A$, there exists $\xi_t \in \mathcal{T}$ satisfying $|\xi_t(t)| = \sup\{|\chi(t)| \mid \chi \in \mathcal{T}\}$. Let $\mathcal{W}$ be the set of L-algebra homomorphisms from B to K whose restrictions to A belong to $\mathcal{T}$. On B we put $\|x\| = \sup\{|\chi(x)| \mid \chi \in \mathcal{W}\}$. Let $y \in B$ and let $P = irr(y, A)$. Then $\mathrm{S}(P, \| \,.\, \|) = \|y\|$. Moreover there exists $\zeta \in \mathcal{W}$ such that $|\zeta(y)| = \|y\|$.*

Proof: By Lemma 5.4, the function $\| \,.\, \|$ defined on B is a L-algebra semi-multiplicative semi-norm. Therefore, by Lemma 29.1 we have $\mathrm{S}(P, \| \,.\, \|) \geq \|y\|$.

We will prove the opposite inequality. Let $P(X) = \sum_{j=0}^{q} a_j X^j \in A[X]$ and $h \in \{0, \ldots, q-1\}$ be such that $\mathrm{S}(P, \| \,.\, \|_{si}^{A}) = \sqrt[q-h]{\|a_h\|^A}$. By hypothesis there exists $\xi \in \mathcal{T}$ such that $|\xi(a_h)| = \|a_h\|$. Let $f(X) = \sum_{j=0}^{q} \xi(a_j) X^j \in L[X]$. Then, of course, $\mathrm{S}(f, | \,.\, |) = \mathrm{S}(P, \| \,.\, \|)$ because $|\xi(a_j)| \leq \|a_j\| \ \forall j = 0, \ldots, q$. Let $\alpha_1, \ldots, \alpha_q$ be the zeros of f in K. By Lemma 29.3 we have $\max_{1 \leq j \leq q} |\alpha_j| = \mathrm{S}(f, | \,.\, |)$. Let β be a zero of f in K such that $|\beta| = \mathrm{S}(f, | \,.\, |)$. Let ϕ be the homomorphism from $A[X]$ into K defined as

$$\phi\left(\sum_{j=0}^{q} c_j X^j\right) = \sum_{j=0}^{q} \xi(c_j)\beta^j$$

and let θ be the canonical surjection from $A[X]$ onto $A[y]$. Since $P(X)A[X] \subset Ker(\phi)$, ϕ factorizes in the form $\widetilde{\xi} \circ \theta$, and then $\widetilde{\xi}$ is a L-algebra homomorphism from $A[y]$ into K such that $\widetilde{\xi}(y) = \beta$. Then by Lemma 1.18 $\widetilde{\xi}$ has continuation to a L-algebra homomorphism $\widehat{\xi}$ from B to K. Thus, $\widehat{\xi}$ satisfies $\mathrm{S}(P, \| \,.\, \|) = |\widehat{\xi}(y)|$, and of course $|\widehat{\xi}(y)| \leq \|y\|$, hence $\mathrm{S}(P, \| \,.\, \|) = |\widehat{\xi}(y)|$, which finishes proving our claim. □

Corollary 29.5: *Let B be finite over A and let A be integrally closed and provided with a semi-multiplicative norm $\| \,.\, \|$ such that for each $t \in A$ there exists a L-algebra homomorphism χ from A into K satisfying $\|t\| = |\chi(t)|$. Then the A-polnorm of B is a semi-multiplicative norm.*

Theorem 29.6: *Let F be an algebraic extension of L. The L-polnorm of F is a L-algebra semi-multiplicative norm.*

Proof: Let $\widehat{L}$ be the completion of L with respect to its absolute value and let Ω be an algebraic closure of $\widehat{L}$. By Theorem 3.8, Ω is provided with the unique absolute value that extends this of $\widehat{L}$. Then by Theorem 3.9 the completion of Ω with respect to its absolute value is algebraically closed. Now let $x \in F$, let $P = irr(x, L)$ and let $\alpha_1, \dots, \alpha_q$ be the zeros of P in Ω. By Lemma 29.3 we have $\max_{1\leq j\leq q} |\alpha_j| = \mathtt{S}(P, |\,.\,|)$. Let G be the Gallois group of F over L. Thus, considering F as a subfield of Ω, we have $\mathtt{S}(P, |\,.\,|) = \max_{g\in G} |g(x)|$, and then we can easily check that this is a semi-multiplicative L-algebra norm. □

Theorem 29.7: *Let L be perfect and let Ω be its algebraic closure provided with its L-polnorm. Then Ω is L-productal.*

Proof: Let E be a finite field extension of L and let Tr be the trace function of E over L. Then Tr is obviously continuous with respect to the L-polnorm on Ω. Since L is perfect, E is a separable extension of L, hence for every $a \in E$, $a \neq 0$, there exists $b \in E$ such that $Tr(ab) \neq 0$. Now consider the L-linear mapping ϕ from E into L defined by $\phi(x) = Tr(xb)$. We have $\phi(a) \neq 0$, hence by Corollary 4.2, Ω is L-productal. □

Proposition 29.8: *Let E be an algebraic extension of L and let F be an algebraic extension of E. If the L-polnorm on F is multiplicative, E being normed with this absolute value, then the L-polnorm and the E-polnorm coincide on F.*

Proof: Let $\|\,.\,\|_{pol}^{L}$ (resp. $\|\,.\,\|_{pol}^{E}$) be the L-polnorm (resp. E-polnorm) on F. Since $\|\,.\,\|_{pol}^{E}$ is obviously a semi-multiplicative L-algebra norm, by Lemma 29.2 we have $\|x\|_{pol}^{L} \geq \|x\|_{pol}^{E}$ $\forall x \in F$. Next, $\|\,.\,\|_{pol}^{L}$ induces on E the L-polnorm of E because given $x \in E$, its minimal polynomial over L is the same, whether we consider x as an element of E or as an element of F. Consider now the L-polnorm of F: since it is a L-algebra norm, it satisfies $\|xy\|_{pol}^{L} \leq \|x\|_{pol}^{L}\|y\|_{pol}^{L}$, $\forall \lambda \in L$, $x \in F$, hence $\|\,.\,\|_{pol}^{L}$ is a mapping ψ from F to $\mathbb{R}_+$ satisfying $\psi(x+y) \leq \max(\psi(x), \psi(y))$, $\psi(xy) \leq \psi(x)\psi(y)$ and $\psi(x^n) = \psi(x)^n$ $\forall n \in \mathbb{N}^*$, $\forall x, y \in F$. Consequently, by Lemma 29.2, we have $\|x\|_{pol}^{E} \geq \|x\|_{pol}^{L}$. □

Proposition 29.9: *Let L have characteristic $p > 0$. For every $n \in \mathbb{N}$, the L-polnorm of $L^{\frac{1}{p^n}}$ is an absolute value extending this of L.*

Proof: Since the L-polnorm is a L-algebra norm, we just have to check that it is multiplicative. On the other hand, by induction, it is sufficient to show the claim when $n = 1$. Let x, $y \in L^{\frac{1}{p}}$. By hypothesis, $irr(x, L)$ is of the form $X^p - a$, $irr(y, L)$ is of the form $X^p - b$, with a, $b \in L \setminus L^{\frac{1}{p^n}}$. Thus we have $\|x\|_{pol}^{L} = \sqrt[p]{|a|}$, $\|y\|_{pol}^{L} = \sqrt[p]{|b|}$. Now, $(xy)^p = ab$, hence $irr(xy, L)$ divides $X^p - ab$, and is of the form $X^p - c$. Hence $c = ab$, and thereby, $\|xy\|_{pol}^{L} = \sqrt[p]{|ab|} = \|x\|_{pol}^{L}\|y\|_{pol}^{L}$. □

Theorem 29.10: *Let L have characteristic $p > 0$. If $L^{\frac{1}{p}}$ is L-productal with respect to the L-polnorm, then so is the algebraic closure of L.*

Proof: Let $L^{\infty} = \bigcup_{n=1}^{\infty} L^{\frac{1}{p^n}}$, and let Ω be an algebraic closure of L containing L^{∞}. Then by Theorem 29.1 L^{∞} is perfect, and by Proposition 29.9 the L-polnorm is an absolute value that continues this of L. By Theorem 29.7, since L^{∞} is perfect, Ω is L^{∞}- productal with respect to the L^{∞}-polnorm. But by Proposition 29.8 the L^{∞}-polnorm is equal to the L-polnorm on Ω. Consequently Ω is L^{∞}-productal with respect to the L-polnorm which is an absolute value on L^{∞}. Hence by Lemma 4.4, Ω is L-productal with respect to the L-polnorm. $\square$

Chapter 30

Definition of Affinoid Algebras

Affinoid algebras were introduced by John Tate in [49] who called them *algebras topologically of finite type.* As this last name suggests, such an algebra is the completion of an algebra of finite type for a certain norm.

Definitions and notation: In this chapter, L is a complete ultrametric field, A, B are commutative ultrametric Banach L-algebras with unity.

Recall that $U_L = d(0,1)$, $M_L = d(0,1^-)$ and the residue class field of L is $\mathcal{L}$. Here we put $V = \{x \in A \mid \|x\| \leq 1\}$.

Let $(E, \| \ . \ \|)$ be a normed L-vector space which also is a A-module. The norm of E will be called a *A-quasi-algebra-norm* if it satisfies $\|af\| \leq \|a\|\|f\|$ $\forall a \in A$, $\forall f \in E$. If E is provided with such a A-quasi-algebra-norm, $(E, \| \ . \ \|)$ will be called a *A-quasi-algebra-normed A-module.*

A U_L-submodule E_1 of A will be said *to define the topology of E* if it is a bounded neighborhood of zero i.e. if there exist $r,\ s \in]0, +\infty[$ with $r < s$ such that $\{x \in E | \ \|x\| \leq r\} \subset E_1 \subset \{x \in E | \ \|x\| \leq s\}$.

We will denote by E_0 the U_L-submodule of the $x \in E$ (resp. $x \in A$) such that $\|x\| \leq 1$.

We will recall definitions given in [49] about algebras topologically of finite type, now mainly called affinoid algebras.

Proposition 30.1: *Let E be a normed L-vector space which also is a A-module. Let $t \in M_L$. Let E_1 be a A_0-submodule defining the topology of E. Assume that there exists $e_1, \dots, e_q \in E_1$ such that $E_1 = \sum_{i=1}^q A_0 e_i + tE_1$. Let ϕ be the mapping from A^q into E defined as $\phi(t_1, \dots, t_q) = \sum_{i=1}^q t_i e_i$. Then ϕ is surjective and E is finite over A.*

Proof: Let $y \in E_1$. We can write y in the form $\sum_{i=1}^{q} f_{1,i}e_i + ty_1$, with $f_{1,i} \in A_0$ and $y_1 \in E_1$, and in the same way, we can write $y_1 = \sum_{i=1}^{q} f_{2,i}e_i + ty_2$, with $f_{2,i} \in A_0$, and $y_2 \in E_1$. By induction, it clearly appears that for every $m \in \mathbb{N}^*$, we obtain $y = \sum_{i=1}^{q}(\sum_{j=1}^{m} f_{m,i}t^{m-1})e_i + t^m y_m$. And since all terms $f_{j,i}$ lie in A_0, and all y_j lie in E_1, each series $\sum_{j=1}^{m} f_{m,i}t^{m-1}$ converges in the Banach algebra A to an element f_i. Then, since each set $t^m E_1$ is closed, $y - \sum_{i=1}^{q} f_i e_i$ lies in $t^m E_1$ for all $m \in \mathbb{N}^*$. Therefore, $y = \sum_{i=1}^{q} f_i e_i$. Consequently $\phi(A^q)$ contains E_1 and therefore ϕ is surjective on E. Moreover, E is finite over A. □

Proposition 30.2: *Let F be a A-quasi-algebra-normed A-module whose completion is finite over A. Then F is complete.*

Proof: Let E be the completion of F, and let $e_1, \ldots, e_q \in F$ be such that $E = \sum_{i=1}^{q} Ae_i$. Let ϕ be the canonical surjection from A^q onto E : $\phi(a_1, \ldots, a_q) = \sum_{i=1}^{q} a_i e_i$. On A^q we can define the product norm of A^q:

$$|||(a_1, \ldots, a_q)||| = \max_{1 \le i \le q} \|a_i\|.$$

Putting then $c = \max_{1 \le i \le q} \|e_i\|$, we have

$$|||(a_1, \ldots, a_q)||| \ge c\|\phi(a_1, \ldots, a_q)\|.$$

Thus ϕ is continuous, hence $Ker(\phi)$ is closed and therefore E is provided with the quotient norm $\| \,.\, \|'$ of $||| \,.\, |||$. Then, E is complete for the norm $\| \,.\, \|'$. On the other hand, given $f \in A^q$, we have

$$\|\phi(f)\|' = \inf_{x \in Ker(\phi)} |||f + x||| \ge c \inf_{x \in Ker(\phi)} \|\phi(f + x)\|.$$

But since $\phi(f + x) = \phi(f)$, we check that $\|\phi(f)\|' \ge c\|\phi(f)\|$. Thus E is complete for both norms $\| \,.\, \|$ and $\| \,.\, \|'$ and therefore by Banach's Theorem (Theorem 3.4), the two norms are equivalent. By definition, ϕ is open with respect to the norm $\| \,.\, \|'$ on E. Consequently $\sum_{i=1}^{q} A_0 e_i$ is a A_0-submodule E_1 that defines the topology of E. Since F is dense in E, each e_i is of the form $f_i + th_i$, with $h_i \in E_1$. Therefore E_1 is included in $\sum_{i=1}^{q} A_0 f_i + tE_1$, and then by Proposition 30.1, we have $E = \sum_{i=1}^{q} A_0 f_i \subset F$. □

Corollary 30.3: *Let A be noetherian and let F be a complete A-quasi-algebra-normed finite A-module. Then every A-submodule of F is closed.*

Proof: Indeed, since A is Noetherian, and since F is finite, F is Noetherian, hence any A-submodule E has a finite closure in F, hence by Proposition 30.2 is closed.

Corollary 30.4: *Let A be noetherian. Every ideal of A is closed.*

Proof: Indeed, the closure of an ideal I is of finite type, and therefore, I is closed. □

Definitions and notation: Let us recall the notation given in Chapter 8: Let E be a commutative ultrametric normed L-algebra with identity. Let $n \in \mathbb{N}$ and let $\| \,.\, \|_0$ be the norm of E. The E-algebra of polynomials in n variables $E[X_1, \ldots, X_n]$ is provided with the *Gauss norm* $\| \,.\, \|$ defined as $\| \sum_{i_1,,i_n} a_{i_1,\ldots,i_n} X_1^{i_1} \ldots X_n^{i_n} \| = \sup_{i_1,\ldots,i_n} \|a_{i_1,\ldots,i_n}\|_0$. By Proposition 6.18 this norm is a L-algebra norm. In particular, if the norm of E is multiplicative, so is the norm $\| \,.\, \|$.

We denote by $E\{X_1, \ldots, X_n\}$ the set of power series in n variables $\sum_{i_1,,i_n} a_{i_1,\ldots,i_n} X_1^{i_1} \ldots X_n^{i_n}$ such that $\lim_{i_1+\ldots+i_n \to \infty} a_{i_1,\ldots,i_n} = 0$. The elements of such an algebra are called *the restricted power series in n variables, with coefficients in E*. Hence by definition $E[X_1, \ldots, X_n]$ is dense in $E\{X_1, \ldots, X_n\}$. By Proposition 6.18, if E is complete, $E\{X_1, \ldots, X_n\}$ is just the completion of $E[X_1, \ldots, X_n]$.

More generally, consider a subring F of E, and elements t_j ($1 \leq j \leq q$) of $E\{Y_1, \ldots, Y_n\}$ such that $\|t_j\| \leq 1 \ \forall j = 1, \ldots, q$. Consider the canonical mapping ϕ from $F[Y_1, \ldots, Y_q]$ to $F[t_1, \ldots, t_q]$ defined as $\phi(P) = P(t_1, \ldots, t_q)$. Then ϕ is clearly continuous and has continuation to $F\{Y_1, \ldots, Y_q\}$: to each restricted power series in $Y_1, \ldots, Y_q$ we can associated a converging power series in $t_1, \ldots, t_q$. The image of $F\{Y_1, \ldots, Y_q\}$ by ϕ is denoted by $F\{t_1, \ldots, t_q\}$.

Given $n \in \mathbb{N}$, the algebra $L\{X_1, \ldots, X_n\}$ is called *a topologically pure extension of L of dimension n* and is denoted by T_n. By definition, an algebra T_m is included in T_n for all $n > m$ and the Gauss norm of T_m is induced by this of T_n. Such a Gauss norm on T_n will be just denoted by $\| \,.\, \|$.

Let B be a commutative Banach A-algebra with unity. B is said to be a *A-affinoid algebra* if it is isomorphic to a quotient of any algebra of the form $A\{X_1, \ldots, X_n\}$ by one of its ideals.

Let $t_1, \ldots, t_n \in B$ such that $\|t_j\| \leq 1$ and let ϕ be the L-algebra homomorphism from $A\{X_1, \ldots, X_n\}$ into B defined as $\phi(F(X_1, \ldots, X_n)) = F(t_1, \ldots, t_n)$. We will denote by $A\{t_1, \ldots, t_n\}$ its image i.e. the set of sums of series in $t_1, \ldots, t_n$ whose coefficients tend to 0 along the filter of complementals of finite subsets of $\mathbb{N}^n$.

Remarks: (1) By definition, given $f = \sum_{i_1,\ldots,i_n} a_{i_1,\ldots,i_n} X_1^{i_1} \ldots X_n^{i_n} \in T_n$, there are finitely many coefficients $a_{i_1,\ldots,i_n}$ such that $|a_{i_1,\ldots,i_n}| = \|f\|$. Consequently, $\|f\|$ lies in $|L|$ for every $f \in T_n$.

(2) Let $f(X) = \sum_{i=0}^{\infty} a_i X^i \in U_L\{X\} = (T_1)_0$ and let $q = N^+(f,0)$. Consider the residue class $\mathcal{L}$-algebra $\overline{A} = \frac{(T_1)_0}{M_L(T_1)_0}$. In $\overline{A}$ we denote by $\overline{f}$ the residue class of $f(X)$. Then $\overline{f}$ is the polynomial $\sum_{i=0}^{q} \overline{a}_i X^i$.

(3) Since the Gauss norm on T_n is multiplicative, it is obviously equal to the associated semi-norm $\| \,.\, \|_{si}$.

Lemma 30.5: *Let $L = K$, let $f_1, \ldots, f_m \in T_1$. There exists $\alpha \in U_K$ such that $|f_j(\alpha)| = \|f_j\|$ $\forall j = 1, \ldots, m$.*

Proof: Indeed, for each $j = 1, \ldots, m$ the equality $|f_j(x)| = \|f_j\|$ holds in all classes of U_K except in finitely many ones. Consequently, all equalities $|f_j(x)| = \|f_j\|$ hold in all classes of U_L except in finitely many. □

Proposition 30.6: *Let A, B be L-affinoid algebras such that B is A-affinoid. Suppose that $\frac{B_0}{M_L B_0}$ is finite over $\frac{A_0}{M_L A_0}$. Then B is finite over A.*

Proof: By hypothesis, A is the quotient of certain L-algebra $A\{T_1, \ldots, T_r\}$, and therefore is of the form $A\{t_1, \ldots, t_r\}$. Let $\lambda \in M_L$, $\lambda \neq 0$. For each $t \in B_0$, here we denote by $\tilde{t}$ the residue class of t in $\frac{B_0}{\lambda B_0}$. Thus, we have

$$\frac{B_0}{\lambda B_0} = \frac{A_0}{\lambda A_0}[\tilde{t}_1, \ldots, \tilde{t}_r].$$

Now, by hypothesis for each $i = 1, \ldots, r$, there exists a polynomial $P_i(T) = T^{q_i} + f_{q_i - 1} T^{q_i - 1} + \ldots + f_0$, with $f_i \in A_0$, and $P_i(t_i) \in M_L B_0$. Let $P_i(t_i) = x_i g_i$, with $x_i \in M_L$, and $g_i \in B_0$. Now, we can choose $l_i \in \mathbb{N}$ such that $x_i^{l_i} \in \lambda A_0$, and thus $P_i(t_i)^{l_i}$ lies in λB_0, hence $\widetilde{P_i}(\tilde{t}_i)^{l_i} = 0$. Therefore, $\frac{B_0}{\lambda B_0}$ is finite over $\frac{A_0}{\lambda A_0}$, and finally, by Proposition 30.1, B is finite over A. □

Lemma 30.7: *Let A be a L-affinoid algebra which is a quotient of a topologically pure extension $L\{X_1, \ldots, X_n\}$ of L by an ideal. Let θ be the canonical surjection, and for each $i = 1, \ldots, n$, we put $t_i = \phi(X_i)$. Then $U_L\{t_1, \ldots, t_n\}$ defines the topology of A.*

Proof: By definition, $U_L\{t_1, \ldots, t_n\}$ is included in B_0. So we just have to check that $U_L\{t_1, \ldots, t_n\}$ is a neighborhood of 0 in B. Let $\lambda \in]0,1[$ and let $f \in B$ satisfy $\|f\| \leq \lambda$. Since $\lambda < 1$, by definition of the quotient norm of B, there exist $F \in U\{X_1, \ldots, X_n\}$ such that $\phi(F) = f$. Hence $f = F(t_1, \ldots, t_n)$ lies in $U_L\{t_1, \ldots, t_n\}$, and therefore $U_L\{t_1, \ldots, t_n\}$ is a neighborhood of 0 in B. □

Theorem 30.8: *Suppose the topology of A is defined by a U_L-algebra A_1. Let $y_1, \ldots, y_s \in A_1$ be such that their residue classes in $\frac{A_1}{M_L A_1}$ be algebraically independent over $\mathcal{L}$. Then the canonical homomorphism ϕ*

from $L\{Y_1,\ldots,Y_s\}$ into a A defined as $\phi(Y_j) = y_j$, $(1 \le j \le s)$ is an isomorphism from $L\{Y_1,\ldots,Y_s\}$ onto a closed subalgebra of A.

Proof: Let $F(Y_1,\ldots,Y_s) \in L\{Y_1,\ldots,Y_s\}$ be such that $F(y_1,\ldots,y_s) \in A_1$. Suppose that $F(Y_1,\ldots,Y_s) \notin U\{Y_1,\ldots,Y_s\}$. Since $\|F\|$ lies in $|L|$, there exists $\lambda \in M_L$ such that $\|\lambda F\| = 1$. Let $P = \lambda F$. Then the residue class $\overline{P}$ of P in $\mathcal{L}[X_1,\ldots,X_s]$ is different from 0, but satisfies $\overline{P}(\overline{y}_1,\ldots,\overline{y}_s) = 0$ in $\frac{A_1}{M_L A_1}$, a contradiction. Consequently, $F(Y_1,\ldots,Y_s)$ belongs to $U_L\{Y_1,\ldots,Y_s\}$, and therefore ϕ is open. Hence, ϕ is a bicontinuous isomorphism from $L\{Y_1,\ldots,Y_s\}$ onto a closed L-subalgebra of A. □

Proposition 30.9: *A is a L-affinoid algebra if and only if it is a finite extension of a topologically pure extension of L. Particularly, if A is of the form $L\{X_1,\ldots,X_n\}[y_1,\ldots,y_q]$ with $y_1,\ldots y_q$ integral over $L\{X_1,\ldots,X_n\}$ and $\max_{1\le j\le q}\|y_j\| \le 1$, then A is isomorphic to a quotient of*

$$L\{X_1,\ldots,X_n,Y_1,\ldots,Y_q\}.$$

Proof: First suppose that A is a finite extension of a topologically pure extension of L, say $A = L\{X_1,\ldots,X_n\}[y_1,\ldots,y_q]$. Without loss of generality, we may clearly assume that $\max(|y_1|,\ldots,|y_q|) \le 1$. Then the canonical homomorphism ϕ from $L[X_1,\ldots,X_n,Y_1,\ldots,Y_q]$ into A defined by $\phi(X_i) = X_i$ $\forall i = 1,\ldots,n$, and $\phi(Y_j) = y_j$ $\forall j = 1,\ldots,q$, is clearly continuous with respect to the Gauss norm on $L[X_1,\ldots,X_n,Y_1,\ldots,Y_q]$ and extends by continuity to a homomorphism $\overline{\phi}$ from $L\{X_1,\ldots,X_n,Y_1,\ldots,Y_q\}$ into A. Then $\overline{\phi}$ is surjective by construction, so A is a quotient of $L\{X_1,\ldots,X_n,Y_1,\ldots,Y_q\}$.

Conversely, we now assume that A is a L-affinoid algebra, and suppose that A is a quotient of a topologically pure extension $L\{X_1,\ldots,X_n\}$ of L. Let θ be the canonical surjection, and for each $i = 1,\ldots,n$, let $t_i = \phi(X_i)$. Then by Lemma 30.7 the U_L-algebra $B_1 = U_L\{t_1,\ldots,t_n\}$ defines the topology of B. Now, consider the residue algebra $\frac{B_1}{M_L B_1}$, and for each $i = 1,\ldots,n$, let τ_i be the residue class of t_i in $\frac{B_1}{M_L B_1}$. Then $\frac{B_1}{M_L B_1} = \mathcal{L}[\tau_1,\ldots,\tau_n]$. By Theorem 1.16, we can choose $\xi_1,\ldots\xi_s \in \frac{B_1}{M_L B_1}$ such that $\frac{B_1}{M_L B_1}$ be finite over $\mathcal{L}[\xi_1,\ldots,\xi_s]$. For each $i = 1,\ldots,s$, we can choose $y_i \in B_1$ whose residue class is ξ_i. Let $E = L\{y_1,\ldots,y_s\}$. By Proposition 30.8 E is a topologically pure extension of L. Now, putting $E_0 = U\{y_1,\ldots,y_s\}$, we can see that $\frac{B_1}{M_L B_1}$ is finite over $\frac{E_0}{M_L E_0}$, and therefore, by Proposition 30.6 B is finite over E. □

Chapter 31

Algebraic Properties of Affinoid Algebras

We shall recall the proofs of many classical algebraic properties of K-affinoid algebras such as noetherianness, finite codimension of maximal ideals and other easy spectral properties [49].

Notation: As usual, K is an algebraically closed complete ultrametric field whose valuation is not trivial. We denote by U the disk $d(0,1)$ and by U_n the unit ball K^n : $\{(x_1,\dots x_n) \mid \max_{1\leq j\leq n} |x_j| \leq 1\}$. We denote by M the disk $d(0,1^-)$, and by $\mathcal{K}$ the residue class field of K.

Throughout the chapter, A will be a K-affinoid algebra.

Theorem 31.1: *A is Noetherian, every ideal is closed, and each maximal ideal of A has codimension* 1.

Proof: We will prove by induction that topologically pure extensions are noetherian. This is trivial for K. Suppose this has been proven for $K\{X_1,\dots,X_s\}$ whenever $s < r$, and consider $B = K\{X_1,\dots,X_r\}$. Let $f \in B$, $f \neq 0$ and let $B' = \frac{B}{fB}$. In B the Gauss norm satisfies $\|f\| \in |K|$, so we can find $\lambda \in K$ such that $\|\lambda f\| = 1$ and therefore, without loss of generality, we can assume $\|f\| = 1$. Since the Gauss norm is multiplicative, now the mapping ϕ from B to B defined as $\phi(y) = fy$ is an isometry from B onto fB. Therefore, fB is complete, and closed in B. Consequently, B', provided with its quotient norm, is a Banach K-algebra, and therefore is a K-affinoid algebra. Let θ be the canonical surjection from B onto B', and for each $i = 1,\dots,r$, let $x_i = \theta(X_i)$. Then, $B' = K\{x_1,\dots,x_r\}$. Moreover in B' we have $f(x_1,\dots,x_r) = 0$. Now, for every $x \in B'$, we denote by $\overline{x}$ the residue class of x. Let $\overline{f} \in \mathcal{K}[Y_1,\dots Y_r]$ be the residue class of f.

Since $\|f\| = 1$, we have $\overline{f} \neq \overline{0}$. However, we notice that $\overline{f}(\overline{x}_1, \ldots, \overline{x}_r) = \overline{0}$. Consequently, $\{\overline{x}_1, \ldots, \overline{x}_r\}$ is not algebraically free over $\mathcal{K}$. Hence, by Theorem 1.16 we can find $\tau_1, \ldots, \tau_s$ in $\frac{B'}{MB'}$, with $s < r$, such that $\tau_1, \ldots, \tau_s$ are algebraically free over $\mathcal{K}$ and such that $\frac{B'}{MB'}$ is finite over $\mathcal{K}[\tau_1, \ldots, \tau_s]$. Therefore, taking $t_1, \ldots, t_s$ in B' such that $\overline{t}_i = \tau_i$ $(1 \leq i \leq s)$, by Proposition 30.6 $K\{t_1, \ldots, t_s\}$ is a topologically pure extension of K and B' is finite over $K\{t_1, \ldots, t_s\}$. But by induction hypothesis $K\{t_1, \ldots, t_s\}$ is noetherian, and so is B' which is finite over $K\{t_1, \ldots, t_s\}$. This is true for every $f \in B$, and therefore by Proposition 1.5 B also is noetherian.

Consequently, by Corollary 30.3 every ideal is closed. Finally, consider a maximal ideal $\mathcal{M}$ of B. Then, $\frac{B}{\mathcal{M}}$ is a field F and a K-affinoid algebra. Thus, F is finite over a topologically pure extension $K\{X_1, \ldots, X_q\}$, and then by Proposition 1.4 $K\{X_1, \ldots, X_q\}$ is a field. But such a ring is never a field except when $q = 0$. Hence, F is a finite field extension of K, and then, since K is algebraically closed, we have $F = K$. □

Corollary 31.2: *The quotient of a topologically pure extension by an ideal is a K-affinoid algebra.*

Corollary 31.3: *Let A be a K-affinoid algebra and let B be a A-affinoid algebra. Then B is a K-affinoid algebra.*

Corollary 31.4: *Let $B = T_n[y_1, \ldots, y_q]$ be finite over a topologically pure extension T_n. Then B is isomorphic to a K-affinoid algebra of the form $\frac{T_n\{Y_1,\ldots,Y_q\}}{I}$ where I is an ideal of $T_n\{Y_1, \ldots, Y_q\}$ such that $T_n \cap I = \{0\}$.*

Corollary 31.5: *In A there are finitely many minimal prime ideals.*

Corollary 31.6: *Let A admit an idempotent u. Then uA is a K-affinoid algebra admitting u for unity.*

Theorem 31.7: *For all $\alpha \in U_n$ and $f \in T_n$, we put $\chi_\alpha(f) = f(\alpha)$. The mapping ϕ from U_n into $\mathcal{X}(T_n)$ defined as $\phi(\alpha) = \chi_\alpha$ is a bijection.*

Proof: Indeed ϕ is an injection from U_n into $\mathcal{X}(T_n)$. Now, let $\gamma \in \mathcal{X}(T_n)$, and let $T_n = K\{X_1, \ldots, X_n\}$. By Theorem 31.1 γ takes values in K. For every $j = 1, \ldots, n$, let $\alpha_j = \varphi(X_j)$, and let $\alpha = (\alpha_1, \ldots, \alpha_n)$. Then we see $\gamma = \chi_\alpha$. □

Corollary 31.8: *Let $\chi \in \mathcal{X}(T_{n-1})$ and let $\alpha \in U$. There exists $\widehat{\chi} \in \mathcal{X}(T_n)$ such that $\widehat{\chi}(h) = \chi(h) \forall h \in T_{n-1}$, and $\widehat{\chi}(X_n) = \alpha$.*

Proposition 31.9: *The following three properties are equivalent:*

(*i*) $|\chi(f)| < 1 \ \forall \chi \in \mathcal{X}(A)$;
(*ii*) *In* $A\{Y\}$, $1 - fY$ *is invertible*;
(*iii*) $\|f\|_{si} < 1$.

Proof: Let us show that (*i*) implies (*ii*). By Corollary 31.3 $A\{Y\}$ is a K-affinoid algebra. Next, by definition, we have $\|Y\| = 1$. Let $\chi \in \mathcal{X}(A\{Y\})$. Then $|\chi(fY)| < 1$, hence $\chi(1 - fY) \neq 0$. Consequently, by Theorem 31.1 $1 - fY$ does not belong to any maximal ideal of $A\{Y\}$, which proves (*ii*).

Let us show that (*ii*) implies (*iii*). Since $1-fY$ is invertible, its inverse is a series $\sum_{n=0}^{\infty} g_n Y^n$, with $g_n \in A$ and $\lim_{n\to\infty} g_n = 0$. Since $(1-fY)g = 1$, we check that $g_0 = 1$, and $g_{n+1} = fg_n$, $\forall n \in \mathbb{N}$. Consequently, $g_n = f^n$, and $\lim_{n\to\infty} f^n = 0$, hence by Theorem 6.9, $\|f\|_{si} < 1$.

Finally, we check that (iii) implies (i): by Theorem 6.19 we have $|\chi(f)| \leq \|f\|_{si}$. □

Theorem 31.10: *Every K-affinoid algebra owns Property* (*p*).

Proof: Suppose a K-affinoid algebra A does not own Property p), and let $f \in A$ be such that $\|f\|_{sa} < \|f\|_{si}$. Taking $\lambda \in K$ such that $\|f\|_{sa} < |\lambda| < \|f\|_{si}$ we have $\|\frac{f}{\lambda}\|_{sa} < 1 < \|\frac{f}{\lambda}\|_{si}$, hence Condition (i) is satisfied but Condition (iii) is not, a contradiction. □

Lemma 31.11: *Let $\mathcal{I}$ be an ideal of T_n, and let $A = \frac{T_n}{\mathcal{I}}$. Let θ be the canonical surjection from T_n onto A. Let $\psi \in Mult(T_n, \| \, . \, \|)$. There exists $\phi \in Mult(A, \| \, . \, \|)$ such that $\psi = \phi \circ \theta$ if and only if $\mathcal{I} \subset Ker(\psi)$.*

Proof: Let $\| \, . \, \|$ be the Gauss norm on T_n, and let $\| \, . \, \|_q$ be the quotient K-algebra norm of A. Let $f \in T_n$. Suppose that $\mathcal{I} \subset Ker(\psi)$. For every $t \in \mathcal{I}$ we have $\psi(f) = \psi(f + t) = \|f + t\|$, and therefore, $\psi(f) \leq \|\theta(f)\|_q$. Consequently, $\phi \circ \theta$ belongs to $Mult(A, \| \, . \, \|)$. Conversely, if ψ is of the form $\phi \circ \theta$, with $\phi \in Mult(A, \| \, . \, \|)$, then $Ker(\psi)$ obviously contains $\mathcal{I}$. □

Theorem 31.12: *For all $x \in T_n$ there exists $\chi \in \mathcal{X}(T_n)$ such that $|\chi(x)| = \|x\|_{si}$.*

Proof: Since the Gauss norm on T_n satisfies $\| \, . \, \| = \| \, . \, \|_{si}$, we are reduced to show that for every $f \in T_n$ there exists $\chi \in \mathcal{X}(T_n)$ such that $|\chi(f)| = \|f\|$. We will proceed by induction. When $n = 1$, by Theorem 17.7, there exists $\alpha \in U$ such that $|f(\alpha)| = \|f\|$. Consequently, the mapping

$\chi_\alpha \in \mathcal{X}(T_1)$ defined as $\chi_\alpha(g) = g(\alpha)$ $(g \in T_1)$ satisfies $|\chi_\alpha(f)| = \|f\|_{si}$. Now, suppose we have already proven the following property:

$(\mathcal{Q}_n)$: For every $n \leq q$, given any $f_1, \dots, f_m \in T_n$, there exists $\chi \in \mathcal{X}(T_n)$ such that $|\chi(f_j)| = \|f_j\|\ \forall j = 1, \dots, m$.

Consider $f_1, \dots, f_m \in T_{q+1}$. We can write each f_h in the form $\sum_{k=0}^{\infty} g_{k,h} X_{q+1}^k$, with $g_{k,h} \in T_q$. There obviously exists $N \in \mathbb{N}$ such that $\|g_{k,h}\| < \|f_h\|$ whenever $k \geq N$, whenever $h = 1, \dots, m$. Now by the induction hypothesis we can find $\chi \in \mathcal{X}(T_q)$ such that

$$\left|\chi\left(\prod_{0\leq k\leq N,\ 1\leq h\leq m} g_{k,h}\right)\right| = \left\|\prod_{0\leq k\leq N,\ 1\leq h\leq m} g_{k,h}\right\|.$$

And since both χ, $\|\cdot\|$ are multiplicative, and obviously satisfy $|\chi(g_{k,h})| \leq \|g_{k,h}\|\ \forall k = 0, \dots, N$, $\forall h = 1, \dots, m$, it is clear that $|\chi(g_{k,h})| = \|g_{k,h}\|\ \forall k = 0, \dots, N$, $\forall h = 1, \dots, m$. For each $h = 1, \dots, m$, consider $\phi_h(X) = \sum_{k=0}^{\infty} \chi(g_{k,h}) X^k$ which obviously lies in T_1. Since the property holds for $n = q$, there exists $\chi \in \mathcal{X}(T_1)$ such that $\chi(\phi_h) = \|\phi_h\|$. Putting $\alpha = \chi(X)$, we have $|\phi_h(\alpha)| = \|\phi_h\|\ \forall h = 1, \dots, m$. Therefore, we can clearly give χ an extension to an element $\widehat{\chi} \in \mathcal{X}(T_{q+1})$ defined as $\widehat{\chi}(\sum_{k=0}^{\infty}(g_k)X_{q+1}^k) = \sum_{k=0}^{\infty} \chi(g_k)\alpha^k$, and this homorphism $\widehat{\chi}$ satisfies $\|f_h\| = |\widehat{\chi}(f_h)|\ \forall h = 1, \dots, m$. Thus, we have proven the property when $n = q + 1$, and therefore our claim holds for all K-algebra T_n. □

Notation: In this chapter and in the following, given $f(X_1, \dots, X_n) \in T_n$, we will write f in the form $\sum_{(i)\in\mathbb{N}^n} a_{(i)} X^{(i)}$, where $(i) = (i_1, \dots, i_n)$, and where $X^{(i)}$ means $X_1^{i_1} \dots X_n^{i_n}$. The set of indices $\mathbb{N}^n$ is provided with this external law from $\mathbb{N} \times \mathbb{N}^n$ to $\mathbb{N}^n$: given $q \in \mathbb{N}$ and $(i) = (i_1, \dots, i_n) \in \mathbb{N}^n$, $q.(i)$ will mean $(qi_1, \dots, qi_n)$. We will denote by (u) the unit index $(i_1, \dots, i_n)$ when $i_1 = \dots = i_n = 1$, and by (0) the zero of $\mathbb{N}^n$.

Lemma 31.13: *Let $f \in T_n$. Then $sp(f)$ is the disk $d(f((0)), \|f - f((0))\|)$.*

Proof: Without loss of generality, we can assume that $f((0)) = 0$. Let $r = \|f\|$. Of course $sp(f) \subset d(0, r)$. Then f is of the form $\sum_{(j)\in\mathbb{N}^n} b_{(j)} X^{(j)}$, and there exists $(k) \in \mathbb{N}^n$ such that $\|b_{(k)}\| = \|f\|$. Since $f((0)) = 0$, at least one of the coordinates of (k) is not 0. For convenience we can assume that (k) is of the form $(k_1, \dots, k_n)$, with $k_n > 0$. Now, we can write T_n in the form $T_{n-1}\{Y\}$, and put $f = \sum_{n=0}^{\infty} a_n Y^n$. Let $l = k_n$. So, we have $\|f\| = \|a_l\|$. Let $\alpha \in d(0, r)$. Since $l > 0$, we have $\|f - \alpha\| = \|f\| = \|a_l\|$. By Theorem 31.12 there exists $\chi \in \mathcal{X}(T_{n-1})$ such that $|\chi(a_l)| = \|a_l\|$.

Let $P(Y) = \sum_{n=0}^{\infty} \chi(a_n)Y^n - \alpha \in K\{Y\}$. Clearly $\|P\| = |\chi(a_l)| = r$, hence by Theorem 17.2 P has its zeros in $d(0,r)$. Now, by Corollary 31.8 there exists $\widehat{\chi} \in \mathcal{X}(T_n)$ such that $\widehat{\chi}(h) = \chi(h)\ \forall h \in T_{n-1}$, and $\widehat{\chi}(Y) = \alpha$. Then $\widehat{\chi}(f) = \alpha$. □

Theorem 31.14: *Let $f \in T_n$. Then f is invertible in T_n if and only if $|f((0))| > \|f - f((0))\|$.*

Proof: If $|f((0))| > \|f - f((0))\|$, f is invertible by Theorem 6.13. Now suppose $|f((0))| \leq \|f - f((0))\|$. Since $sp(f) = d(f(0), \|f - f(0)\|)$, it is seen that 0 belongs to $sp(f)$, hence f is not invertible. □

Chapter 32

Jacobson Radical of Affinoid Algebras

Notation: Throughout the Chapter, A is a K-affinoid algebra.

In A we will show that $Mult_m(A, \| \cdot \|)$ is dense in $Mult(A, \| \cdot \|)$. It will help us show that the nilradical of A is equal to the Jacobson radical, and more generally that A is a Jacobson ring.

Proposition 32.1: *Let $A = T_n[x_1, \ldots x_q]$ be a K-affinoid algebra without non zero divisors of zero. Let $\phi \in Mult'(T_n, \| \cdot \|)$. There exists $\overline{\phi} \in Mult'(A, \| \cdot \|)$ whose restriction to T_n is ϕ.*

Proof: Let F be the completion of the field of fractions of T_n with respect to ϕ. Then $x_1, \ldots, x_q$ are algebraic over F. Let $A' = F[x_1, \ldots, x_q]$. Then A is isomorphic to a subring of A' through an isomorphism which induces the identity on T_n. By Theorem 3.8 there exists a unique absolute value $\overline{\phi}$ extending that of F to A' and then, considering A as included in A', $\overline{\phi}$ defines on A an absolute value $\widetilde{\phi}$ extending ϕ. Let $m = \max_{1 \leq i \leq q} \widetilde{\phi}(x_i)$, let $\lambda \in K$ be such that $|\lambda| \leq \frac{1}{m}$, and let $y_i = \lambda x_i$ $(1 \leq i \leq q)$. Let θ be the canonical surjection from $T_n\{X_1, \ldots, X_q\}$ onto A inducing the identity on T_n and such that $\theta(X_i) = y_i$ $(1 \leq i \leq q)$. Since $\widetilde{\phi}(y_i) \leq 1$, and since ϕ belongs to $Mult(T_n, \| \cdot \|)$, $\widetilde{\phi} \circ \theta$ clearly satisfies

$$\widetilde{\phi} \circ \theta(Y) \leq \|Y\| \ \forall Y \in T_n\{X_1, \ldots, X_q\}. \tag{32.1}$$

Now, let $t \in A$. By definition $\|t\| = \inf\{\|Y\| \mid \theta(Y) = t\}$. So, given $Y \in T_n\{X_1, \ldots, X_q\}$ such that $\theta(Y) = t$, we have $\widetilde{\phi}(t) = \widetilde{\phi} \circ \theta(Y)$, which obviously does not depend on the Y such that $\theta(Y) = t$. So, by (32.1), considering such a Y we obtain $\widetilde{\phi}(t) \leq \inf\{\|Y\| \mid \theta(Y) = t\} = \|t\|$, and therefore $\widetilde{\phi}$ is continuous and belongs to $Mult(A, \| \,.\, \|)$. □

Remark: Such a continuous absolute value extending ϕ to A is not unique, due to the fact that elements which are conjugate over the field of fractions E of A are not always conjugate over F. For example suppose that $A = T_n[x, y]$ and that x and y are conjugate over E but not over F. Thus, we can have several isomorphism from A onto subrings of A', so we can have $\overline{\phi}(x) \neq \overline{\phi}(y)$ and then ϕ may admit 2 different expansions to A: ϕ_1, ϕ_2, satisfying $\phi_2(x) = \phi_1(y)$. Such a situation will be illustrated later by an example of Krasner-Tate algebra.

Corollary 32.2: *A K-affinoid algebra without non zero divisors of zero, admits continuous absolute values.*

Corollary 32.3: *Every prime ideal of A is the kernel of at least one continuous multiplicative semi-norm of A.*

Remark: As recalled in Theorem 19.15 there exist Krasner algebras $H(D)$ without non zero divisors of zero, admitting no continuous absolute values, and therefore in certain Banach ultrametric commutative K-algebras with unity, certain closed prime ideals are not the kernel of any continuous multiplicative semi-norm. Nevertheless, as long as Krasner algebras $H(D)$ are concerned, it is uneasy to construct a counter-example of an algebra $H(D)$ without non zero divisors of zero admitting no continuous absolute value. Here, we see that such counter-examples don't exist among affinoid K-algebras.

Theorem 32.4 (B. Guennebaud): *$Mult_m(A, \| \,.\, \|)$ is dense inside $Mult(A, \| \,.\, \|)$. Let B be a K-algebra of finite type. Then $Mult(B)$ is locally compact and $Mult_m(B)$ is dense inside $Mult(B)$.*

Proof: Let $\psi \in Mult(B)$ and let $F = \{x_1, \dots x_n\}$ be a finite generating subset of B over K such that $\psi(x_j) < 1 \ \forall j = 1, \dots, n$. Let θ be the canonical surjection from $K[X_1, \dots, X_n]$ onto B such that $\theta(X_j) = x_j$, $(1 \leq j \leq n)$. Then θ has continuation to a continuous morphism $\widehat{\theta}$ from $K\{X_1, \dots, X_n\}$ onto a K-affinoid algebra $\widehat{B}$ quotient of $K\{X_1, \dots, X_n\}$, and $\widehat{B}$ is a Banach K-algebra with respect to the quotient norm that we denote by $\| \,.\, \|_F$. Let $\mathcal{T}(\psi)$ be the set of finite subsets of generators F of B such that $\psi(x) \leq 1 \ \forall x \in F$.

Until the end of the proof, given a subset S of $Mult(B, \| \,.\, \|_F)$, $\overline{S}$ will denote the closure of S in $Mult(B, \| \,.\, \|_F)$. We shortly check that the family of compact subsets $\{\overline{Mult_m}(B, \| \,.\, \|_F),\ F \in \mathcal{T}(\psi)\}$ is a base of a filter because given $F,\ G \in \mathcal{T}(\psi)$, $(\overline{Mult_m}(B, \| \,.\, \|_F)) \cap (\overline{Mult_m}(B, \| \,.\, \|_G))$ contains $Mult_m(B, \| \,.\, \|_{F \cup G})$. Let $Y = \bigcap_{F \in \mathcal{T}(\psi)} \overline{Mult_m}(B, \| \,.\, \|_F)$. Then Y is not empty. For every $x \in \widehat{B}$, we put $\nu(x) = \sup\{\varphi(x) |\ \varphi \in Y\}$. Then we have $\nu(x) \leq \inf\{\|x\|_F \mid F \in \mathcal{T}(\psi)\}\ \forall x \in \widehat{B}$. But $\inf\{\|x\|_F \mid F \in \mathcal{T}(\psi)\} = \psi(x)\ \forall x \in \widehat{B}$. Consequently, we have

$$\nu(x) \leq \psi(x)\ \forall x \in \widehat{B}. \tag{32.2}$$

Now, we fix $x \in \widehat{B}$ such that $\nu(x) < 1$. Since Y is compact, we can easily construct in $Mult(B)$ an open neighborhood W of Y such that $\varphi(x) < 1\ \forall \varphi \in W$. Suppose that for every $F \in \mathcal{T}(\psi)$, $\overline{Mult_m}(B, \| \,.\, \|_F)$ is not included in W. Then the family of compact sets $\overline{Mult_m}(B, \| \,.\, \|_F) \setminus W$ is a base of a filter because $(\overline{Mult_m}(B, \| \,.\, \|_F) \setminus W) \cap (\overline{Mult_m}(B, \| \,.\, \|_G) \setminus W)$ contains $\overline{Mult_m}(B, \| \,.\, \|_{F \cup G}) \setminus W$. But since W is open, each set $\overline{Mult_m}(B, \| \,.\, \|_F)$ $(F \in \mathcal{T}(\psi))$ is a compact and therefore the intersection S is not empty. And since W is open, we have $S \cap W = \emptyset$, a contradiction. Hence, there exists $J \in \mathcal{T}(\psi)$ such that $\overline{Mult_m}(B, \| \,.\, \|_J) \subset W$. Consequently, $\|x\|_J < 1$, and therefore, $\psi(x) < 1$. Thus $\nu(x) < 1$ implies $\psi(x) < 1$. Consequently, since, by Lemma 6.1, ψ and ν are two semi-multiplicative semi-norms, we can conclude that $\psi(x) \leq \nu(x)$. So, by (32.2) the two semi-norms are equal. As a consequence, ψ lies in Y. Then, for every $F \in \mathcal{T}(\psi)$, ψ lies in $\overline{Mult_m}(B, \| \,.\, \|_F)$. Since $Mult_m(B, \| \,.\, \|_F)$ is a subset of $Mult_m(B)$, this finishes proving that $Mult_m(B)$ is dense in $Mult(B)$.

Now, since A is a quotient of a topologically pure extension $T_n = K\{X_1, \ldots, X_n\}$ by an ideal J, we have a canonical homomorphism θ from T_n onto A. Putting $x_j = \theta(X_j)$, $1 \leq j \leq n$, the K-subalgebra of finite type $B = K[x_1, \ldots x_n]$ is dense in A. Putting $F = \{x_1, \ldots x_n\}$, the norm of A is equivalent to $\| \,.\, \|_F$. We now take $\phi \in \underline{Mult(A, \| \,.\, \|)}$. As we just saw above, since ϕ is an element of $Mult(B, \| \,.\, \|_F)$, it lies in $\overline{Mult_m}(B, \| \,.\, \|_F)$. But by Theorem 6.8 $Mult(B, \| \,.\, \|_F)$ and $Mult(A, \| \,.\, \|_F)$ are homeomorphic. Therefore, since every maximal ideal of B and A has codimension one, $Mult_m(B, \| \,.\, \|_F)$ and $Mult_m(A, \| \,.\, \|_F)$ also are homeomorphic. Thus, all elements of $\overline{Mult_m}(B, \| \,.\, \|_F)$ have continuation to elements of $Mult_m(A, \| \,.\, \|_F)$, and finally ϕ lies in $\overline{Mult_m}(A, \| \,.\, \|)$. □

Theorem 32.5: *The nilradical of A is equal to the Jacobson radical.*

Proof: Let $\mathcal{R}$ be the Jacobson radical and let $x \in \mathcal{R}$. Let $\mathcal{J}$ be a prime ideal of A, let $B = \frac{A}{\mathcal{R}}$ and let θ be the canonical surjection from A onto B.

Since B has no non zero divisors of zero different from 0, by Corollary 32.2 it admits a continuous absolute value ϕ, hence $\phi \circ \theta$ lies in $Mult(A, \| \, . \, \|)$. Let $\psi = \phi \circ \theta$. Since $Mult_m(A, \| \, . \, \|)$ is dense in $Mult(A, \| \, . \, \|)$, for all $\epsilon > 0$ we can find $\zeta_\epsilon \in Mult_m(A, \| \, . \, \|)$ such that $|\zeta_\epsilon(x) - \psi(x)|_\infty \leq \epsilon$. But since x belongs to $\mathcal{R}$, by hypothesis we have $\zeta_\epsilon(x) = 0$. Consequently $\psi(x) = 0$. But since ϕ is an absolute value of B, $Ker(\psi)$ is equal to $\mathcal{J}$. Thus, x belongs to $\mathcal{J}$, and therefore $\mathcal{R}$ is equal to the nilradical of A. □

Corollary 32.6: *The semi-norm* $\| \, . \, \|_{si}$ *is a norm if and only if* A *is reduced.*

Theorem 32.7: *A is a Jacobson ring.*

Proof: Let $\mathcal{J}$ be a prime ideal of A and let θ be the canonical homomorphism from A onto the quotient algebra $B = \frac{A}{\mathcal{J}}$. Since B has no non zero divisors of zero, its radical is $\{0\}$, and since it is a K-affinoid algebra, the intersection of all its maximal ideals also equals $\{0\}$. But the maximal ideals of B are the images by θ of the maximal ideals of A containing $\mathcal{J}$. Consequently, we have

$$\mathcal{J} = \theta^{-1}(0) = \theta^{-1}\left(\bigcap_{\mathcal{H} \in Max(B)} \mathcal{H}\right)$$

$$= \bigcap_{\mathcal{H} \in Max(B)} \theta^{-1}(\mathcal{H}) = \bigcap_{\mathcal{J} \subset \mathcal{M}, \mathcal{M} \in Max(A)} \mathcal{M}.$$

Thus $\mathcal{J}$ is equal to the intersection of the maximal ideals that contain it. □

Chapter 33

Salmon's Theorems

The chapter is designed to show that topologically pure extensions are factorial, a theorem due to Paolo Salmon [45].

Definitions and notation: Recall that $U = d(0,1)$, $M = d(0,1^-)$ and the residue class field of K is $\mathcal{K}$. Here, T_n denotes the topologically pure extension $L\{X_1, \ldots, X_n\}$ and we put $V_n = \{x \in T_n \mid \|x\| \leq 1\}$.

Let $f(X_1, \ldots, X_n) \in T_n$. We denote by $\overline{f}$ the residue class of f in $\frac{V_n}{MV_n}$. Then f is said to be *k-regular* if it is of the form cg with $c \in K$ and $g \in V_n$ such that $\overline{g}$ is monic as a polynomial in $\overline{X}_k$.

Lemma 33.1: *Let P, $Q \in V_n[Y]$ be monic polynomials associated to each other in $T_n\{Y\}$. Then $P = Q$.*

Proof: By hypothesis there exists u invertible in $T_n\{Y\}$ such that $Q = uP$. By Theorem 31.14 u is of the form $l + h$ with $l \in K$, $h \in T_n\{Y\}$ and $\|h\| < |l|$. Since the Gauss norm is multiplicative, and since both P, Q are monic, we have $\|P\| = \|Q\| = 1$ and then $\|u\| = |l| = 1$. Suppose $P \neq Q$. Since both $\overline{P}$, $\overline{Q}$ are monic and associated in $\frac{V_n}{MV_n}[\overline{X_n}]$, they have same degree, and therefore we check that $\overline{l} = \overline{1} = \overline{u}$. Consequently, u is of the form $1 + f$, with $f \in T_n\{Y\}$ and $\|f\| < 1$. Therefore $Pf = Q - P$. Let $\lambda \in K$ be such that $\|\lambda(P - Q)\| = 1$. Then the degree in Y of $\overline{\lambda(P - Q)}$ is at most $q - 1$. But since P is monic, and since $\|\lambda f P\| = 1$, the degree in Y of $\overline{\lambda P f}$ is at least q, a contradiction. Hence $P = Q$. □

Lemma 33.2: *Let $f \in T_n$ be n-regular. There exists one and only one monic polynomial $P \in V_{n-1}[X_n]$ associated to f in T_n.*

Proof: Since f is n-regular of order q, it is of the form $\sum_{j=0}^{\infty} a_j X_n^j$, with $a_j \in V_{n-1}$ and a_q invertible in V_{n-1}. Hence by Corollary 7.11 f is associated in T_n to a monic polynomial in X_n. Suppose we can find two monic polynomials P and Q associated to f in T_n. In particular, Q is associated to P in T_n. Without loss of generality we can clearly assume that both P, Q belong to V_n. Then by Lemma 33.1 they are equal. □

Definition: Let $f \in T_n$ be k-regular. The unique monic polynomial $P \in U_K\{X_1, \dots, X_{k-1}, X_{k+1}, \dots, X_n\}[X_k]$, associated to f in T_n will be called *the k-canonical associate to f.*

Let $f \in T_n$ be k-regular and let $f = cg$ with g the k-canonical associate to f. By definition, the degree q of $\overline{g}$ in $\overline{X}_k$ is such that $|a_{0,\dots,0,q,0,\dots,0} - 1| < 1$, $|a_{j_1,\dots,j_{k-1},q,j_{k+1},\dots,j_n}| < 1 \ \forall (j_1, \dots, j_{n-1}) \neq (0, \dots, 0)$, and $|a_{j_1,\dots,j_n}| < 1$ whenever $j_n > q$. The integer q will be called *the k-order of f.*

Theorem 33.3: *Let $f \in T_n$ be k-regular of k-order q. There exists a unique monic polynomial $P(X_k) \in K\{X_1, \dots, X_{k-1}, X_{k+1}, \dots, X_n\}[X_k]$ which is associated to f in T_n. Moreover, the degree of P in X_k is q.*

Proof: Without loss of generality, we can obviously assume that $k = n$, what we do for convenience. We put $T_{n-1} = E$ and $Y = X_n$. Without loss of generality, we can also assume that the coefficient $a_{0,\dots,0,q}$ of Y in f is 1. Let q be the n-order of f. In $E\{Y\}$ f is of the form $\sum_{j=0}^{\infty} f_j Y^j$ with $\lim_{j\to\infty} f_j = 0$, $\|f_j\| \leq 1 \ \forall j \in \mathbb{N}$, $f_q = 1$ and $\|f_j\| < 1 \ \forall j > q$. Then $\overline{f}$ is a monic polynomial Π of degree q, hence by Corollary 7.11 there exists $Q \in E[Y]$ such that $\overline{Q} = \overline{f}$ and $h \in E\{Y\}$ such that $\overline{h} = 1$ and $f = Qh$. Since $\overline{Q} = \overline{f}$, the coefficient l of Y^q in Q is of the form $1 + \epsilon$ with $\|\epsilon\| < 1$, therefore by Theorem 6.13 Q is invertible in the Banach K-algebra T_n. In the same way, since $\overline{h} = 1$, h is invertible in T_n. Consequently, the q-degree monic polynomial $P = l^{-1}Q$ is associated to f in T_n.

Now, we will show such a factorization is unique. Let wS be another factorization of f, with w invertible in T_n and S a monic polynomial in $E[Y]$. Thus in $E[Y]$ P and S are two monic polynomials associated to each other. Consequently, by Lemma 33.1 they are equal and this shows the uniqueness of the polynomial P. □

Theorem 33.4: *Let $f \in T_n$ be k-regular of k-order q and factorize in T_n in the form $f_1 f_2$. Then, each f_j is k-regular of order q_j ($j = 1, 2$), such that $q = q_1 + q_2$.*

Proof: Without loss of generality, we can assume that f, f_1, f_2 lie in V_n and that $\overline{f}$ is monic of degree q. Then $\overline{f} = \overline{f}_1 \overline{f}_2$. Putting

$q_j = \deg(\overline{f_j(X_k)})$, $(j = 1, 2)$, we have $q = q_1 + q_2$. Since $\overline{f}$ is monic, each $\overline{f}_j$ has a coefficient of degree q_j in $\overline{X}_k$ which is constant, hence each f_j is k-regular, of k-order q_j. □

Theorem 33.5: *Let $f \in V_{n-1}[X_n]$ be monic and irreducible in $T_{n-1}[X_n]$. Then it is irreducible in $T_{n-1}\{X\}$.*

Proof: Suppose f is not irreducible in T_n. By Theorem 33.4, there exists n-regular elements f_1, $f_2 \in T_n$ such that $f = f_1 f_2$, and therefore by Theorem 33.3 there exists a unique monic polynomial $g_1(X_n) \in T_{n-1}[X_n]$ associated to f_1 in T_n and a unique monic polynomial $g_2(X_n) \in T_{n-1}[X_n]$ associated to f_2 in T_n. Then $g_1 g_2$ is a monic polynomial in X_n and is associated to f hence it is equal to f, a contradiction to the hypothesis: f irreducible in $T_{n-1}[X_n]$. □

Notation: In $\mathbb{N}^n$ we will denote by $\mathcal{O}_n$ the order relation defined as it follows: given $(i_1, \ldots, i_n)$, $(j_1, \ldots, j_n) \in \mathbb{N}^n$, we put $(i_1, \ldots, i_n) < (j_1, \ldots, j_n)$ if for certain $t \leq n$ we have $i_{t-1} < j_{t-1}$ and $i_m = j_m$ $\forall m = t, \ldots, n$, and we put $(i_1, \ldots, i_n) \leq (j_1, \ldots, j_n)$ if either $(i_1, \ldots, i_n) < (j_1, \ldots, j_n)$ or $(i_1, \ldots, i_n) = (j_1, \ldots, j_n)$. Then this relation is a total order relation in $\mathbb{N}^n$.

Proposition 33.6: *Let $h \in V_n$ be such that in $\overline{h}$, the coefficient of maximal index, with respect to $\mathcal{O}_n$ is equal to 1. There exists a K-automorphism ψ of T_n preserving V_n, such that $\overline{\psi(h)}$ is a monic polynomial in $\overline{X}_n$.*

Proof: Let $(k_1, \ldots, k_n)$ be the maximum index (with respect to $\mathcal{O}_n$) whose coefficient in $\overline{h}$ is not 0. We can clearly find $t_2, \ldots, t_n \in \mathbb{N}$ such that $h(X_1, X_2 + X_1^{t_2}, \ldots, X_n + X_1^{t_n})$ is a monic polynomial in X_1 of degree $k_1 + k_2 t_2 + \ldots + k_n t_n$. Let ϕ be the K-algebra endomorphism defined in $K[X_1, \ldots X_n]$ as $\phi(f) = f(X_1, X_2 + X_1^{t_2}, \ldots, X_n + X_1^{t_n})$. Then we can check that ϕ is a K-algebra automorphism of $K[X_1, \ldots X_n]$. Moreover, it also induces on $\mathcal{K}[\overline{X}_1, \ldots, \overline{X}_n]$ a $\mathcal{K}$-algebra automorphism, so it satisfies $\|\phi(f)\| = \|f\|$. Therefore ϕ has continuation to T_n and preserves V_n. Now, let ω be the K-algebra automorphism of T_n defined as $\omega(f(X_1, \ldots, X_n)) = f(X_n, X_2, \ldots, X_{n-1}, X_1)$. Then $\omega \circ \phi$ is a continuous K-algebra automorphism ψ of T_n such that $\overline{\psi(h)}$ is a monic polynomial in $\overline{X}_n$. □

Theorem 33.7 (P. Salmon [45]): *T_n is factorial for every $n \in \mathbb{N}$.*

Proof: We assume the theorem true for every $n < m$, and will prove it when $n = m$. For convenience we put $E = T_{n-1}$. Let g be an irreducible element of T_n. We will prove that gT_n is a prime ideal of T_n. Without loss of generality we can obviously assume that g belongs to V_n and that in $\overline{g}$ the coefficient of maximal index, with respect to $\mathcal{O}_n$, is equal to 1. Consequently, by Proposition 33.6 there exists a K-algebra automorphism ψ of T_n preserving V_n, such that $\overline{\psi(g)}$ is a monic polynomial in $\overline{X}_n$. Thus, without loss of generality, we can also assume that $\overline{g}$ is a monic polynomial in $\overline{X}_n$. Then, by Theorem 33.2, g is associated to a monic polynomial $P(X_n) \in V_{n-1}[X_n]$. Let $q = \deg(P)$. Suppose that P admits a factorization in $V_{n-1}[X_n]$ in the form Q_1Q_2, with $Q_i(X_n) = \sum_{j=0}^{q_i} a_{j,i}X_n^j$ $(i = 1,2)$. Then $q_1 + q_2 = q$ and $a_{q_1}a_{q_2} = 1$. Since g is irreducible in T_n, so is P, hence one of the Q_i is invertible in T_n. Both polynomials $\frac{Q_i}{a_{q_i}}$ are monic and $\frac{Q_1Q_2}{a_{q_1}a_{q_2}}$ divides P in T_n. Now, we consider T_n as $E\{X_n\}$. By Theorem 33.3 there exists only one monic polynomial in X_n associated to a n-regular series such as P. Hence one of the polynomials $\frac{Q_i}{a_{q_i}}$ is equal to 1, and therefore P is irreducible in $E[X_n]$.

Now, suppose that P divides a product f_1f_2 in $E\{X_n\}$ with $f_1,\ f_2 \in E\{X_n\}$. And according to Theorem 7.5, let R_i be the rest of the Euclidean division of f_i by P $(i = 1,2)$ in $E\{X_n\}$. Then P divides R_1R_2 in B. But by Theorem 7.5 the Euclidean division in $E[X_n]$ is induced by the one in $E\{X_n\}$. Consequently, P divides R_1R_2 in $E[X_n]$. And since, by induction hypothesis, E is factorial, so is $E[X_n]$. Consequently P must divide one of the R_i, say R_1, in $E[X_n]$. But since $\deg(R_1) < q$, of course $R_1 = 0$, hence P divides f_1, and therefore generates a prime ideal in $E[X_n]$ and so does g in $E\{X_n\}$. Thus every irreducible element of T_n generates a prime ideal, and consequently, by Lemma 1.2 we know that $E\{X_n\}$ is factorial, hence the theorem holds when $m = n$. $\square$

Chapter 34

Separable Fields

When the field K is separable (in the topological meaning) we can prove that, given a K-affinoid algebra A, the topology of simple convergence on $Mult(A, \| \,.\, \|)$ is metrizable, and we construct an equivalent metric, a study first made by N. Mainetti [39]. Of course, such a metric defines a topology weaker than this defined by δ in the case of one variable, as it was seen in Chapter 13.

Notation: In this chapter, the field K is supposed to be separable, i.e. to have a dense countable subset that we will denote by S, and we will denote by $S[Z_1, \dots Z_q]$ the set of polynomials in q variables with coefficients in K. A will be a K-affinoid algebra. As in previous chapters, we will denote by $\| \,.\, \|$ the Gauss norm on an algebra of polynomials $K[Y_1, \dots, Y_q]$ and on $K\{Y_1, \dots, Y_q\}$. In order to avoiding any confusion, $\| \,.\, \|_a$ will denote the norm of K-affinoid algebra of A.

We will denote by $K[x_1, \dots, x_q]$ a K-algebra of finite type which is not necessarily a K-algebra of polynomials, i.e. the x_j are not supposed to be algebraically independent.

Remark: A field which is separable is weakly valued. But a weakly valued field is not necessarily separable. For instance, it is known that the spherical completion of a $\mathbb{C}_p$ is not separable, but is weakly valued as $\mathbb{C}_p$.

Lemma 34.1 and 34.2 are immediate:

Lemma 34.1: *Let $f \in K[x_1, \dots, x_q]$. The mapping ν defined on $(Mult(K[x_1, \dots, x_q]))^2$ by $\nu(\phi, \psi) = \min(|\phi(f) - \psi(f)|_\infty, 1)$ is a semi-distance.*

Lemma 34.2: *There exists a sequence $(P_n)_{n\in\mathbb{N}^*}$ in $S[Z_1,\ldots,Z_q]$ satisfying $P_n = Z_n$ $\forall n \le q$, such that, given $Q \in K[x_1,\ldots,x_q]$ and $\epsilon > 0$, there exists $m \in \mathbb{N}$ such that $\|Q - P_m\| < \epsilon$ and $\deg_h(Q) = \deg_h(P_m)$ $\forall h = 1,\ldots,q$.*

Lemma 34.3: *A admits a dense countable subset of the form $S[x_1,\ldots,x_q]$, with $\|x_j\|_a \le 1$ $\forall$ $j = 1,\ldots,q$.*

Proof: Since A is a K-affinoid algebra, it is of the form $T_m[y_1,\cdots,y_t]$ with $T_m = K\{Y_1,\ldots,Y_m\}$. Consequently, the K-algebra of finite type $K[Y_1,\ldots,Y_m,y_1,\ldots y_t]$ is dense in A, and of course $S[X_1,\ldots,X_m,y_1,\ldots,y_t]$ is dense in $K[Y_1,\ldots,Y_m,y_1,\ldots,y_t]$, hence in A. By hypothesis the Y_j satisfy $\|Y_j\|_a = 1$ $\forall$ $j = 1,\ldots,m$ and of course we can take the y_j such that $\|y_j\|_a = 1$ $\forall$ $j = 1,\ldots,t$. $\square$

Definitions and notation: In $K[Z_1,\ldots,Z_q]$, we consider a sequence $(P_n(Z_1,\ldots,Z_q))_{n\in\mathbb{N}^*}$ in $S[Z_1,\ldots,Z_q]$ satisfying $P_n = Z_n$ $\forall n \le q$, such that, given $Q \in K[Z_1,\ldots,Z_q]$ and $\epsilon > 0$, there exists $m \in \mathbb{N}$ such that

$$\|Q(Z_1,\ldots,Z_q) - P_m(Z_1,\ldots,Z_q)\| < \epsilon$$

(with respect to the Gauss norm in $K[Z_1,\ldots,Z_q]$) and

$$\deg_h(Q(Z_1,\ldots,Z_q)) = \deg_h(P_m(Z_1,\ldots,Z_q))\ \forall h = 1,\ldots,q.$$

Such a sequence will be called a *a S-appropriate sequence of* $K[Z_1,\ldots,Z_q]$.

On $Mult(K[x_1,\ldots,x_q])$, for $\phi,\psi \in Mult(K[x_1,\ldots,x_q])$ we put

$$\mathcal{D}(\phi,\psi) = \sum_{n=1}^{\infty} \frac{1}{n^2}\min(|\phi(P_n) - \psi(P_n)|_\infty, 1).$$

Then $\mathcal{D}$ will be called *the metric associated to the sequence $(P_n)_{n\in\mathbb{N}^*}$ on* $Mult(K(x_1,\ldots,x_q])$.

In the same way, in A, we consider a dense K-subalgebra of finite type of the form $K[x_1,\ldots,x_q]$ and an appropriate sequence of $K[Z_1,\ldots,Z_q]$. Then by Theorem 6.8 there is a natural injective mapping from $Mult(A,\|\ .\ \|_a)$ into $Mult(K[x_1,\ldots,x_q])$ which lets us consider $Mult(A,\|\ .\ \|_a)$ as a subset of $Mult(K[x_1,\ldots,x_q])$. Consequently the metric on $Mult(K[x_1,\ldots,x_q])$ associated to the S-appropriate sequence $(P_n)_{n\in\mathbb{N}^*}$ is defined on $Mult(A,\|\ .\ \|_a)$, and then, as a semi-distance defined on $Mult(A,\|\ .\ \|_a)$, it will be called *the metric associated to the S-appropriate sequence $(P_n)_{n\in\mathbb{N}^*}$ on $Mult(A,\|\ .\ \|_a)$.*

In $Mult(K[x_1,\ldots,x_q])$ and in $Mult(A,\|\ .\ \|_a)$ an open ball of center ϕ and diameter r will be denoted by $B(\phi,r)$.

Theorem 34.4: *Let $K[x_1,\ldots,x_q]$ be a dense K-subalgebra of A. Let $(P_n)_{n\in\mathbb{N}^*}$ be a S-appropriate sequence of $K[Z_1,\ldots,Z_q]$. Let $\mathcal{D}$ be the metric associated to the S-appropriate sequence $(P_n)_{n\in\mathbb{N}^*}$ on $Mult(K[x_1,\ldots,x_q])$ (resp. on $Mult(A,\|\ .\ \|_a)$). Then $\mathcal{D}$ is a distance on $Mult(K[x_1,\ldots,x_q])$ (resp. on $Mult(A,\|\ .\ \|_a)$) which defines the topology of simple convergence.*

Proof: Let $\phi\in Mult[Z_1,\ldots,Z_q]$, (resp. let $\phi\in Mult(A,\|\ .\ \|_a)$), and let $\epsilon\in]0,1[$. In order to show the statement, we will first show that $B(\phi,\epsilon)$ contains a certain neighborhood $\mathcal{V}(\phi,Q_1,\ldots,Q_s,\eta)$. Let $s\in\mathbb{N}$ be such that $\sum_{n=s+1}^{\infty}\frac{1}{n^2}<\frac{\epsilon}{2}$ and let $\eta=\frac{3\epsilon}{\pi^2}$. Let $\psi\in\mathcal{V}(\phi,P_1,\ldots,P_s,\eta)$. Clearly, we have

$$\begin{aligned}\sum_{n=1}^{s}\frac{1}{n^2}\min(|\phi(P_n)-\psi(P_n)|_\infty,1)&=\sum_{n=1}^{s}\frac{1}{n^2}|\phi(P_n)-\psi(P_n)|_\infty\\&\le\eta\sum_{n=1}^{s}\frac{1}{n^2}\le\frac{\eta\pi^2}{6}\\&=\frac{\epsilon}{2}.\end{aligned}$$

And

$$\sum_{n=s+1}^{\infty}\frac{1}{n^2}\min(|\phi(P_n)-\psi(P_n)|_\infty,1)\le\sum_{n=s+1}^{\infty}\frac{1}{n^2}<\frac{\epsilon}{2}.$$

Consequently, $\mathcal{D}(\phi,\psi)<\epsilon$ therefore $\mathcal{V}(\phi,P_1,\ldots,P_s,\eta)$ is included in $B(\phi,\epsilon)$. Thus, a ball of center ϕ, with respect to $\mathcal{D}$, is a neighborhood of ϕ with respect to the topology of simple convergence, and therefore is an open set with respect to the topology of simple convergence, so the topology defined by the semi-distance $\mathcal{D}$ is weaker (in the large sense) than the topology of simple convergence (we notice that the reasoning holds in the same way in $K[x_1,\ldots,x_q]$ and in A).

Conversely, we now consider a neighborhood $\mathcal{V}(\phi,Q_1,\ldots,Q_t,\epsilon)$, with $Q_j\in K[x_1,\ldots,x_t]$ (resp. $Q_j\in A$)) and $\epsilon\in]0,1[$ and we will look for a ball $B(\phi,\eta)$ included in $\mathcal{V}(\phi,Q_1,\ldots,Q_t,\epsilon)$.

We first consider the case of $Mult(K[x_1,\ldots,x_q])$. For each $j=1,\ldots,t$, let u_j be the total degree of Q_j and let $\rho>1+\max_{1\le h\le q}\phi(x_h)$. On the K-algebra of polynomials $K[Z_1,\ldots,Z_q]$, we can consider the Gauss norm $\|\ .\ \|$, and then, with respect to this norm, for each $j=1,\ldots,t$ there exists

$n_j \in \mathbb{N}^*$ such that

$$\|P_{n_j}(Z_1,\ldots,Z_q) - Q_j(Z_1,\ldots,Z_q)\|\rho^{u_j} < \frac{\epsilon}{3} \tag{34.1}$$

and

$$\deg_h(Q_j) = \deg_h(P_{n_j})\ \forall h = 1,\ldots,q. \tag{34.2}$$

Let $l = \max(n_1,\ldots,n_t,q)$ and let $\eta = \frac{\epsilon}{3l^2}$ (hence $l^2\eta < 1$). Let $\psi \in B(\phi,\eta)$. Then we have

$$\sum_{n=1}^{l} \frac{1}{n^2} \min(|\phi(P_n) - \psi(P_n)|_\infty, 1) < \eta.$$

In particular for each $n = 1,\ldots,l$ we obtain

$$\frac{1}{n^2} \min(|\phi(P_n) - \psi(P_n)|_\infty, 1) < \eta \tag{34.3}$$

and therefore $\min(|\phi(P_n) - \psi(P_n)|_\infty, 1) < n^2\eta = \frac{n^2\epsilon}{3l^2} < 1\ \forall n = 1,\ldots,l$. Consequently, $\min(|\phi(P_n) - \psi(P_n)|_\infty, 1) = |\phi(P_n) - \psi(P_n)|_\infty$, hence by (34.3) we have

$$|\phi(P_n) - \psi(P_n)|_\infty < n^2\eta = \frac{n^2\epsilon}{3l^2} \leq \frac{\epsilon}{3}\ \forall n = 1,\ldots,l.$$

Thus, we have proven

$$|\phi(P_n) - \psi(P_n)|_\infty < \frac{\epsilon}{3}\ \forall n = 1,\ldots,l. \tag{34.4}$$

Now, consider the obvious inequality

$$|\psi(Q_j) - \phi(Q_j)|_\infty \leq |\psi(Q_j) - \psi(P_{n_j})|_\infty + |\psi(P_{n_j}) - \phi(P_{n_j})|_\infty + |\phi(P_{n_j}) - \phi(Q_j)|_\infty.$$

On one hand, since $\rho > \phi(x_h)\ \forall h = 1,\ldots,t$, by (34.2) we check that

$$\phi(P_{n_j}(Z_1,\ldots,Z_q) - Q_j(Z_1,\ldots,Z_q)) \leq \|P_{n_j} - Q_j\|\rho^{u_j},$$

and therefore by (34.1) we obtain $\phi(P_{n_j} - Q_j) < \frac{\epsilon}{3}$, hence

$$|\phi(P_{n_j}) - \phi(Q_j)|_\infty \leq \phi(P_{n_j} - Q_j) < \frac{\epsilon}{3}. \tag{34.5}$$

On the other hand, since $x_h = P_h\ \forall h = 1,\ldots,q$, and since $\psi \in B(\phi,\eta)$, we have

$$\frac{1}{h^2}|\psi(x_h) - \phi(x_h)|_\infty < \eta, \text{ hence } |\psi(x_h) - \phi(x_h)|_\infty < \frac{\epsilon h^2}{3l^2} < 1.$$

Consequently, since $\rho > \phi(x_h)+1$, we check that $\psi(x_h) < \rho$. Consequently, similarly to (34.5), we obtain

$$\begin{aligned} |\psi(P_{n_j}) - \psi(Q_j)|_\infty &\le \psi(P_{n_j} - Q_j) \\ &\le \|P_{n_j}(Z_1,\dots,Z_q) - Q_j(Z_1,\dots,Z_q)\|\rho^{u_j} \\ &< \frac{\epsilon}{3}. \end{aligned} \tag{34.6}$$

Then, by (34.4), (34.5), (34.6) we can conclude that $|\psi(Q_j)-\phi(Q_j)|_\infty < \epsilon$, hence $\psi \in \mathcal{V}(\phi, Q_1,\dots,Q_t,\epsilon)$, and therefore $B(\phi,\eta)$ is included in $\mathcal{V}(\phi, Q_1,\dots,Q_t,\epsilon)$. This finishes proving that the two topologies are equivalent on $Mult(K[x_1,\dots,x_q])$.

We now consider $Mult(A, \| \,.\, \|_a)$ and therefore assume that $Q_j \in A$. By hypothesis, the sequence $(P_n(Z_1,\dots,Z_q))_{n\in\mathbb{N}}$ is dense in $K[Z_1,\dots,Z_q]$, with respect to the Gauss norm, hence the sequence $P_n(x_1,\dots,x_q)_{n\in\mathbb{N}}$ is dense $K[x_1,\dots,x_q]$ and in A with respect to the norm $\| \,.\, \|_a$ of A. Consequently, for each $j = 1,\dots,t$ there exists $n_j \in \mathbb{N}^*$ such that

$$\|P_{n_j} - Q_j\|_a < \frac{\epsilon}{3}. \tag{34.7}$$

The beginning of this proof is almost similar to that in the case previously studied of $Mult(K[x_1,\dots,x_q])$. Let $l = \max(n_1,\dots,n_t)$ and let $\eta = \frac{\epsilon}{3l^2}$, (hence $l^2\eta < 1$). Let $\psi \in B(\phi,\eta)$. Then we have $\sum_{n=1}^{l} \frac{1}{n^2}\min(|\phi(P_n) - \psi(P_n)|_\infty, 1) < \eta$. In particular for each $n = 1,\dots,l$ we obtain $\frac{1}{n^2}\min(|\phi(P_n) - \psi(P_n)|_\infty, 1) < \eta$, and therefore

$$\min(|\phi(P_n) - \psi(P_n)|_\infty, 1) < n^2\eta = \frac{n^2\epsilon}{3l^2} < 1 \;\forall n = 1,\dots,l. \tag{34.8}$$

Consequently, $\min(|\phi(P_n) - \psi(P_n)|_\infty, 1) = |\phi(P_n) - \psi(P_n)|_\infty$, hence by (34.8)

$$|\phi(P_n) - \psi(P_n)|_\infty < n^2\eta = \frac{n^2\epsilon}{3l^2} \le \frac{\epsilon}{3} \;\forall n = 1,\dots,l.$$

Thus, we have proven

$$|\phi(P_n) - \psi(P_n)|_\infty < \frac{\epsilon}{3} \;\forall n = 1,\dots,l. \tag{34.9}$$

Then, thanks to (34.7) and (34.9), by Lemma 34.2 we know that $|\psi(Q_j) - \phi(Q_j)|_\infty < \epsilon$, and therefore $B(\phi,\eta)$ is included in $\mathcal{V}(\phi, Q_1,\dots,Q_t,\epsilon)$. Thus the two topologies are equal on $Mult(A, \| \,.\, \|_a)$.

Moreover, in both cases, since the topology of simple convergence is separated, this shows that $\mathcal{D}$ is a distance. □

Corollary 34.5: *$Mult(A, \| \,.\, \|_a)$ is sequentially compact.*

Corollary 34.6: *In $Mult(K[x_1, \ldots, x_q])$, every compact is sequentially compact.*

Corollary 34.7: *Let D be a closed bounded subset of K. The topology of simple convergence on $Mult(H(D), \| \,.\, \|_D)$ is metrizable.*

Notation: By Lemma 34.2 we can assume that $\|x_j\|_a \leq 1 \; \forall \, j = 1, \ldots, q$. Then, in $Mult(A, \| \,.\, \|_a)$ we can also define another distance by considering a sequence $(R_n)_{n \in \mathbb{N}}$ of $(S \cap U)[Z_1, \ldots, Z_q]$ which is dense in $U[Z_1, \ldots, Z_q]$ and then we put $\mathcal{D}'(\phi, \psi) = \sum_{n=1}^{\infty} \frac{1}{n^2} |\phi(R_n) - \psi(R_n)|_\infty$. Indeed, since $\|x_j\|_a \leq 1 \; \forall \, j = 1, \ldots, q$, for each $n \in \mathbb{N}$ we have $\phi(R_n) \leq \|R_n(x_1, \ldots, x_q)\|_a \leq \|R_n\| \leq 1$, $\psi(R_n) \leq \|R_n(x_1, \ldots, x_q)\|_a \leq \|R_n\| \leq 1$, hence the series is convergent. We will denote by $B'(\phi, r)$ the open ball for this metric.

Theorem 34.8: *$\mathcal{D}'$ defines the same topology as this of simple convergence on $Mult(A, \| \,.\, \|_a)$.*

Proof: On one hand it is easily seen that $\mathcal{D}'(\phi, \psi) \leq \mathcal{D}(\phi, \psi)$. Indeed, we can consider the sequence R_m as a subsequence of the sequence P_m, hence $\mathcal{D}(\phi, \psi) \geq \sum_{n=1}^{\infty} \frac{1}{n^2} \min(|\phi(R_n) - \psi(R_n)|_\infty, 1)$. But since $R_n \in U[Z_1, \ldots, Z_q]$, we have $|\phi(R_n)| \leq \|R_n\| \leq 1$, $|\psi(R_n)| \leq \|R_n\| \leq 1$, hence

$$\min(|\phi(R_n) - \psi(R_n)|_\infty, 1) \geq \frac{|\phi(R_n) - \psi(R_n)|_\infty}{2},$$

and therefore $\mathcal{D}(\phi, \psi) \geq \frac{\mathcal{D}'(\phi, \psi)}{2}$, which shows that the topology defined by $\mathcal{D}$, and therefore the topology of simple convergence is stronger than this defined by $\mathcal{D}'$.

On the other hand, we will show that the topology defined by $\mathcal{D}'$ is stronger than this of simple convergence, in a similar way as in Theorem 34.5. Let $Q_1, \ldots, Q_t \in A$, let $N = 1 + \max_{0 \leq j \leq t} \|Q_j\|_a$, and let $\epsilon \in]0, 1[$. Let $l = \max(n_1, \ldots, n_t)$. For each $j = 1, \ldots, t$, we can find $\mu_j \in K$ such that $\|\mu_j Q_j\|_a \leq 1$ and $|\mu_j| < N$. Let $\eta = \frac{\epsilon}{3Ml^2}$, (hence $Nl^2\eta < 1$). Let $\psi \in B(\phi, \eta)$. For each $j = 1, \ldots, t$, there exists $R_{n_j} \in S[Z_1, \ldots, Z_q]$ such that $\|R_{n_j}(x_1,\dots,x_q)-\mu_jQ_j\|_a<\eta$, so we have

$$\phi(R_{n_j}(x_1,\dots,x_q)-\mu_jQ_j)<\eta,\ \psi(R_{n_j}(x_1,\dots,x_q)-\mu_jQ_j)<\eta. \quad (34.10)$$

Now, since $\psi\in B(\phi,\eta)$, we have

$$\sum_{n=1}^{l}\frac{1}{n^2}|\phi(R_n)-\psi(R_n)|_\infty<\eta.$$

In particular for each $n=1,\dots,l$ we obtain $\frac{1}{n^2}|\phi(R_n)-\psi(R_n)|_\infty<\eta$, and therefore

$$|\phi(R_n)-\psi(R_n)|_\infty<n^2\eta=\frac{n^2\epsilon}{3Nl^2}<1\le\frac{\epsilon}{3N}\ \forall n=1,\dots,l. \quad (34.11)$$

Consequently, by (34.10) and (34.11) we obtain $|\psi(\mu_jQ_j)-\phi(\mu_jQ_j)|_\infty<N\epsilon$, therefore $|\psi(Q_j)-\phi(Q_j)|_\infty<\epsilon$ and finally $B'(\phi,\eta)$ is included in $\mathcal{V}(\phi,Q_1,\dots,Q_t,\epsilon)$. This finishes proving that the two topologies are equal on $Mult(A,\|\cdot\|_a)$. □

Chapter 35

Spectral Norm of Affinoid Algebras

In Chapter 31 we saw that if A is reduced, its spectral semi-norm is a norm. Then it is a natural question to ask whether A is uniform. The answer is quite easy and was soon given in [49] when the characteristic p of K is 0. It is much more difficult when $p \neq 0$ (such a study was also made in [3]).

Notation: Throughout the chapter and in the next one we denote by p the characteristic of K. Next, considering $\mathbb{Z}^n$ as a $\mathbb{Z}$-module, for all $(i) = (i_1, \ldots, i_n) \in \mathbb{Z}$, and $q \in \mathbb{Z}$ we put $q(i) = (qi_1, \ldots, qi_n)$. Let

$$f(Y_1, \ldots, Y_n) = a_{i_1,\ldots,i_n} Y_1^{i_1} \ldots Y_n^{i_n} \in T_n.$$

We put $(Y) = (Y_1, \ldots, Y_n)$, $(Y)^{(i)} = (Y_1^{i_1}, \ldots, Y_n^{i_n})$ and $f(Y_1, \ldots, Y_n) = \sum_{(i)\in\mathbb{N}^n} \in a_{(i)}(Y)^{(i)}$.

For each $n \in \mathbb{N}^*$ we denote by L_n the field of fractions of T_n and A is a K-affinoid algebra. By Corollary 31.5 A has finitely many minimal prime ideals $\mathcal{S}_1, \ldots, \mathcal{S}_l$. For each $j = 1, \ldots, l$, we denote by A_j the K-affinoid algebra $\frac{A}{\mathcal{S}_j}$, by θ_j the canonical surjection from A onto A_j, and by $\| \cdot \|_{si,j}$ the spectral norm of A_j.

Theorem 35.1: *Let A be reduced. For every $f \in A$, we have*

$$\|f\|_{si} = \max(\|\theta_j(f)\|_{si,j}).$$

Proof: By Theorem 31.10 we have $\|f\|_{si} = \sup_{\lambda\in sp_A(f)} |\lambda|$, and

$$\|\theta_j(f)\|_{si,j} = \sup_{\lambda\in sp_{A_j}(\theta_j(f))} |\lambda|, \ (j = 1, \dots, l).$$

But by Lemma 1.15 $sp_A(f) = \cup_{j=1}^{l} sp_{A_j}(\theta_j(f))$. So, our claim is obvious. □

Theorem 35.2: *For all $x \in A$ there exists $\chi \in \mathcal{X}(A)$ such that $|\chi(x)| = \|x\|_{si}$. Moreover, if A is an integral domain finite over T_n, then for all $f \in A$, we have $\mathtt{S}(irr(f, T_n)) = \|f\|_{si}$.*

Proof: Let $f \in A$. First, we will show that there exists $\chi \in \mathcal{X}(A)$ such that $|\chi(x)| = \|x\|_{si}$. When A is a topologically pure extension, the statement is proven in Theorem 31.12.

Here we first consider a K-affinoid algebra A without non zero divisors of zero. There exists a topologically pure extension T_n such that A is a finite over T_n. By Lemma 1.18 every $\chi \in \mathcal{X}(T_n)$ admits at least one extension to an element of $\mathcal{X}(A)$. By Theorem 31.10 the spectral norm $\| \cdot \|_{si}$ of A satisfies $\|f\|_{si} = \sup\{|\chi(f)| \mid \chi \in \mathcal{X}(\mathcal{A})\}$. On the other hand, by Proposition 29.4 we have $\sup\{|\chi(f)| \mid \chi \in \mathcal{X}(\mathcal{A})\} = \mathtt{S}(irr(f, T_n))$ and there exists $\chi \in \mathcal{X}(A)$ such that $|\chi(f)| = \sup\{|\chi(f)| \mid \chi \in \mathcal{X}(\mathcal{A})\}$. Consequently, there exists $\chi \in \mathcal{X}(A)$ such that $|\chi(f)| = \|f\|_{si}$. Thus, our claim is now proven when A is an integral domain.

We now consider the case when A is just reduced. By Theorem 35.1 we have $\max_{1\le j\le l}(\|\theta_j(f)\|_{si,j}) = \|f\|_{si}$. Thus $\|f\|_{si}$ lies in $|K|$. Let $\lambda \in K$ be such that $|\lambda| = \|f\|_{si}$ and let $g = \frac{f}{\lambda}$. So, $\|g\|_{si} = 1$. By Theorem 31.7, if $|\chi(g)| < 1 \ \forall\chi \in \mathcal{X}(A)$ then $\|g\|_{si} < 1$, hence there exists $\chi \in \mathcal{X}(A)$ such that $|\chi(g)| = 1$, and therefore $|\chi(f)| = \|f\|_{si}$.

We can immediately generalize this property when A is not reduced. Indeed, by Theorem 31.10 we have $\sup\{|\chi(f)| \ |\chi \in \mathcal{X}(A)\} = \|f\|_{si} \ \forall f \in A$. Now, consider the radical $\mathcal{R}$ of A, let $B = \frac{A}{\mathcal{R}}$, and let ϕ be the canonical surjection of A onto B. Then, we have $\sup\{|\chi(f)| \ |\chi \in \mathcal{X}(A)\} = \sup\{|\chi \circ \phi(f)| \ |\chi \in \mathcal{X}(B)\}$. Since B is reduced, the property holds for B, and therefore also holds for A. □

Theorem 35.3 *The algebraic closure of L_n is L_n-productal with respect to $\| \cdot \|_{pol}^{L_n}$.*

Proof: If $p = 0$, the claim comes from Theorems 29.1 and 29.7. So, we suppose $p \neq 0$. By Theorem 29.10, it is sufficient to prove that $L_n^{\frac{1}{p}}$ is L_n-productal. For each $k = 1, \dots, n$, we put $Y_k = \sqrt[p]{X_k}$ and $Y = (Y_1, \dots, Y_n)$.

So, we have $Y^{p.(u)} = X$, and more generally $Y^{jp.(u)} = X^{j.(u)}$. Let $g \in L^{\frac{1}{p}}$. Then g is of the form $\sum_{j=0}^{p-1} \frac{a_j}{b_j} Y^{j.(u)}$, with $a_j,\ b_j \in T_n\ \forall j = 1, \ldots, n$. And

$$g^p = \sum_{j=0}^{p-1} \frac{(a_j)^p Y^{jp.(u)}}{(b_j)^p} = \sum_{j=0}^{p-1} \frac{(a_j)^p X^{j.(u)}}{(b_j)^p}.$$

Let $b = \prod_{j=0}^{p-1} b_j$ and for each $j = 0, \ldots, p-1$ we set $c_j = \frac{b}{b_j}$, which belongs to T_n. Then

$$(bg)^p = \sum_{j=1}^{p-1} (a_j c_j)^p X^{j.(u)}.$$

Now, for each $j = 0, \ldots, p-1$, $a_j c_j$ is of the form $\sum_{(i)\in\mathbb{N}^n} \lambda_{(i),j} X^i$, with $\lambda_{(i),j} \in K$, hence $(a_j c_j)^p = \sum_{(i)\in\mathbb{N}^n} \lambda_{(i),j} X^i$ and thereby $b^p g^p = \sum_{(i)\in\mathbb{N}^n,\ 0\le j\le p-1} \lambda_{(i),j}^p X^{p.(i)+j.(u)}$.

Next, it is clear that the mapping ω from $\mathbb{N}^n \times \{0, \ldots, p-1\}$ into $\mathbb{N}^n$ defined as $\omega((i), j) = p.(i) + j.(u)$ is an injection. Consequently, we have $\|b^p g^p\| = \max\{|\lambda_{(i),j}^p \mid (i) \in \mathbb{N}^n,\ 0 \le j \le p-1\}$. On the other hand,

$$\begin{aligned} \max\{|\lambda_{(i),j}^p \mid (i) \in \mathbb{N}^n,\ 0 \le j \le p-1\} &= \max_{0\le j\le p-1} (\max\{|\lambda_{(i),j}^p \mid (i) \in \mathbb{N}^n\}) \\ &= \max_{0\le j\le p-1} \|a_j c_j\|^p \end{aligned}$$

hence $\|g^p\| = \max_{0\le j\le p-1}\left(\frac{\|a_j c_j\|^p}{\|b\|^p}\right)$. Since $\|\ .\ \|$ is an absolute value on T_n, finally we have

$$\|g^p\| = \max_{0\le j\le p-1} \left(\left\|\frac{a_j}{b_j}\right\|\right)^p.$$

But $\|g\|_{pol}^{L_n} = \sqrt[p]{\|g^p\|}$, hence

$$\|g\|_{pol}^{L_n} = \max_{0\le j\le p-1} \left(\left\|\frac{a_j}{b_j}\right\|\right).$$

Thus, $L_n^{\frac{1}{p}}$ is L_n-productal with respect to $\|\ .\ \|_{pol}^{L_n}$. □

Theorem 35.4 *Let A be reduced. Then the norm of Banach K-algebra of A is equivalent to the norm $\|\ .\ \|_{si}$.*

Proof: First we suppose that A is an integral domain. Let T_n be a topologically pure extension of K such that A is finite over T_n. As a finite

T_n-module, A admits a basis $\{e_1, \dots, e_q\}$. Putting $M = \max_{1\le j\le q} \|e_j\|$, an element $f = \sum_{j=1}^{q} a_j e_j \in A$, with $a_j \in T_n$ $\forall j = 1, \dots, q$, satisfies

$$\|f\| \le M \max_{1\le j\le q} \|a_j\|. \tag{35.1}$$

Let L_n be the field of fractions of T_n. Of course $\{e_1, \dots, e_q\}$ is also a basis of $L_n[x_1, \dots, x_q]$ as a L_n-vector space. By Theorem 35.3 the algebraic closure of L_n is L_n-productal with respect to $\| \,.\, \|_{pol}^{L_n}$. So we have a constant $V > 0$ such that

$$\|f\|_{pol}^{L_n} \ge V \max_{1\le j\le q} \|a_j\|. \tag{35.2}$$

Thus, by (35.1), (35.2) and by Theorems 35.2 we obtain

$$V \max_{1\le j\le q} \|a_j\| \le \|f\|_{pol}^{L_n} = \|f\| = \|f\|_{si} \le M \max_{1\le j\le q} \|a_j\|.$$

Thereby, $\| \,.\, \|_{si}$ is clearly equivalent to $\| \,.\, \|$ on A.

We now consider the general case, when A is reduced but not necessarily an integral domain. Let $B = \prod_{j=1}^{m} \frac{A}{\mathcal{S}_j}$. Let ϕ be the K-algebra homomorphism from A into B defined as $\phi(x) = (\theta_1(x), \dots, \theta_m(x))$. Since A is reduced, the intersection of all prime ideals is null, and therefore, $\bigcap_{j=1}^{m} \mathcal{S}_j = \emptyset$. Thus, ϕ is injective and consequently, B is provided with a structure of finite A-module whose external law is defined as follows: given $g \in A$, $(f_1, \dots, f_m) \in B$, then $g(f_1, \dots, f_m)$ is defined as $(\theta_1(g)f_1, \dots, \theta_m(g)f_m)$. Then, as a A-submodule, A itself is isomorphic to the A-submodule of B consisting of the $g(f_1, \dots, f_m)$ where $f_j = 1$ $\forall j = 1, \dots, m$. Each K-algebra A_j is a K-affinoid algebra without non zero divisors of zero, hence is complete for its own spectral norm $\| \,.\, \|_{si,j}$, and of course, B is obviously complete for the product K-algebra norm defined as $\|(f, \dots, f_m)\|_B = \max_{1\le j\le m} \|f_j\|_{si,j}$.

On the other hand, the norm $\| \,.\, \|^B$, B is provided with, is a A-quasi-algebra-norm because for each $j = 1, \dots, m$, we have

$$\|\theta_j(g)f_j\|_{si,j} \le \|\theta_j(g)\|_{si,j}\|f_j\|_{si,j} \le \|g\|_{si}\|f_j\|_{si,j} \le \|g\|\|f_j\|_{si,j}.$$

Consequently, since B is finite over A, by Corollary 30.3 every A-submodule of B is closed with respect to its norm $\| \,.\, \|^B$. In particular, as a A-submodule, A is closed in B with respect to $\| \,.\, \|^B$.

Now, given $f \in A$, we have $\|f\|_{si} \ge \max_{1\le j\le m} \|\theta_j(f)\|_{si,j}$. The inequality actually is an equality. Indeed, according to Theorem 35.2, we can find $\chi \in \mathcal{X}(A)$ such that $|\chi(f)| = \|f\|_{si}$, and there does exist an index

h such that $\mathcal{S}_h \subset Ker(\chi)$, so χ factorizes in the form $\chi_h \circ \theta_h$, hence $\|\theta(f)\|_{si,h} \geq \|f\|_{si}$, so we get $\|f\|_{si} = \max_{1\leq j\leq m} \|\theta_j(f)\|_{si,j}$.

Finally, it is easily seen that the spectral norm $\| \,.\, \|_{si}^{B}$ of B is just the product norm B has been provided with, because $\mathcal{X}(B) = \prod_{j=1}^{m} \mathcal{X}(A_j)$. Thus, we have $\max_{1\leq j\leq m} \|\theta_j(f)\|_{si,j} = \|(\theta_1(f),\dots,\theta_m(f))\|_{si}^{B}$, and therefore, the spectral norm $\| \,.\, \|_{si}$ of A is induced by that of B. Consequently, A is complete for both norm $\| \,.\, \|_{si}$ and $\| \,.\, \|$, and therefore, by Theorem 3.4 the two norms are equivalent. □

By Theorems 15.14, 32.4 and 35.4 we can state Corollary 35.5:

Corollary 35.5: *Let A be reduced and let $f,\ g \in A$. Then for every $\phi \in Mult(A, \| \,.\, \|)$, we have $\delta(f^*(\phi), g^*(\phi)) \leq \|f - g\|$. Moreover, if $\varphi_{\mathcal{F}} = f^*(\phi)$, $\varphi_{\mathcal{G}} = g^*(\phi)$, and if $\phi(f-g) < diam(\sup(\mathcal{F},\mathcal{G}))$ then $f^*(\phi) = g^*(\phi)$.*

Theorem 35.6: *Let A be reduced and let $f \in A$ be such that $\widetilde{sp(f)} = U$. Then the closure of $K[f]$ in A is equal to $K\{f\}$. Let $g \in A$ be such that $\|f - g\|_{si} < 1$. Then $\widetilde{sp(g)} = U$. Moreover, if $g \in K\{f\}$, then $K\{f\} = K\{g\}$.*

Proof: Suppose first that A is reduced, and therefore admits $\| \,.\, \|_{si}$ as its norm. The restriction of the norm to $K[f]$ is the Gauss norm. Indeed, let $P(X) \in K[X]$, let $\| \,.\, \|$ be the Gauss norm on $K[X]$ and let $\mathcal{F}$ be the circular filter of center 0 and diameter 1. Then by Lemma 12.1 we have $\|P\| = \lim_{\mathcal{F}} |P(x)|$. On the other hand $\|P(f)\|_{si} = \sup_{\lambda\in sp(f)} |\lambda|$. But since $\widetilde{sp(f)} = U$, $\mathcal{F}$ is secant with $\widetilde{sp(f)}$, therefore $\lim_{\mathcal{F}} |P(x)| = \sup_{\lambda\in sp(f)} |\lambda|$, thereby we get the equality. Consequently, the closure of $K[f]$ in A is equal to $K\{f\}$. Now, consider $g \in A$ such that $\|f - g\|_{si} < 1$. Then $sp(g)$ is included in U, as $sp(f)$. And since $diam(sp(f)) = 1$, we have $diam(sp(g)) = 1$, hence of course $\widetilde{sp(g)} = U$.

Now, suppose $g \in K\{f\}$, hence g is of the form $\sum_{m=0}^{\infty} b_m f^m$, with $|b_0 - 1| < 1$, $|b_m| < 1\ \forall m \in \mathbb{N}$. Without loss of generality, we can clearly assume that $b_0 = 1$. Thus, g is of the form $f + h(f)$, with $h(Y) \in K\{Y\}$, and $\|h\| < 1$. Let $l(Y) = Y + h(Y)$. Since $K\{Y\}$ is identical to $H(U)$, and since $\|h\| < 1$, it clearly satisfies the hypothesis of Theorem 17.12, hence $l(Y)$ is a strictly injective analytic element in U, making a bijection from U onto U. Consequently we can apply Theorem 17.14 showing that, if we put $Z = l(Y)$, then Y is a strictly injective function in Z, hence of the form $Z + t(Z)$, with $t \in K\{Z\}$ and $\|t\| < 1$. Consequently, applying this to f, we see that f belongs to $K\{g\}$. □

for such that [illegible], is [illegible] of the form [illegible], hence [illegible].

Finally, it is easily seen that the spectral norm $\|\cdot\|_\sigma^B$ of B is just the product norm B has been provided with, because [illegible]. Thus, we have [illegible], and therefore the spectral norm $\|\cdot\|_\sigma$ of A is induced by that of B. Consequently A is complete for both norms $\|\cdot\|_\sigma$ and $\|\cdot\|$, and therefore, by Theorem [illegible], the two norms are equivalent. $\square$

By Theorems 15.12, 22.4 and 15.1 we can state Corollary 22.5.

Corollary 22.5. *Let A be [illegible] and let [illegible]. Then for every [illegible] [illegible] [illegible]*

Theorem 22.6. *Let K be [illegible] and let [illegible] [illegible] [illegible]. Moreover [illegible] [illegible].*

Proof. Suppose first that [illegible] is [illegible] and therefore [illegible] is complete. The restriction of the norm of $K[X]$ to the Gauss norm. Indeed, let [illegible] be the Gauss norm on $K[X]$ and let $\mathcal{S}$ be the [illegible] of [illegible]. Then by Lemma 12.7 we have [illegible]

[illegible] clearly, except the [illegible] is separable, the element [illegible] is equal to [illegible]. Now [illegible] such that [illegible]. Then [illegible] is integral [illegible] and therefore [illegible] equal [illegible] hence [illegible].

Now suppose [illegible] is of the form [illegible] with [illegible]. Without loss of generality, we can clearly assume that [illegible] of the form [illegible] and [illegible]. Since [illegible] is [illegible] of [illegible] and since [illegible] satisfies the hypotheses of Theorem 17.14, there [illegible] the element [illegible] in [illegible] than [illegible]. Consequently, we can apply Theorem 17.11 showing that there [illegible] is a [illegible] function in [illegible] of the form [illegible] with [illegible] and [illegible]. Consequently, [illegible], we see that [illegible] belongs to [illegible]. $\square$

Chapter 36

Spectrum of an Element of an Affinoid Algebra

In [40] Y. Morita proved that the spectrum $sp(x)$ of an element x of an affinoid algebra is an affinoid subset of K. However, the proof involved considerations on Tate's acyclicity Theorem [49], and spherical completion of the field K. Here we give a more simple explanation to show that such a spectrum is an affinoid set.

Notation: Keeping notation introduced in Chapter 32, here we put $\|(i_1,\dots,i_n)\|_\infty = \max(|i_1|_\infty,\dots,|i_n|_\infty)$. In K^n, (0) denotes $(0,\dots,0)$.

Proposition 36.1: *Let $F(X) = \sum_{i=0}^q f_i X^i \in T_n[X]$ be monic and irreducible in $T_n[X]$. The set of $\lambda \in K$ such that $F(\lambda)$ is not invertible is an affinoid subset of K.*

Proof: For each $i = 0,\dots,q$, we put $f_i = \sum_{(j)\in\mathbb{N}^n} a_{(j),i} T^{(j)}$. So, $F(X)$ is of the form $\sum_{(j)\in\mathbb{N}^n} b_{(j)}(X)T^{(j)}$, whereas each $b_{(j)}$ lies in $K[X]$ and satisfies $\deg(b_{(j)}) < q\ \forall (j) \neq (0)$, and $\deg(b_{(0)}) = q$ because F is monic. Moreover, since F is irreducible, we notice that there exist no $\alpha \in K$ such that $b_{(j)}(\alpha) = 0\ \forall (j) \in \mathbb{N}^n$. Further, since $\lim_{\|(j)\|_\infty\to+\infty} a_{(j),i} = 0\ \forall i = 0,\dots,q$, we have $\lim_{\|(j)\|_\infty\to+\infty} \|b_{(j)}\| = 0$. For every $\lambda \in K$ we put $B(\lambda) = \sup_{(j)\in\mathbb{N}^n} |b_{(j)}(\lambda)|$. By Theorem 31.14 $F(\lambda)$ is not invertible in T_n if and only if $|b_{(0)}(\lambda)| \leq \sup_{(j)\neq(0)} |b_{(j)}|$. Let $D = \{\lambda \in K \mid |b_{(0)}(\lambda) \leq \sup_{(j)\neq(0)} |b_{(j)}(\lambda)]\}$ and $s = \sup_{(j)\neq(0)} \|b_{(j)}\|$. We first notice that D is bounded. Indeed, let $d(0,\rho)$ be a disk containing all zeros of $b_{(0)}$, and let $r = \max(\rho, s, 1)$. Let $\lambda \in K$ be such that $|\lambda| > r$. Since all zeros of $b_{(0)}$ lie in $d(0,r)$ and since $b_{(0)}$ is monic, by Corollary 12.8 we have $|b_{(0)}(\lambda)| = |\lambda|^q$. On the other hand, for each $(j) \neq (0)$, if $t = \deg b_{(j)}$, we have the basic inequality $|b_{(j)}(\lambda)| \leq \|b_{(j)}\|.|\lambda|^t \leq \|b_{(j)}\|.|\lambda|^{q-1}$ because $|\lambda| > 1$. Consequently, $|b_{(j)}(\lambda)| \leq s|\lambda|^{q-1} < |\lambda|^q$, hence $\lambda \notin D$. This shows that $D \subset d(0, r)$.

Now, we will show that there exists a finite subset S of $\mathbb{N}^n \setminus (0)$ such that $\sup_{(j)\neq(0)} |b_{(j)}(\lambda)| = \sup_{(j)\in S} |b_{(j)}(\lambda)| \ \forall \ \lambda \in D$. Indeed, suppose this is not true. There exist injective sequences $(\lambda_m)_{m\in\mathbb{N}}$ in D and $(k_m)_{m\in\mathbb{N}}$ in $\mathbb{N}^n$ such that $\sup_{(j)\neq(0)} |b_{(j)}(\lambda_m)| = |b_{(k)_m}(\lambda_m)| \ \forall \ m \in \mathbb{N}$. Since D is bounded, by Proposition 10.7 we can extract from the sequence (λ_m) a subsequence thinner than a circular filter $\mathcal{F}$. Thus, without loss of generality, we can directly assume for convenience that the sequence (λ_m) itself is thinner than $\mathcal{F}$. Since the sequence $((k)_m)$ is injective, by definition of T_n we have $\lim_{m\to\infty} \|b_{(k)_m}\| = 0$, therefore $\varphi_{\mathcal{F}}(b_{(0)}) = 0$, hence $\varphi_{\mathcal{F}}$ is not an absolute value on $K[X]$. Consequently, by Theorem 13.1 $\mathcal{F}$ is the filter of neighborhoods of a point $a \in D$, hence $b_{(0)}(a) = 0$. But as we noticed above there exists $(h) \in \mathbb{N}^n$ such that $b_{(h)}(a) \neq 0$, hence the inequality $|b_{(h)}(x)| > |b_{(0)}(x)|$ holds in a neighborhood of a, a contradiction to the hypothesis: $\sup_{(j)\neq(0)} |b_{(j)}(\lambda_m)| = |b_{(k)_m}(\lambda_m)| \ \forall \ m \in \mathbb{N}$. Thus we have proven this existence of S.

Now, for each $(j) \in S$, we set $D_{(j)} = \{\lambda \in K \mid |b_{(0)}(\lambda)| \leq |b_{(j)}(\lambda)|\}$. Then $D = \cup_{(j)\in S} D_{(j)}$. Since $\deg(b_{(0)}) > \deg(b_{(j)}) \ \forall (j) \neq (0)$, by Theorem 12.10 each set $D_{(j)}$ is affinoid and therefore by Lemma 8.14 so is D. □

Theorem 36.2: *For all $x \in A$ $sp(x)$ is an affinoid subset of K.*

Proof: Assume that A is finite over T_n and let $x \in A$. First suppose that A is an integral domain and set $F(X) = \sum_{i=0}^q f_i X^i = irr(x, T_n)$. For each $i = 0, \ldots, q$, we put

$$f_i = \sum_{(j)\in\mathbb{N}^n} a_{(j),i} Y^{(j)}, P(X) = \sum_{i=0}^{q} a_{(0),i} X^i \text{ and } Q(X) = F(X) - P(X).$$

Now let $\lambda \in K$ and $G_\lambda(X) = irr(x-\lambda, T_n)$. One checks that $G_\lambda(0) = F(\lambda)$, so by Lemma 1.7 $x - \lambda$ is invertible in A if and only if $F(\lambda)$ is invertible in T_n. Thus, $sp(x)$ is equal to the set of $\lambda \in K$ such that $F(\lambda)$ is not invertible, and then by Proposition 36.1 this set is affinoid.

We now suppose that A is not necessarily an integral domain. By Corollary 31.5 A admits finitely many minimal prime ideals, $\mathcal{P}_1, \ldots, \mathcal{P}_k$. For each $j = 1, \ldots, k$, let $A_j = \frac{A}{\mathcal{P}_j}$ and let θ_j be the canonical surjection from A to A_j. Each algebra A_j is a K-affinoid algebra without non zero divisors of zero, hence $sp(\theta_j(x))$ is an affinoid set. Then by Theorem 1.15 we have

$sp(x) = \cup_{j=1}^{k} sp(\theta_j(x))$, and therefore by Lemma 8.14 $sp(x)$ is an affinoid set. So our claim is proven in the general case. $\square$

Theorem 36.3: *Let A be a reduced K-affinoid algebra without non trivial idempotents, whose spectral norm is denoted by $\| \cdot \|$, let $y \in A$, let $a \in sp_A(y)$, let $r \in |K| \cap]0, \|y - a\|]$ and let $S = \{y - b \mid b \in d(a, r^-)\}$. Let $c \in C(0, r)$, and let $B = \frac{A\{Z\}}{(c-(y-a)Z)A\{Z\}}$. Then A is canonically identified with a K-subalgebra of the K-affinoid algebra B, $S^{-1}A$ is dense in B, $d(a, r^-)$ is a hole of $sp_B(y)$ and for every $\chi \in \mathcal{X}(A)$ satisfying $|\chi(y)| \geq r$, χ has a unique extension to an element of $\mathcal{X}(B)$. Moreover, if $r < \|y - a\|$, then for every $\phi \in Mult(A, \| \cdot \|)$ satisfying $\phi(y) \geq r$, ϕ has a unique extension to an element of $Mult(B, \| \cdot \|)$.*

Proof: Without loss of generality we can obviously assume that $a = 0$. Let $D = sp_A(y)$. Suppose first $D \cap d(0, r^-) = \emptyset$. Then y is invertible in A and

$$\sup_{\chi \in \mathcal{X}(A)} \left|\chi\left(\frac{1}{y}\right)\right| \leq \frac{1}{r} \leq \phi\left(\frac{1}{y}\right) \leq \left\|\frac{1}{y}\right\|.$$

But by Theorem 35.2 we have

$$\sup_{\chi \in \mathcal{X}(A)} \left|\chi\left(\frac{1}{y}\right)\right| = \left\|\frac{1}{y}\right\|,$$

hence

$$\sup_{\chi \in \mathcal{X}(A)} \left|\chi\left(\frac{1}{y}\right)\right| = \frac{1}{r}, \text{ i.e. } \inf_{\chi \in \mathcal{X}(A)} |\chi(y)| = r$$

and therefore $d(0, r^-)$ is a hole of $sp_A(y)$. Consequently, we can just take $A = B$.

Now, suppose that $D \cap d(0, r^-) \neq \emptyset$. Let $\| \cdot \|^B$ be the norm of B, let $\| \cdot \|_{si}^B$ be the spectral semi-norm of B and let θ be the canonical surjection from $A\{Z\}$ to B. Then B is a K-affinoid algebra and the restriction of θ to A is an injection. Since y is clearly invertible in B we have $\theta(\frac{c}{y}) = \theta(Z)$, hence in B we have

$$\left\|\theta\left(\frac{c}{y}\right)\right\|^B = \left\|\theta\left(\frac{c}{y}\right)\right\|^B \leq 1,$$

hence $B = A\{Z\} = A\{\frac{c}{y}\}$. For each $b \in d(0, r^-)$, we see that the series $\sum_{n=0}^{\infty} \frac{b^n}{y^{n+1}}$ converges to $\frac{1}{y-b}$. So $S^{-1}A$ is included in B, and consequently $d(0, r^-) \cap sp_B(y) = \emptyset$. On the other hand, since $B = A\{\frac{c}{y}\}$, $S^{-1}A$ is obviously dense in B and more precisely, if we denote by S_0 the set of $y^n, n \in \mathbb{N}$, then $S_0^{-1}A$ is dense in B.

Consequently, if an element ϕ of $Mult(A, \| \,.\, \|)$ has continuation to an element $\widetilde{\phi}$ of $Mult(B, \| \,.\, \|^B)$, such a continuation is unique, so we can consider $Mult(B, \| \,.\, \|^B)$ as a subset of $Mult(A, \| \,.\, \|^B)$. In the same way, since all characters are continuous, if an element χ of $\mathcal{X}(A)$ has continuation to an element $\widetilde{\chi}$ of $\mathcal{X}(B)$ it is unique, so we can consider $\mathcal{X}(B)$ as a subset of $\mathcal{X}(A)$.

Let $\chi \in \mathcal{X}(A)$ satisfy $|\chi(y)| \geq r$ and let $\alpha = \chi(y)$. Since $|\alpha| \geq r$, we can define in $A\{Z\}$, an extension $\overline{\chi}$ of χ defined as $\overline{\chi}(\sum_{j=0}^{\infty} a_j Z^j) = \sum_{j=0}^{\infty} a_j \frac{c}{\alpha^j}$. Clearly $Ker(\theta)$ is included in $Ker(\chi)$, hence $\overline{\chi}$ factorizes in the form $\widetilde{\chi} \circ \theta$. Consequently, $\widetilde{\chi}$ is an extension of χ to B. Since A has no idempotents, D is an infraconnected affinoid set, hence since $r \in |K|$, all classes of $C(0,r)$, except maybe finitely many, are included in D, and therefore, are included in $sp_B(y)$. Since we have seen that $d(0,r^-) \cap sp_B(y) = \emptyset$, $d(0,r^-)$ is a hole of $sp_B(y)$.

Now, assume that $r < \|y\|$. Let $\phi \in Mult(A, \| \,.\, \|)$ satisfy $\phi(y) > r$. Let $f \in S_0^{-1}A$ and let us show that $\phi(f) \leq \|f\|^B$ with respect to the norm $\| \,.\, \|^B$. By definition, f is of the form $\frac{h}{y^q}$ $q \in \mathbb{N}$. Let $\epsilon \in]0, \min(\phi(y) - r, \phi(h))[$. By Theorem 32.4 there exists $\psi \in Mult_m(A, \| \,.\, \|)$ satisfying $|\phi(y) - \psi(y)|_\infty < \epsilon$, $|\phi(h) - \psi(h)|_\infty \leq \epsilon$, hence

$$\frac{\phi(h)}{\phi(y)^q} \leq \frac{(\psi(h) + \epsilon)}{(\psi(y) - \epsilon)^q} \leq \frac{\psi(h)}{\psi(y)^q}\Big(\frac{r}{r - \epsilon}\Big)^q \Big(\frac{\psi(h) + \epsilon}{\psi(h)}\Big).$$

But it is easily checked that

$$\frac{\psi(h) + \epsilon}{\psi(h)} \leq \frac{\phi(h) + 2\epsilon}{\phi(h) - \epsilon},$$

so we obtain

$$\frac{\phi(h)}{\phi(y)^q} \leq \frac{\psi(h)}{\psi(y)^q}\Big(\frac{r}{r - \epsilon}\Big)^q \frac{\phi(h) + 2\epsilon}{\phi(h) - \epsilon}. \tag{36.1}$$

On the other hand, since $\epsilon < \phi(y) - r$, it is clear that $\psi(y) > r$, hence ψ is of the form $|\chi|$, with $\chi \in \mathcal{X}(A)$ and then, as we just saw, χ has continuation to an element $\widetilde{\chi}$ of $\mathcal{X}(B)$. Putting $\widetilde{\psi} = |\widetilde{\chi}|$, $\widetilde{\psi}$ is an element of $Mult(B, \| \,.\, \|^B)$

which extends ψ to B. Therefore, in B we have

$$\frac{\widetilde{\psi}(h)}{\widetilde{\psi}(y)^q} = \widetilde{\psi}(f) \leq \|f\|^B,$$

hence by (36.1) we obtain

$$\frac{\phi(h)}{\phi(y)^q} \leq \|f\|^B \Big(\frac{r}{r-\epsilon}\Big)^q \frac{\phi(h)+2\epsilon}{\phi(h)-\epsilon}.$$

This last relation is true for all $\epsilon \in]0, \min(\phi(y)-r, \phi(h))[$, and actually shows $\frac{\phi(h)}{\phi(y)^q} \leq \|f\|^B$. Therefore, since $S_0^{-1}A$ is dense in B, ϕ has continuation to an element of $Mult(B, \| \,.\, \|^B)$.

Now, let $\phi \in Mult(A, \| \,.\, \|)$ satisfy $\phi(y) = r$. Since $r < \|y\|$, every neighborhoods of ϕ in $Mult(A, \| \,.\, \|)$ contains elements ψ such that $\psi(y) > r$, therefore such a ψ has extension to $Mult(B, \| \,.\, \|^B)$. Considering $Mult(B, \| \,.\, \|^B)$ as a subset of $Mult(A, \| \,.\, \|)$, we can see that the filter of neighborhoods of ϕ is secant with $Mult(B, \| \,.\, \|^B)$, and therefore has a point of adherence ϕ' in $Mult(B, \| \,.\, \|^B)$. But since the filter converges in $Mult(A, \| \,.\, \|)$, the restriction of ϕ' to A is equal to ϕ, hence ϕ' is an element of $Mult(B, \| \,.\, \|^B)$ which is the continuation of ϕ to B. □

Remark: The hypothesis $r < \|y-a\|$ assumed to proving assertions about continuation of elements of $Mult(A, \| \,.\, \|)$ to elements of $Mult(B, \| \,.\, \|^B)$ does not seem really necessary. The problem just comes from the fact that in the proof of Theorem 36.3, Theorem 32.4 provides us with a $\psi \in Mult_m(A, \| \,.\, \|)$ satisfying $|\phi(y) - \psi(y)|_\infty < \epsilon$, $\phi(h) \leq \psi(h) + \epsilon$, but if $r = \|y - a\|$ we can no longer assure that $\psi(y) \geq \phi(y)$ (which actually here would mean $\psi(y) = \phi(y)$).

Chapter 37

Krasner-Tate Algebras

In this chapter we consider a Banach K-algebra which is isomorphic to both a K-algebra of analytic elements $H(D)$, and a K-affinoid algebra. Such a Banach K-algebra called *a Krasner-Tate algebra* [18]. The issue is to characterize Krasner-Tate algebras among algebras $H(D)$ on one hand, and among affinoid algebras on the other hand.

Notation: Henceforth, for each ideal $\mathcal{J}$ of T_n, we denote by $\Lambda(\mathcal{J})$ the set of maximal ideals of T_n containing $\mathcal{J}$, and similarly, for every $h \in T_n$, we put $\Lambda(h) = \Lambda(hT_n)$.

Theorem 37.1: *Let P, $Q \in K[X]$ be relatively prime polynomials. Then the ideal $(P(X) - YQ(X))K\{Y, X\}$ of $K\{Y, X\}$ is equal to its radical.*

Proof: By Theorem 33.7 $K\{Y, X\}$ is factorial. Suppose its factorization into irreducible factors contains a factor g with a power $q > 1$. Then g divides $\frac{\partial(P(X)-YQ(X))}{\partial Y} = -Q(X)$. Consequently, g divides both P and Q in $K\{Y, X\}$. Let $\mathcal{M}$ be a maximal ideal of $K\{Y, X\}$ such that $g \in \mathcal{M}$ and let $\chi \in \mathcal{X}(K\{Y, X\})$ be such that $Ker(\chi) = \mathcal{M}$. Let $\chi(X) = \alpha$. Then we have $\chi(P(X)) = P(\alpha) = 0$, $\chi(Q(X)) = Q(\alpha) = 0$, a contradiction to the hypothesis "P and Q relatively prime". This shows that $P(X) - YQ(X)$ has no multiple irreducible factors in its decomposition. But since $K\{Y, X\}$ is factorial, then by Lemma 1.3 the ideal $(P(X) - YQ(X))K\{Y, X\}$ of $K\{Y, X\}$ is equal to its radical. $\square$

Theorem 37.2: *Let D be affinoid. Let P, $Q \in K[X]$ be relatively prime polynomials such that* $\deg(P) > \deg(Q)$, $\frac{P}{Q} \in R(D)$, $\frac{P}{Q}(D) = U$,

$\left(\frac{P}{Q}\right)^{-1}(U) = D$. *Then the K-algebra $H(D)$ is isomorphic to the K-affinoid algebra*

$$\frac{K\{Y,X\}}{(P(X) - YQ(X))K\{Y,X\}}$$

which is also isomorphic to $K\{h\}[x]$, with $h = \frac{P(x)}{Q(x)}$.

Proof: Since $\deg(P) > \deg(Q)$, in $H(D)$ x is integral over $K[h]$ and satisfies $P(x) - hQ(x) = 0$. Now, since $sp_{H(D)}(h) = h(D) = U$, the norm $\| \cdot \|_D$ induces the Gauss norm on $K[h]$. So, by Theorem 35.6 the closure of $K[h]$ in $H(D)$ is isometrically isomorphic to T_1 that we can denote here by $K\{h\}$. Without loss of generality we can obviously assume that $\|x\|_D \leq 1$. Thus in $H(D)$, x is integral over $K\{h\}$ and satisfies $P(x) - hQ(x) = 0$. Let $B = K\{h\}[x]$. By Corollary 31.4, B is a K-affinoid algebra isomorphic to a quotient of $K\{Y,X\}$ by an ideal $\mathcal{I}$ that contains $P(X) - YQ(X)$ and the canonical surjection θ from $K\{Y,X\}$ onto B satisfies $\theta(Y) = h$, $\theta(X) = x$. On the other hand, by definition B is a K-subalgebra of $H(D)$. For every $a \in K \setminus D$ since $|h(a)| > \|h\|_D$, by Theorem 6.13 $h(a) - h$ is invertible in B, hence $R(D) \subset B \subset H(D)$. In particular $sp_B(x) = sp_{H(D)}(x) = D$. We notice that $H(D)$ is reduced, hence so is B. Consequently, $\mathcal{I}$ is equal to its radical and therefore, by Theorem 35.4, B is complete for its spectral norm. But since $sp_B(x) = sp_{H(D)}(x)$ we can easily deduce that the spectral norm of B is induced by the norm $\| \cdot \|_D$. Indeed, given $\chi \in \mathcal{X}(B)$, since $\chi(x) \in D$, χ has continuation by continuity to $H(D)$, and therefore belongs to $\mathcal{X}(H(D))$, hence for all $f \in B$, the spectral norm $\| \cdot \|_{si}^B$ of B satisfies $\|f\|_{si}^B = \|f\|_D$. Therefore, since $R(D) \subset B$, we have $B = H(D)$.

Since $P(x) - YQ(X) \in \mathcal{I}$, of course $\Lambda(\mathcal{I}) \subset \Lambda(P(X) - YQ(X))$. Now, consider a maximal ideal $\mathcal{M}$ which contains $\mathcal{I}$ and let $\chi \in \mathcal{X}(K\{Y,X\})$ be such that $Ker(\chi) = \mathcal{M}$. Let $\chi(X) = \alpha$, $\chi(Y) = \beta$. Since $x - a$ is invertible whenever $a \notin D$, we see that α lies in D, and then $\chi(Y) = h(\alpha)$, thereby $P(\alpha) - h(\alpha)Q(\alpha) = 0$, hence $P(X) - YQ(X)$ belongs to $\mathcal{M}$. Consequently, we have $\Lambda(\mathcal{I}) = \Lambda(P(X - YQ(X))$. Therefore, the ideals $(P(X) - YQ(X))K\{Y,X\}$ and $\mathcal{I}$ have the same radical. Thus, on one hand, by Proposition 37.1 $(P(X) - YQ(X))K\{Y,X\}$ is equal to its radical, on the other hand, since $H(D)$ is reduced, $\mathcal{I}$ also is equal to its radical, hence $(P(X) - YQ(X))K\{Y,X\} = \mathcal{I}$. □

Theorem 37.3: *A K-algebra $H(D)$ is a Krasner-Tate algebra if and only if D is affinoid.*

Proof: Indeed, if $H(D)$ is a Krasner-Tate algebra, then by Theorem 36.2 $sp(x)$, i.e. D, is affinoid. Conversely, suppose that D is affinoid. By Theorem 12.11 there exists $h \in R(D)$ such that $h(D) = U$, $D = h^{-1}(U)$, with $h = \frac{P}{Q}$, P, Q relatively prime and such that $\deg(P) > \deg(Q)$. Consequently, Theorem 37.2 shows that $H(D)$ is a Krasner-Tate algebra isomorphic to $\frac{K\{Y,X\}}{(P-YQ)K\{Y,X\}}$. □

Corollary 37.4: *Let D be affinoid and let $f \in H(D)$. Then $f(D)$ is equal to $sp(f)$ and is affinoid.*

Proof: Indeed by Theorem 37.3 $H(D)$ is a K-affinoid algebra, hence by Theorem 36.2 $sp(f)$ is an affinoid subset of K, but since K-affinoid algebras are noetherian, by Theorem 19.2 we have $sp(f) = f(D)$. □

Another way to describe Krasner-Tate algebras $H(D)$ when the set D is not infraconnected consists of focusing on the infraconnected components of D, in order to show a product of Krasner-Tate algebras.

Theorem 37.5: *Let D be affinoid and let $D_1, \dots, D_q$ be its infraconnected components. For every $j = 1, \dots, q$, let P_j, $Q_j \in K[X]$ be relatively prime and satisfy:*

$$\frac{P_j}{Q_j} \in R(D_j),\ \frac{P_j}{Q_j}(D_j) = U,\ \left(\frac{P_j}{Q_j}(D_j)\right)^{-1}(U) = D_j.$$

Let $\mathcal{I} = \prod_{j=1}^{q}(P_j - YQ_j)K(Y,X)$.

Then $H(D)$ is isomorphic to $\frac{K\{Y,X\}}{\mathcal{I}}$.

Proof: Without loss of generality we can obviously assume that $\|x\|_D \leq 1$. For each $j = 1, \dots, q$, we denote by I_j the ideal $(P_j - YQ_j)K\{Y,X\}$ of $K\{Y,X\}$. By Theorem 19.12 $H(D)$ is isomorphic to the K-algebra $H(D_1)\times \dots \times H(D_q)$. Each set D_j is an infraconnected set, hence by Theorem 37.2 $H(D_j)$ is isomorphic to the Krasner-Tate algebra without non zero divisors of zero $\frac{K\{Y,X\}}{I_j}$. On the other hand, since $D_k \cap D_l = \emptyset\ \forall k \neq l$, we have $\Lambda(I_k + I_l) = \emptyset$, therefore $I_k + I_l = K\{Y,X\}\ \forall k \neq l$. Consequently, by Lemma 1.1 the direct product

$$\frac{K\{Y,X\}}{(P_1 - YQ_1)K\{Y,X\}} \times \dots \times \frac{K\{Y,X\}}{(P_q - YQ_q)K\{Y,X\}}$$

is isomorphic to $\frac{K\{Y,X\}}{I}$, which completes the proof. □

Remark: Given a Krasner-Tate algebra $\frac{K\{Y,X\}}{(P-YQ)K\{Y,X\}}$, the form shown in Theorem 37.3 is not the only one possible. Indeed, consider the

K-affinoid algebra

$$A = \frac{K\{Y,Z\}}{(Z^3 + 4Z^2Y + 5ZY^2 + 2Y^3 + Z + 1)K\{Y,Z\}}.$$

Now, let $X = Z + Y$. Then A becomes

$$\frac{K\{Y,X\}}{[(X^3 + X + 1 - Y(-X^2 + 1)]K\{Y,X\}}.$$

Thus, A appears as a Krasner-Tate algebra $H(D)$ where $D = h^{-1}(U)$ with $h = \frac{x^3+x+1}{-x^2+1}$.

Now, consider the affinoid algebra $B = \frac{K\{Y,Z\}}{(Z^2-YZ)K\{Y,Z\}}$. Apparently, it looks like a Krasner-Tate algebra, except that the polynomials $P = Z^2$ and $Q = Z$ are not relatively prime. Of course, B has non zero divisors of zero. Putting $X = Y - Z$, we obtain $B = \frac{K\{Y,X\}}{(XY)K\{Y,X\}}$, so we can check that such a K-algebra has no idempotent different from 0 and 1. Suppose B is a Krasner-Tate algebra $H(D)$. Since B has n non trivial idempotent, D is infraconnected, hence by Theorem 19.2 we know that B not noetherian. Consequently, it can't be a Krasner-Tate algebra.

Lemma 37.6: *Let $A = \frac{K\{Y_1,\ldots,Y_n\}}{I}$ be a K-affinoid algebra without non zero divisors of zero such that the field of fractions of A is a pure degree one transcendental extension $K(t)$ of K. Let θ be the canonical surjection from $K\{Y_1,\ldots,Y_n\}$ onto A. For each $j = 1,\ldots,n$, let $y_j = \theta(Y_j) = h_j(t)$, with $h_j(t) \in K(t)$. If t belongs to A, then $sp(t)$ contains no poles of h_j and no points a such that $h'_j(a) = 0\ \forall j = 1,\ldots,n$.*

Proof: Suppose that $t \in A$ and let $D = sp(t)$. Suppose that for some $k \in \{1,\ldots,n\}$, h_k admits a pole b in D. Let $h_k = \frac{P(t)}{Q(t)}$, with P, Q relatively prime and let $\chi \in \mathcal{X}(A)$ be such that $\chi(t) = b$. Then $\chi(P(t)) = \chi(y_k)\chi(Q(t)) = \chi(y_k)Q(b) = 0$, a contradiction to the hypothesis: P, Q relatively prime. Consequently no poles of the h_j lie in $sp(t)$, whenever $j = 1,\ldots,n$. On the other hand let $F(Y_1,\ldots,Y_n) \in K\{Y_1,\ldots,Y_n\}$ be such that $t = F(y_1,\ldots,y_n)$. Consider the mapping G from $sp(t)$ into K^n defined as $G(\alpha) = F(h_1(\alpha),\ldots,h_n(\alpha))$. Then G is the identical mapping on $sp(t)$. If there exists $\alpha \in sp(t)$ such that $h'_j(\alpha) = 0\ \forall j = 1,\ldots,n$, then we have $G'(\alpha) = 0$, a contradiction since G is the identical mapping on $sp(t)$. □

Theorem 37.7: *The K-affinoid algebra $A = \frac{K\{X,Y\}}{(Y^2-X^3)K\{X,Y\}}$ is an integral domain but is not integrally closed.*

Proof: Let θ be the canonical surjection from $K\{X,Y\}$ onto A, let $x = \theta(X)$, $y = \theta(y)$. Suppose that A is has non zero divisors of zero. Since by Theorem 33.7 $K\{X,Y\}$ is factorial, there exist non invertible elements f_1, $f_2 \in K\{X\}[Y]$ such that $f_1 f_2 = Y^2 - X3$. Then, by Theorem 33.3 each f_j is associated in $K\{X,Y\}$ to a monic polynomial in Y, $P_j(Y)$. Since both P_j are not invertible in $K\{X,Y\}$, each P_j is of the form $Y + S_j(X)$, with $S_j(X) \in K\{X\}$. Consequently, we have $Y^2 - X^3 = Y^2 + Y(S_1(X) + S_2(X)) + S_1(X)S_2(X)$, hence by identification, $S_2 = -S_1$, and $S_1^2 = X^3$. But X^3 is not a square in $K\{X\}$ because it has a zero of order 3 at 0. Consequently we are led to a contradiction proving that $Y^2 - X^3$ is irreducible in $K\{X,Y\}$, and therefore A is an integral domain.

Suppose now that A is integrally closed. Let $B = K[x,y]$. The field of fractions of B is clearly the field $K(u)$ with $u = \frac{y}{t}$. And since u is integral over B, it belongs to A. On the other hand, both x, y are not invertible in A, hence neither is u. But putting $x = g(u)$, $y = h(u)$ we have $g'(u) = h'(u) = 0$, a contradiction with regards to Lemma 37.6. □

Corollary 37.8: *The K-affinoid algebra $A = \frac{K\{X,Y\}}{(Y^2-X^3)K\{X,Y\}}$ is not a Krasner-Tate algebra.*

Proof: Indeed, a Krasner-Tate algebra without non zero divisors of zero is a principal ideal ring. □

Proposition 37.9 will be useful in further consideration in chapter 39:

Proposition 37.9: *Let $r \in]0,1[\cap |K|$, let $E = d(0,r)$, $F = d(1,r)$ and let $D = E \cup F$. Let u be the characteristic function of E in $H(D)$, and $A = \frac{H(D)}{ux^2 H(D)}$. Then A is a non reduced K-affinoid algebra containing a dense Luroth K-algebra.*

Proof: Since D is an affinoid subset of K, $H(D)$ is a K-affinoid algebra, hence so is A. Let θ be the canonical surjection of $H(D)$ onto A. Then, $\theta(ux)$ is a nilpotent element of A, hence A is not reduced. But since $H(D)$ is a Krasner-Tate algebra, it is of the form $K\{h\}[x]$, with $h \in K(x), \deg(h) > 0$, and $h(D) = U$, $D = h^{-1}(U)$, so $K[h,x]$ is a dense K-subalgebra of $H(D)$. Let $\tau = \theta(h)$, $\xi = \theta(x)$. Then $Ker(\theta) \cap K\{h\}$ is equal to $\{0\}$, and so is $Ker(\theta) \cap K[h,x]$. Consequently, $K[\tau, \xi]$ is a Luroth K-subalgebra of A isomorphic to $K[h,x]$, and is dense in A. □

Chapter 38

Universal Generators in Tate Algebras

It is possible to find another way of characterizing Krasner-Tate algebras among K-affinoid algebras, by using universal generators, a study made in [22].

Proposition 38.1: *Let A be a K-affinoid algebra without non zero divisors of zero, of the form $K\{Y\}[x]$, with x integral over $K\{Y\}$, let*

$$\begin{aligned}F(Y,X) &= \sum_{j=0}^{n} f_j(Y)\lambda^j \\ &= X^n + \sum_{i=0}^{n-1}(a_i + b_iY + Y^2\epsilon_i(Y))X^i \\ &= irr(x, K\{Y\}) \in K\{Y\}[X]\end{aligned}$$

and suppose that $\|F\| = 1$.

Let $P(X) = X^n + \sum_{i=0}^{n-1} a_iX^i$ and let $Q(X) = \sum_{i=0}^{n-1} b_iX^i$. Then Conditions (i), (ii), (iii), (iv) are equivalent:

(i) x is a universal generator;

(ii) for every $\lambda \in K$, $\sum_{j=0}^{n} \lambda^j f_j$ is either invertible or irreducible in $K\{Y\}$;

(iii) for every $\lambda \in K$, $\max(|P(\lambda)|, |Q(\lambda)| > \| \sum_{j=0}^{n-1} \lambda^j \epsilon_j(Y)\|$;

(iv) for every $\lambda \in sp(x)$, $|Q(\lambda)| > \| \sum_{j=0}^{n-1} \lambda^j \epsilon_j(Y)\|$ and $|Q(\lambda)| \geq |P(\lambda)|$.

Further, if the above conditions are satisfied, then

$$sp(x) = \{\lambda \in K \mid |P(\lambda)| \leq |Q(\lambda)|\}.$$

Proof: For each $\lambda \in K$, we denote by $G_\lambda(X) = X^n + \sum_{j=0}^{n-1} g_{\lambda,j}(Y)X^j$ the minimal polynomial of $x - \lambda$ over $K\{Y\}$. We can easily check that $g_{\lambda,0} = F(Y,\lambda) = \sum_{j=0}^{n} f_j(Y)\lambda^j$. By Theorem 19.2 $K\{Y\}$ is principal, so the equivalence between (i) and (ii) are immediate. Now, by Theorem 17.6 $F(Y,\lambda)$ is invertible in $K\{Y\}$ if and only if $|P(\lambda)| > \max(|Q(\lambda)|, \sup_{0\leq j\leq n-1}(\|\epsilon_j(Y,\lambda)\|))$. In the same way, $F(Y,\lambda)$ is irreducible in $K\{Y\}$ if and only if $|P(\lambda)| \leq |Q(\lambda)|$ and $|Q(\lambda)| > \sup_{0\leq j\leq n-1}(\|\epsilon_j(Y,\lambda)\|))$. Thus, Condition (iii) is equivalent to "for every $\lambda \in sp(x)$, we have $|P(\lambda)| \leq |Q(\lambda)|$ and $|Q(\lambda)| > \sup_{0\leq j\leq n-1}(\|\epsilon_j(Y,\lambda)\|)$", which shows that Condition (iii) is equivalent to Conditions (ii) and (iv).

Now, suppose the four conditions satisfied. Since every maximal ideal of A has codimension 1, given $\lambda \in K$, it lies in $sp(x)$ if and only if there exists $\chi \in \mathcal{X}(A)$ such that $\chi(x) = \lambda$, i.e. there exists $\mu \in U$ such that $F(\mu,\lambda) = 0$. Let $\Omega = \Lambda(F)$, let $\Omega' = \Lambda(P(X) - YQ(X))$, and let π be the canonical projection from $U \times U$ to U defined as $\pi(\mu,\lambda) = \lambda$. Then $\pi(\Omega) = sp(x)$, and $\pi(\Omega') = \{\lambda \mid |P(\lambda)| \leq |Q(\lambda)|\}$. On the other hand, given $\lambda \in K$, $x - \lambda$ is not invertible if and only if (by (iii)) if $|Q(\lambda)| \geq |P(\lambda)|$, i.e. $\pi(\Omega') = sp(x)$ and this shows the last statement. □

Remark: Proposition 38.1 gives a piece of answer to this general problem: given the quotient of $K\{Y,X\}$ by two different principal ideals, are these quotients be isomorphic when the generators are very close?

Lemma 38.2: *Let A be a K-affinoid algebra without non zero divisors of zero, of the form $K\{Y\}[x]$, with x integral over $K\{Y\}$ and let $F(Y,X) = irr(x, K\{Y\}) \in K\{Y\}[X]$. Let $\mathcal{I}$ be an ideal of $K\{Y,X\}$ containing $F(Y,X)$ and included in infinitely many maximal ideals of A.*

Then $\mathcal{I} = F(Y,X)K\{Y,X\}$.

Proof: Without loss of generality, we can assume that $\|x\|$ is small enough to assure that $\|F(Y,X)\| \leq 1$. Since A is an integral domain, it is isomorphic to $\frac{K\{Y\}[X]}{F(Y,X)K\{Y\}[X]}$, and $F(Y,X)$ is irreducible in $K\{Y\}[X]$. But then, by Theorem 33.5 it is irreducible in $K\{Y,X\}$, hence $F(Y,X)K\{Y,X\}$ is a prime ideal of $K\{Y,X\}$. Since by Theorem 32.7 $K\{Y,X\}$ is a Jacobson ring, then $F(Y,X)K\{Y,X\}$ is equal to $\bigcap_{\mathcal{M}\in\Lambda(F)} \mathcal{M}$. And since $K\{Y,X\}$ is noetherian, the family of infinite sets $\Lambda(\mathcal{J})$ contained in $\Lambda(\mathcal{I})$ obviously admits minimal elements. Let N be such a minimal element, and let $\mathcal{S} = \bigcap_{\mathcal{M}\in N} \mathcal{M}$. Then $\mathcal{S}$ contains $\mathcal{I}$. Suppose that two elements $f,\ g \in K\{Y,X\}$ satisfy $fg \in \mathcal{S}, f \notin \mathcal{S}, g \notin \mathcal{S}$. Then the ideal generated by f and $\mathcal{S}$ strictly contains $\mathcal{P}$, hence $\Lambda(f)$ does not contain L, therefore

$\Lambda(f) \cap N$ is finite. Similarly, $\Lambda(g) \cap N$ is finite, and so is $\Lambda(g) \cap N$. But this is a contradiction to the hypothesis $fg \in \mathcal{S}$. Consequently, $\mathcal{S}$ is a prime ideal of $K\{Y, X\}$ which contains $\mathcal{I}$.

Let $\mathcal{S}' = \mathcal{S} \cap K\{Y\}$. Suppose first $\mathcal{S}' \neq \{0\}$. Since $\mathcal{S}'$ is a prime ideal of $K\{Y\}$ and since $K\{Y\}$ is isomorphic to $H(U)$, by Theorem 19.2 $\mathcal{S}'$ is of the form $(Y - a)K\{Y\}$, with $a \in U$. For every $\mathcal{M} \in N$, the unique $\chi \in \mathcal{X}(K\{Y, X\})$ such that $Ker(\chi) = \mathcal{M}$ must satisfy $F(a, \chi(x)) = 0$, an equation that only has finitely many solutions, and therefore L is finite, a contradiction. Consequently, we have $\mathcal{S}' = \{0\}$. Since $\mathcal{I} \subset \mathcal{S}$, there exists a canonical homomorphism θ from A onto $\frac{K\{Y,X\}}{\mathcal{S}}$. But since $\mathcal{S} \cap K\{Y\} = \{0\}$, $\theta(x)$ is integral over $K\{Y\}$ and then $irr(\theta(x), K\{Y\})$ divides F in $K\{Y\}[X]$. But since F is irreducible in $K\{Y\}[X]$, $irr(\theta(x), K\{Y\})$ is just $F(Y, X)$, hence $\mathcal{S} = F(Y, X)K\{Y, X\}$, and therefore $\mathcal{I} = \mathcal{S} = F(Y, X)K\{Y, X\}$. □

Theorem 38.3: *K is supposed to have characteristic* 0. *Let A be a K-affinoid algebra without non zero divisors of zero of the form $K\{Y\}[x]$ where x is a universal generator of A. Let*

$$F(Y, X) = X^n + \sum_{i=0}^{n-1} (a_i + b_i Y + Y^2 \epsilon_i(Y)) X^i$$
$$= irr(x, K\{Y\}) \in K\{Y\}[X]$$

and suppose that $\|F\| = 1$.

Let $P(X) = X^n + \sum_{i=0}^{n-1} a_i X^i$ let $Q(X) = -\sum_{i=0}^{n-1} b_i X^i$, and let $D = sp(x)$. Then A is isomorphic to both $H(D)$ and $\frac{K\{Y,X\}}{(P(X) - YQ(X))K\{Y,X\}}$.

Proof: Since A is an integral domain, by Theorem 35.4 we can consider that the norm of A is its spectral norm $\| \ . \ \|_{si}$. By Condition (iv) in Proposition 38.1 Q has no zero in D, so $\frac{P}{Q}$ belongs to $R(D)$. Then $Q(x)$ doesn't lie in any maximal ideal of A because if there existed $\chi \in \mathcal{X}(A)$ such that $\chi(Q(x)) = 0$, then we would have simultaneously $Q(\chi(x)) = 0$ and $\chi(x) \in D$. Consequently, $Q(x)$ is invertible in A. Let $Z = \frac{P(x)}{Q(x)}$. For each $\alpha \in D$, we put

$$u_\alpha(Y) = \frac{\sum_{i=0}^{n-1} \alpha^i \epsilon_i(Y)}{Q(\alpha)} \in K\{Y\}.$$

By Relation (iv) in Proposition 38.1 we have

$$\left\| \frac{\sum_{i=0}^{n-1} \alpha^i \epsilon_i(Y)}{Q(\alpha)} \right\| < 1 \ \forall \alpha \in D.$$

Let $f(Y) = \frac{\sum_{i=0}^{n-1} x^i \epsilon_i(Y)}{Q(x)}$. Consider $\chi \in \mathcal{X}(A)$ and let $\chi(x) = \alpha$. Then

$$|\chi(f)| = \Big|\frac{\sum_{i=0}^{n-1} \alpha^i \epsilon_i(\chi(Y))}{Q(\alpha)}\Big| \leq \Big\|\frac{\sum_{i=0}^{n-1} \alpha^i \epsilon_i(Y)}{Q(\alpha)}\Big\| < 1.$$

Consequently $|\chi(f)| < 1 \ \forall \chi \in \mathcal{X}(A)$. Therefore by Proposition 31.9 we have $\|f(Y)\|_{si} < 1$. On the other hand, we have $Z = Y - f(Y)$ hence by Theorem 35.6 $K\{Z\}$ is a topologically pure extension of K and of course its residue class ring is equal to this of $K\{Y\}$. Consequently by Proposition 30.6 A is finite over $K\{Z\}$. Let $B = K\{Z\}[x]$. Since P, Q are relatively prime, by Theorem 37.2 B is Krasner-Tate algebra, and since it has no non zero divisors of zero, it is a principal ring, hence in particular, it is integrally closed. Of course, A is finite over B so, by Theorem 1.19 x is a spectrally injective element of B. As a Krasner-Tate algebra B is semi-simple. Consequently, by Theorem 1.19 we have $A = B$. Now, by Theorem 37.2 $K\{Z\}[x]$ is the Krasner-Tate algebra $H(D')$, with $D' = Z^{-1}(U)$. And by Proposition 38.1 we have $D' = D$. □

Proposition 38.4: *Let A be a K-affinoid algebra without non zero divisors of zero of the form $K\{Y\}[x]$ where x is a spectrally injective element and let $F(Y,X) = irr(x, K\{Y\}) \in K\{Y\}[X]$. If x is not a universal generator of A, then we have $p \neq 0$, $\frac{\partial F}{\partial Y} = 0$, and further, there exists $t \in \mathbb{N}$ such that $K\{Y^{p^t}\}[x]$ is a Krasner-Tate algebra admitting x as a universal generator.*

Proof: Let $F(Y,X) = \sum_{j=0}^{n} f_j X^j$. For all $\lambda \in K$ we put

$$F(Y,\lambda) = \sum_{m=0}^{\infty} a_{\lambda,m} Y^m.$$

Suppose x is not a universal generator. By Proposition 38.1 there exists $\nu \in K$ such that $F(Y,\nu)$ is neither invertible nor irreducible in $K\{Y\}$, therefore, by Theorem 17.2 there exists $q > 1$ such that

$$|a_{\nu,q}| \geq |a_{\nu,j}| \ \forall j < q \tag{38.1}$$

Let $r \in]0, \frac{|a_{\nu,q}|}{|\nu| \max_{0 \leq i \leq n}(\|f_i\|)}[$. Now, for every $j < q$, and for all $\lambda \in d(\nu, r)$, it easily seen that $|a_{\lambda,q}| = |a_{\nu,q}|$ and that $|a_{\lambda,j}| < |a_{\nu,q}|$, hence $F(Y,\lambda)$ is neither invertible, nor irreducible for all $\lambda \in d(\nu,r)$. Let $\lambda \in d(\nu,r)$ and suppose that $F(Y,\lambda)$ admits two different zeros α and β in U. Let χ', $\chi'' \in \mathcal{X}(K\{Y\})$ be such that $\chi'(Y) = \alpha$, $\chi''(Y) = \beta$. Since $A = K\{Y\}[x]$, by Lemma 1.18 we can define $\widetilde{\chi}', \widetilde{\chi}'' \in \mathcal{X}(A)$ such that $\widetilde{\chi}'(g) =$

$\chi'(g)\ \forall g \in K\{Y\}$, and $\chi(x) = \lambda$, and similarly, $\widetilde{\chi}''(g) = \chi''(g)\ \forall g \in K\{Y\}$ and $\chi(x) = \lambda$. Thus, $x - \lambda$ lies in $Ker(\widetilde{\chi}') \cap Ker(\widetilde{\chi}'')$, a contradiction to the hypothesis "x spectrally injective". Consequently, for each $\lambda \in d(\nu, r)$, $F(Y, \lambda)$ admits a unique zero $\alpha(\lambda)$. But then, by (38.1), $\alpha(\lambda)$ is a zero of order $\geq q$ for every $\lambda \in d(\nu, r)$. Consequently, we have $\sum_{j=0}^{n} \lambda^i f_j'(\alpha(\lambda)) = 0\ \forall \lambda \in d(\nu, r)$. Therefore, the set of maximal ideals of $K\{Y, X\}$ which contain both F and $\frac{\partial F}{\partial Y}$ is infinite. But since A is an integral domain, by Lemma 38.2 the ideal generated by $\frac{\partial F}{\partial Y}$ and F in $K\{Y, X\}$ is just equal to $F(Y, X)K\{Y, X\}$. Consequently, F must divide $\frac{\partial F}{\partial Y}$ in $K\{Y, X\}$. Hence, by Corollary 7.6 F must divide $\frac{\partial F}{\partial Y}$ in $K\{Y\}[X]$, which obviously implies $\frac{\partial F}{\partial Y} = 0$ because $\frac{\partial F}{\partial Y}$ is a polynomial of degree $q - 1$. Hence, we have $f_i' = 0\ \forall i = 0, \ldots, n$, and therefore each f_i is of the form g_i^p, with $g \in K\{Y\}$. Now, putting $Y_1 = Y^p$, and $F_1(Y_1, X) = F(Y, X)$, by Theorem 1.19 x is a spectrally injective element of $K\{Y_1\}[x]$. Thus, if x is not a universal generator of $K\{Y_1\}[x]$ we are lead to the same problem with Y_1 instead of Y, and F_1 instead of F. Finally, since F is not identically 0, by Theorem 16.10 we can certainly find a power p^t such that, when putting $Y_t = Y^{p^t}$ and $F_t(Y_t, X) = F(Y, X)$, we obtain $\frac{\partial F_t}{\partial Y_t} \neq 0$. Consequently, x is a universal generator of $K\{Y_t\}[x]$, and thereby this K-algebra is a Krasner-Tate algebra by Theorem 38.3. □

Theorem 38.5: *The field K is supposed to have characteristic* 0. *Let A be a K-affinoid algebra without non zero divisors of zero, finite over a topologically pure extension $K\{Y\}$, admitting a spectrally injective element x, and let $D = sp(x)$. Then A is a Krasner-Tate algebra equal to $K\{Y\}[x]$ and isomorphic to $H(D)$ and x is a universal generator.*

Theorem 38.6: *Let A be a K-affinoid algebra without non zero divisors of zero, finite over a topologically pure extension $K\{Y\}$, admitting a universal generator x and let $D = sp(x)$. Then A is a Krasner-Tate algebra equal to $K\{Y\}[x]$ and isomorphic to $H(D)$.*

Proof of Theorems 38.5 and 38.6 (general considerations): Without loss of generality, we can obviously assume that $D \subset U$. Then putting $F(X) = irr(x, K\{Y\})$ we have $\|F\| = 1$. Let $B = K\{Y\}[x]$. By Theorem 1.19 x is a spectrally injective element of B. □

Proof of Theorem 38.5: Since K has characteristic 0, by Proposition 38.4, x is a universal generator of B. Hence by Theorem 38.3 B is a Krasner-Tate algebra and therefore is isomorphic to $H(D)$. Then since K has characteristic 0, by Theorem 1.19 we have $A = B$. □

Proof of Theorem 38.6: Let $D = sp_A(x)$ and let $B = H(D)$. Since A is an integral domain, and since A is complete for its spectral norm, B is clearly the completion of the K-subalgebra $R(D)$ of A. Suppose $B \neq A$ and let $g \in A \setminus B$. By Theorem 1.19 the characteristic p of K is different from 0, and g is of the form $\sqrt[p^t]{f}$, with $f \in A$ and $t \in \mathbb{N}^*$. Now, let $\alpha \in D$. Since x is a universal generator, and since $g - g(\alpha)$ vanishes at α, clearly f admits α as a zero of order p^t. Consequently, f' is identically zero, hence by Theorem 16.10 $\sqrt[p]{f}$ belongs to B. Putting $f_1 = \sqrt[p]{f}$, we have $g = \sqrt[p^{t-1}]{f_1}$, and for the same reason, we can see that every point $\alpha \in D$ is a zero of order p^{t-1} of f_1, hence $\sqrt[p]{f_1}$ also belongs to B. Thus, by induction, we can see that g belongs to B. □

Theorem 38.5 lets us find a sufficient condition to assure that the inverse mapping of an injective analytic element is an analytic element, an old and general problem which usually has no solution as shows the classic example of the injective function $(1+x)^q$ defined in $d(0, 1^-)$, with values in the same disk, whereas the inverse mapping of which does not belong to $H(d(0, 1^-))$.

Theorem 38.7: *K is supposed to have characteristic 0. Let D be an infraconnected affinoid subset of K and let $f \in H(D)$ be injective in D. Then $f(D)$ is affinoid and f^{-1} belongs to $H(f(D))$.*

Proof: Since $H(D)$ is Krasner-Tate algebra, $f(D)$ is affinoid by Theorem 36.2. Since f is an injective function, it is a spectrally injective element of the Krasner-Tate algebra $H(D)$. Then, since K has characteristic 0, by Theorem 38.4 f is a universal generator of $H(D)$. On the other hand, as a Krasner-Tate algebra, $H(D)$ is of the form $K\{Y\}[x]$, where $K\{Y\}$ is a topologically pure extension of K and x is the identical mapping on D. Now, since all maximal ideals of $H(D)$ have codimension 1, by Theorem 19.2 we know that $sp(f) = f(D)$, so by Theorem 38.5 we have $H(D) = K\{Y\}[f]$, so $H(D)$ is isomorphic to $H(f(D))$ because $sp(f) = f(D)$. Then, the homomorphism ϕ from $H(f(D))$ to $H(D)$ defined as $\phi(h) = h \circ f$ is an isomorphism. Consequently, by Corollary 17.10 we know that f^{-1} belongs to $H(f(D))$. □

Chapter 39

Mappings from $H(D)$ to the Tree $Mult(K[x])$

We show that every element $f \in H(D)$ has continuation to a mapping f^* from $Mult(H(D), \| \, . \, \|_D)$ to $Mult(K[x])$. Given a circular filter $\mathcal{F} \in \phi(D)$ the mapping that associates to each $f \in H(D)$ the multiplicative semi-norm $f^*(\mathcal{F})$ is uniformly continuous with respect to the norm of $H(D)$ and the metric δ on $Mult(K[x])$.

Lemma 39.1: *Let $h \in K(x)$ and let $\mathcal{F}$ be a circular filter of center a and diameter r such that $d(a, r^-)$ contains no pole of h. Then $h(\mathcal{F})$ is a base of the circular filter of center $h(a)$ and diameter $\varphi_{\mathcal{F}}(h - h(a))$.*

Proof: Let $s = \varphi_{\mathcal{F}}(h - h(a))$, let $b = h(a)$ and let $\mathcal{H}$ be the circular filter of center $h(a)$ and diameter s. Let $\mathcal{F}'$ be the filter admitting for base $h(\mathcal{F})$, let $A \in \mathcal{F}$ be an infraconnected affinoid set and let $B = h(A)$. There obviously exists $r' \in]0, r[$ and $r'' > r$ such that h has neither any zero nor any poles in $\Gamma(a, r', r) \cup \Gamma(a, r, r'')$. Suppose first that $r \notin |K|$. Then, h has neither any poles nor any zeros inside $\Gamma(a, r', r'')$, and therefore by Corollary 17.5 $h(\Gamma(a, r', r'')$ is an annulus $\Gamma(b, s', s'')$, with $s' < s < s''$. So, B belongs to $\mathcal{H}$.

Now, suppose that $r \in |K|$. Then in each annulus $\Gamma(a, r', r)$ and $\Gamma(a, r, r'')$, $h(x) - b$ is equal to a Laurent series whose constant coefficient is 0. Then by Corollary 17.5 $h(\Gamma(a, r', r))$ is an annulus of the form $\Gamma(b, s', s)$ with $s' < s$, hence B contains $\Gamma(b, s', s)$. We will show that in each class $d(c, s^-)$ of $d(b, s)$ there exists $\alpha \in B$. Since $r \in |K|$, by Lemma 12.1 we know that $s \in |K|$, and therefore without loss of generality we can assume that $r = s = 1$, and of course $a = b = 0$. Suppose that a class $d(c, 1^-)$ of U is such that $d(c, 1^-) \cap B = \emptyset$. Since $a = b = 0$ and $r = s = 1$, h is of the form $\frac{P(x)}{Q(x)}$ with $(P,Q)=1$ and $\|P\|=\|Q\|=1$. Let $\mathcal{K}$ be the residue class field of K and for all $\lambda \in U$, let $\overline{\lambda}$ be the class of λ in $\mathcal{K}$. Similarly, for every $P \in K[x]$, we denote by $\overline{P}$ the residue class of P in $\mathcal{K}[x]$. Then the equation $\overline{P}(x) = \overline{c}\overline{Q}(x)$ has no solutions, a contradiction to the fact that $\mathcal{K}$ is algebraically closed. Consequently, $B \cap d(c, s^-)$ is not empty whenever $c \in d(b,s)$. In the same way, through a change of variable of the form $u = \frac{1}{x-a}$ we can show that $B \cap (K \setminus d(b,s)) \neq \emptyset$. Consequently, by Lemma 10.2 B lies in $\mathcal{H}$, and therefore $\mathcal{F}' = \mathcal{H}$. □

Proposition 39.2: *Let $\mathcal{F}$ be a large circular filter on K and let $h \in K(x)$. Then $h(\mathcal{F})$ is a base of a circular filter denoted by $h_*(\mathcal{F})$, such that $\varphi_{h_*(\mathcal{F})}(f) = \varphi_{\mathcal{F}}(f \circ h)\ \forall f \in K(x)$.*

Proof: Let $r = diam(\mathcal{F})$. Suppose first that $\mathcal{F}$ has a center a. Without loss of generality we can assume that h has no poles in $d(a, r^-)$ because there does exist center b such that $d(b, r^-)$ contains no poles of h. Then by Lemma 39.1 $h(\mathcal{F})$ is a base of the circular filter $\mathcal{G}$ of center $h(a)$ and diameter $\varphi_{\mathcal{F}}(h - h(a))$. Then for every $f \in K[x]$, we have $\lim_{\mathcal{G}} |f(u)| = \lim_{h(\mathcal{F})} |f(u)| = \lim_{\mathcal{F}} |f(h(x))|$. Now, suppose $\mathcal{F}$ has no center and let $(D_n)_{n\in\mathbb{N}}$ be a canonical base of $\mathcal{F}$. Then $(h(D_n))_{n\in\mathbb{N}}$ is a decreasing sequence of disks and is a base of a filter $\mathcal{F}'$ thinner than a circular filter $\mathcal{H}$. But the adherence of a circular filter is empty, hence so is the adherence of $\mathcal{H}$, and so much the more the adherence of $\mathcal{F}'$, hence $\cap_{n=0}^{\infty} h(D_n) = \emptyset$. Therefore, the sequence $(h(D_n))_{n\in\mathbb{N}}$ is a base of a circular filter with no center, thinner than $\mathcal{H}$, hence it is $\mathcal{H}$. □

Lemma 39.3: *Let $h \in K(x)$ have no zero and no poles in an annulus $\Gamma(a,r,s)$ and be equal to a Laurent series $\sum_{-\infty}^{+\infty} a_n(x-a)^n\ \forall x \in \Gamma(a,r,s)$. Assume that there exists an integer $m \in \mathbb{Z}^*$ such that $|a_m|\rho^m > |a_n|\rho^n\ \forall n \neq m$, $\forall \rho \in]r,s[$. Let $\mathcal{F}$ (resp. $\mathcal{G}$) be the circular filter of center a and diameter r (resp. s). Then, $h_*(\mathcal{F})$ (resp. $h_*(\mathcal{G})$) is the circular filter of center 0 and diameter $|a_m|r^m$ (resp. $|a_m|s^m$).*

Proof: By Lemma 39.1 $h(\mathcal{F})$ is a base of a circular filter $\mathcal{F}'$. Let $(a_n)_{n\in N}$ be a sequence of $\Gamma(a,r,s)$ such that $|a_n - a| > |a_{n+1} - a|\ \forall n \in \mathbb{N}$ and $\lim_{n\to\infty} |a_n| = r$. Then the sequence $h(a_n)_{n\in\mathbb{N}}$ is thinner than $\mathcal{F}'$. But since $h(x)$ is equal to the Laurent series $\sum_{-\infty}^{+\infty} a_n(x-a)^n\ \forall x \in \Gamma(a,r,s)$ and since there exists an integer $m \in \mathbb{Z}^*$ such that $|a_m|\rho^m > |a_n|\rho^n\ \forall n \neq m$, $\forall \rho \in]r,s[$, we can see that the sequence $|h(a_n)|_{n\in\mathbb{N}}$ is strictly decreasing and tends to a limit $r' = |a_m| f^m$. Consequently, $\mathcal{F}'$ is the circular filter of center 0 and diameter r'. Similarly, $h(\mathcal{G})$ is a base of a circular filter $\mathcal{G}'$. By considering a sequence $(b_n)_{n\in\mathbb{N}}$ of $\Gamma(a,r,s)$ such that $|b_n - a| < |b_{n+1} - a|$ and $\lim_{n\to\infty} |b_n| = s$, we can show that $\mathcal{G}'$ is the circular filter of center 0 and diameter $s' = \lim_{n\to\infty} |h(b_n)| = |a_m|s^m$. □

Corollary 39.4: *Let $h \in K(x)$ have no poles in an annulus $\Gamma(a,r,s)$ and be equal to a Laurent series $\sum_{-\infty}^{+\infty} a_n(x-a)^n$ $\forall x \in \Gamma(a,r,s)$. Assume that $|a_m|\rho^m > |a_n|\rho^n$ $\forall n \neq 0, m$, $\forall \rho \in]r,s[$. Let $\mathcal{F}$ (resp. $\mathcal{G}$) be the circular filter of center a and diameter r (resp. s). Then, $h_*(\mathcal{F})$ (resp. $h_*(\mathcal{G})$) is the circular filter of center a_0 and diameter $|a_m|r^m$ (resp. $|a_m|s^m$).*

Lemma 39.5: *Let $\mathcal{F}$, $\mathcal{G}$ be two circular filters having a center a, such that $\mathcal{F} \prec \mathcal{G}$, let Ξ be the class of $\mathcal{Q}(\mathcal{G})$, $\mathcal{F}$ is secant with. Let $h \in K(x)$ have no zeros and no poles in $\Xi \setminus \mathcal{Q}(\mathcal{F})$ and be such that $|h(x)|$ is not constant in $\Xi \setminus \mathcal{Q}(\mathcal{F})$. Let w be the absolute degree of h. Then*

$$\left|1 - \frac{diam(h_*(\mathcal{G}))}{diam(h_*(\mathcal{F}))}\right|_\infty \leq \left|1 - \left(\frac{diam(\mathcal{G})}{diam(\mathcal{F})}\right)^w\right|_\infty.$$

Moreover $h_*(\mathcal{F})$ and $h_*(\mathcal{G})$ are comparable for the order $\preceq$.

Proof: Let $r = diam(\mathcal{F})$ and let $s = diam(\mathcal{G})$. In $\Gamma(a,r,s)$, $h(x)$ is equal to a Laurent series $\sum_{-\infty}^{+\infty} a_n(x-a)^n$. Since h has no zeros and no poles in $\Gamma(a,r,s)$, by Lemma 17.4 there exists a unique integer $m \in \mathbb{Z}$ such that $|a_m|\rho^m > |a_n|\rho^n$ $\forall n \neq m$, $\forall \rho \in]r,s[$, and since $|h(x)|$ is not constant in $\Xi \setminus \mathcal{Q}(\mathcal{F})$, by Lemma 12.7 we have $m \neq 0$. Consequently, by Corollary 39.4, $h_*(\mathcal{F})$ is the circular filter of center a_0 and diameter $|a_m|r^m$, and $h_*(\mathcal{G})$ is the circular filter of center a_0 and diameter $|a_m|s^m$. Particularly, $h_*(\mathcal{F})$ and $h_*(\mathcal{G})$ are comparable. Next, we have

$$\frac{diam(h_*(\mathcal{G}))}{diam(h_*(\mathcal{F}))} = \left(\frac{diam(\mathcal{G})}{diam(\mathcal{F})}\right)^m.$$

Now, given a number $t > 1$, we notice that $|1-t|_\infty > |1-\frac{1}{t}|_\infty$, and consequently, we have

$$\left|1 - \frac{diam(h_*(\mathcal{G}))}{diam(h_*(\mathcal{F}))}\right|_\infty \leq \left|1 - \left(\frac{diam(\mathcal{G})}{diam(\mathcal{F}}\right)^m\right|_\infty. \tag{39.1}$$

On the other hand, by Corollary 12.8 m is the difference between the number of zeros and the number of poles of h inside $\mathcal{Q}(\mathcal{F})$. Hence $|m|_\infty \leq w$, and therefore by (39.1) we obtain

$$\left|1 - \frac{diam(h_*(\mathcal{G}))}{diam(h_*(\mathcal{F}))}\right|_\infty \leq \left|1 - \left(\frac{diam(\mathcal{G})}{diam(\mathcal{F}}\right)^w\right|_\infty.$$

□

Lemma 39.6: *Let $\mathcal{F}$ be a circular filter of center a and diameter r. Let $h \in K(x)$ and let w be the absolute degree of h. Then there exists $s < r$ such that every circular filter $\mathcal{G}$ of center a and diameter $\rho \in [s, r[$ satisfies*

$$\left|1 - \frac{diam(h_*(\mathcal{G}))}{diam(h_*(\mathcal{F}))}\right|_\infty \leq \left|1 - \left(\frac{diam(\mathcal{G})}{diam(\mathcal{F})}\right)^w\right|_\infty.$$

Moreover $\mathcal{F}$ and $\mathcal{G}$ are comparable for the order $\preceq$.

Proof: We can find $v \in]0, r[$ such that h has no poles in $\Gamma(a, v, r)$. Hence, inside $\Gamma(a, v, r)$, $f(x)$ is equal to a Laurent series $\sum_{-\infty}^{+\infty} a_n(x-a)^n$. Let $f(x) = h(x) - a_0$ and let $f(x) = \sum_{-\infty}^{+\infty} a'_n(x-a)^n$. There exists $s \in]v, r[$ such that f has no zeros inside $\Gamma(a, s, r)$. Since $a'_0 = 0$, $|f(x)|$ is not constant inside $\Gamma(a, s, r)$, hence we can apply Lemma 39.5 to f and then for every circular filter $\mathcal{G}$ of center a and diameter $\rho \in [s, r[$, we have

$$\left|1 - \frac{diam(h_*(\mathcal{G}))}{diam(h_*(\mathcal{F}))}\right|_\infty \leq \left|1 - \left(\frac{diam(\mathcal{G})}{diam(\mathcal{F})}\right)^w\right|_\infty$$

and $\mathcal{F}$ and $\mathcal{G}$ are comparable for the order $\preceq$.

On the other hand, since $f_*(\mathcal{F})$ and $f_*(\mathcal{G})$ have center 0, then by Corollary 39.4 $h_*(\mathcal{F})$ and $h_*(\mathcal{G})$ have center a_0. □

Theorem 39.7 ([11]): *Let D be a closed bounded subset of K and let $f \in H(D)$. Let $\mathcal{F} \in \Phi(D)$. There exists a unique circular filter $f_*(\mathcal{F})$ less thin than the filter admitting for base $f(\mathcal{F} \cap D)$ and satisfying $\varphi_{f_*(\mathcal{F})} = f^*(\varphi_\mathcal{F})$. Further, if D' is a closed bounded set containing $f(D)$, then*

$$\varphi_{f_*(\mathcal{F})} \in Mult(H(D'), \|\ .\ \|_{D'})$$

and for all $g \in H(D')$, it satisfies

$$\varphi_{f_*(\mathcal{F})}(g) = \varphi_\mathcal{F}(g \circ f).$$

Proof: For every $P \in K[x]$, we put $\psi(P) = \varphi_\mathcal{F}(P \circ f)$. Then ψ belongs to $Mult(K[x])$ and therefore is of the form $\varphi_\mathcal{G}$. Thus we can put $f^*(\varphi_\mathcal{F}) = \psi$, and $f_*(\mathcal{F}) = \mathcal{G}$ because if $f \in R(D)$, we find again the previous definition of f^* and f_*. Let $\mathcal{F}_D$ be the intersection of $\mathcal{F}$ with D. Let $\mathcal{F}'$ be the filter on K admitting for base $f(\mathcal{F}_D)$. Suppose first that $\mathcal{F}'$ is not less thin than $\mathcal{G}$. Then we can find a sequence thinner than $\mathcal{F}'$ which is either a monotonous distances sequence, or an equal distances sequence or a cauchy sequence, and then this sequence is thinner than another circular filters $\mathcal{H}$. Consequently, $\mathcal{F}'$ is secant with another circular filters $\mathcal{H}$. Hence we have $\psi(P) = \varphi_{\mathcal{G}}(P) = \varphi_{\mathcal{H}}\ \forall P \in K[x]$, hence by Theorem 13.1 $\mathcal{H} = \mathcal{G}$, therefore $\mathcal{F}'$ is thinner than $\mathcal{G}$.

Let D' be a closed bounded set containing $f(D)$. By Proposition 17.8, for all $g \in H(D')$, $g \circ f$ belongs to $H(D)$, and we check that $\varphi_{f_*(\mathcal{F})}$ belongs to $Mult(H(D'), \|\ .\ \|_{D'})$. More precisely, for all $g \in H(D')$, we have $\varphi_{f_*(\mathcal{F})}(g) = \lim_{f(\mathcal{F})} |g(u)| = \lim_{\mathcal{F}} |g(f(x))| = \varphi_{\mathcal{F}}(g \circ f)$. Moreover, by definition of the mapping $\mathcal{MG}_A$ in Chapter 15, we can see that $\varphi_{f_*(\mathcal{F})} = f^*(\varphi_{\mathcal{F}})$. □

By Theorem 15.14, we obtain have Corollary 39.8:

Corollary 39.8: *Let D be a closed bounded subset of K. Let $\mathcal{F} \in \Phi(D)$. Given $f, g \in H(D)$ such that $f_*(\mathcal{F}) \neq g_*(\mathcal{F})$, then $diam(\sup(f_*(\mathcal{F}), g_*(\mathcal{F})) = \|f - g\|_D$.*

Corollary 39.9: *Let D be a closed bounded subset of K. Let $\mathcal{F} \in \Phi(D)$. Given $f, g \in H(D)$, then $\delta(g_*(\mathcal{F}), f_*(\mathcal{F})) \leq \|f - g\|_D$. Further, if $\|f - g\|_D < diam(\sup(f_*(\mathcal{F}), g_*(\mathcal{F}))$, then $f_*(\mathcal{F}) = g_*(\mathcal{F})$.*

Remark: In [3] it is shown that when two elements f and g of $H(D)$ are such that $\|f - g\|_D$ is small enough, then $f_*(\mathcal{F}) = g_*(\mathcal{F})$.

Notation: For every $\psi \in Mult(H(D), \|\ .\ \|_D)$, we denote by Z_ψ the mapping from $H(D)$ to $Mult(K[x])$ defined as $Z_\psi(f) = f^*(\psi)$.

Corollary 39.10: *Let D be a closed bounded subset of K. Then Z_ψ is continuous with respect to the norm on $H(D)$ and to both topologies on $Mult(K[x])$ and it is uniformly continuous with respect to the metric δ on $Mult(K[x])$. Moreover, the family of mappings Z_ϕ, ($\phi \in Mult(H(D), \|\ .\ \|_D)$) is uniformly equicontinuous with respect to the metric δ on $Mult(K[x])$.*

Corollary 39.11: *Let D be a closed bounded subset of K and for every $\psi \in Mult(H(D), \|\ .\ \|_D)$, Then Z_ψ is continuous with respect to the norm on $H(D)$ and to the topologies of simple convergence on $Mult(K[x])$.*

Chapter 40

Continuous Mappings on *Mult*($K[x]$)

Every analytic element on a closed bounded set D has expansion to a function from $Mult(H(D), \| \cdot \|_D)$ to $Mult(K[x])$ which is continuous with respect to both topologies on $Mult(K[x])$. The family of functions $Z_{\mathcal{F}}$ defined by circular filters secant with D as $Z_{\mathcal{F}}(f) = f^*(\mathcal{F})$ is a uniformly equicontinuous family with respect to the δ-topology on $Mult(K[x])$. Results apply to meromorphic functions in K which define open continuous function with respect to the topology of simple convergence. Such a mapping is increasing with respect to the order of the tree $Mult(K[x])$ if and only if it is an entire function. Now, let $D = h^{-1}(U)$. We have a natural mapping from $H(U)$ into $H(D)$, and another from $Mult(H(D), \| \cdot \|_D)$ into $Mult(H(U), \| \cdot \|_U)$. We show that the Gauss norm $\| \cdot \|_U$ on $H(U)$ has a number of continuations to the Krasner-Tate algebra $H(D)$ which is equal to the cardinal of $\Sigma(D)$.

Theorem 40.1 ([11]): *Let D be a closed bounded subset of K and let $f \in H(D)$. Let $\mathcal{F} \in \Phi(D)$. The mapping f^* from $Mult(H(D), \| \cdot \|_D)$ to $Mult(K[x])$ defined as $f^*(\varphi_{\mathcal{F}}) = \varphi_{f_*(\mathcal{F})}$ is continuous with respect to the topology of simple convergence and is uniformly continuous with respect to the δ-topology.*

Proof: f^* is continuous with respect to the topology of simple convergence by Theorem 15.11. Here, we will show that the mapping f^* is uniformly continuous for the δ-topology. Thanks to the inequality $\delta(f_*(\mathcal{F}), g_*(\mathcal{G})) \leq \|f - g\|_D$ we can clearly reduce us to the case when $f \in R(D)$ without loss of generality, what we will assume. Let $\tau = \|f\|_D$. Since $R(D)$ is dense inside $H(D)$, we can assume that $f \in R(D)$ without loss of generality. Let w be the absolute degree of f. Let ϵ be > 0. Since f is uniformly continuous in D, we can find $\nu \in]0, r[$ such that $|x - y| \le 2\nu$ implies $|f(x) - f(y)| \le \frac{\epsilon}{2}$. Now, let $\eta \in]0, \frac{\nu}{4}[$ satisfy

$$\tau\Big(|1 - (\frac{\nu + 2\eta}{\nu - 2\eta})^w|_\infty + \big(1 + |1 - (\frac{\nu + 2\eta}{\nu - 2\eta})^w|_\infty\big)\big(|1 - (\frac{\nu + 4\eta}{\nu - 4\eta})^w|_\infty\big)\Big) < \epsilon. \tag{40.1}$$

Consider now $\mathcal{F}, \mathcal{G} \in \Phi(D)$ such that $\delta(\mathcal{F}, \mathcal{G}) \le \eta$ and let $\mathcal{S} = \sup(\mathcal{F}, \mathcal{G})$. Suppose first that $diam(\mathcal{F}) \le \nu$, $diam(\mathcal{G}) \le \nu$. Then $diam(\mathcal{S}) \le \nu + \eta \le 2\nu$. Therefore, $\mathcal{Q}(\mathcal{S})$ is a disk $d(a, \rho)$ with $\rho \le 2\nu$. such that both $\mathcal{F}, \mathcal{G}$ are secant with $d(a, \rho)$. We notice that given $a \in D$, every set A included in $d(a, \nu)$ is such that $A \subset d(f(a), \frac{\epsilon}{2})$. Therefore, we have $\delta(f(a), f(\mathcal{F})) \le \frac{\epsilon}{2}$ and finally $\delta(f_*(\mathcal{F}), f_*(\mathcal{G})) \le \delta(f_*(\mathcal{F}), f(a)) + \delta(f(a), f_*(\mathcal{G})) \le \epsilon$.

Now, suppose that $diam(\mathcal{F}) > \nu$. Consider the ball $\mathcal{B}_\delta$ of center $\varphi_\mathcal{F}$ and radius η with respect to the distance δ. Then, $\mathcal{B}_\delta$ is clearly included in the ball $\mathcal{B}_{\delta'}$ of center $\varphi_\mathcal{F}$ and radius 2η with respect to the distance δ'.

Let $\varphi_\mathcal{G} \in \mathcal{B}$ and let $\mathcal{S} = \sup(\mathcal{F}, \mathcal{G})$. Let $r = diam(\mathcal{F})$, $s = diam(\mathcal{G}), t = diam(\mathcal{S})$. Then $t - r + t - s \le 2\eta$. In particular, $t - r \le 2\eta$. Let $\mathcal{G}' = h_*(\mathcal{G})$, $\mathcal{S}' = h_*(\mathcal{S})$, let $r' = diam(\mathcal{F}')$, let $s' = diam(\mathcal{G}')$, $t' = diam(\mathcal{S}')$. By Lemma 3.8 $\mathcal{S}'$ is comparable to both $\mathcal{F}'$ and to $\mathcal{G}'$ hence we obtain $\delta(\mathcal{S}', \mathcal{F}') = |diam(\mathcal{S}') - diam(\mathcal{F}')|_\infty = r'|1 - (\frac{t'}{r'})|_\infty$ and by Lemma 39.6 we check that

$$\left|1 - \left(\frac{t'}{r'}\right)\right|_\infty \le \left|1 - \left(\frac{t}{r}\right)^w\right|_\infty \le \left|1 - \left(\frac{r + 2\eta}{r - 2\eta}\right)^w\right|_\infty,$$

therefore

$$\delta(\mathcal{S}', \mathcal{F}') \le r' \left|1 - \left(\frac{r + 2\eta}{r - 2\eta}\right)^w\right|_\infty. \tag{40.2}$$

Similarly, we obtain $\delta(\mathcal{S}', \mathcal{G}') = |diam(\mathcal{S}') - diam(\mathcal{G}')|_\infty = t'|1 - (\frac{s'}{t'})|_\infty$. Now by the hypothesis on s' and t' we check that

$$\left|1 - \left(\frac{s'}{t'}\right)\right|_\infty \le \left|1 - \left(\frac{t'}{s'}\right)\right|_\infty \le \left|1 - \left(\frac{t + 2\eta}{t - 2\eta}\right)^w\right|_\infty \le \left|1 - \left(\frac{r + 4\eta}{r - 4\eta}\right)^w\right|_\infty,$$

therefore we obtain

$$\delta(\mathcal{S}', \mathcal{G}') \le t' \left|1 - \left(\frac{r + 4\eta}{r - 4\eta}\right)^w\right|_\infty. \tag{40.3}$$

But by (40.2) we have $|t' - r'| \leq r'|1 - (\frac{r+2\eta}{r-2\eta})^w|_\infty$, therefore by (40.3) we obtain $\delta(\mathcal{S}', \mathcal{G}') \leq r'\big(1 + |1 - (\frac{r+2\eta}{r-2\eta})^w|_\infty\big)(|1 - (\frac{r+4\eta}{r-4\eta})^w|_\infty)$, and consequently, by (40.2) again, we have

$$\delta(\mathcal{F}', \mathcal{G}') \leq r'\left(\left|1 - \left(\frac{r+2\eta}{r-2\eta}\right)^w\right|_\infty \right.$$
$$\left. + \left(1 + \left|1 - \left(\frac{r+2\eta}{r-2\eta}\right)^w\right|_\infty\right)\left(\left|1 - \left(\frac{r+4\eta}{r-4\eta}\right)^w\right|_\infty\right)\right).$$

But by hypothesis, $r' \leq \tau$ and $\nu < r$, hence

$$\delta(\mathcal{F}', \mathcal{G}') \leq \tau\left(\left|1 - \left(\frac{\nu+2\eta}{\nu-2\eta}\right)^w\right|_\infty \right.$$
$$\left. + \left(1 + \left|1 - \left(\frac{\nu+2\eta}{\nu-2\eta}\right)^w\right|_\infty\right)\left(\left|1 - \left(\frac{\nu+4\eta}{\nu-4\eta}\right)^w\right|_\infty\right)\right).$$

Consequently, by (40.1), we obtain $\delta'(\mathcal{F}', \mathcal{G}') \leq \epsilon$, and so much the more $\delta(\mathcal{F}', \mathcal{G}') \leq \epsilon$. Thus we have proven in every cases that $\delta(\mathcal{F}, \mathcal{G}) \leq \eta$ implies $\delta(\mathcal{F}', \mathcal{G}') \leq \epsilon$. □

Theorem 40.2: *Assume K is separable, let D be a closed bounded subset of K and let $f \in H(D)$. The mapping f^* from $Mult(H(D), \| \,.\, \|_D)$ to $Mult(K[x])$ defined as $f^*(\varphi_{\mathcal{F}}) = \varphi_{f_*(\mathcal{F})}$ is uniformly continuous with respect to any metric defining the topology of simple convergence.*

Proof: Since K is topologically separable, by Theorem 34.4 and Corollary 34.8 the topology of simple convergence on $Mult(K[x])$ and on $Mult(H(D), \| \,.\, \|_D)$ is metrizable. Then, since $Mult(H(D), \| \,.\, \|_D)$ is compact, f^* is uniformly continuous with respect to any metric defining the topology of simple convergence. □

Remark: In particular Theorem 40.2 applies to the field $\mathbb{C}_p$. On the other hand, when the field K is topologically separable, we can also ask whether mappings Z_ϕ are uniformly continuous, and whether the family of mappings Z_ϕ, $(\phi \in Mult(H(D), \| \,.\, \|_D))$ is uniformly equicontinuous with respect to the metric of simple convergence. Since each Z_ϕ is uniformly continuous with respect to the metric δ, it is likely uniformly continuous with respect to the metric of simple convergence, and similarly, the family of mappings Z_ϕ, $(\phi \in Mult(H(D), \| \,.\, \|_D))$ should be uniformly equicontinuous with respect to this metric. But a proof is not easily written, due to the complexity to define distances such as $\mathcal{D}$. And yet, the answer might depend on the distance we consider defining the topology of simple convergence. Anyway, we can state Theorem 40.3.

Theorem 40.3 *Assume K is separable and let D be a closed bounded subset of K. The family of functions Z_ϕ, $\phi \in Mult(H(D), \| \, . \, \|_D)$ is uniformly equicontinuous on every ball of $H(D)$ with respect to any metric defining the topology of simple convergence on $Mult(H(D), \| \, . \, \|_D)$.*

Proof: Let $\widehat{d}$ be a distance defining the topology of simple convergence on $Mult(H(D), \| \, . \, \|_D)$. Suppose that the family $\{Z_\phi \mid \phi \in Mult(H(D), \| \, . \, \|_D)\}$ is not uniformly equicontinuous on certain ball $\{f \in H(D) \mid \|f\|_D \leq m\}$. There exists a sequence $(\varphi_{\mathcal{F}_n})_{n\in\mathbb{N}}$ of $Mult(H(D), \| \, . \, \|_D)$, $t > 0$ and a sequence $(f_n, g_n)_{n\in\mathbb{N}}$ of pairs of elements of $H(D)$ such that $\lim_{n\to\infty} \|g_n - f_n\|_D = 0$ and $\widehat{d}(f_n^*(\varphi_{\mathcal{F}_n}), g_n^*(\varphi_{\mathcal{F}_n})) \geq t \; \forall n \in \mathbb{N}$. For each $n \in \mathbb{N}$, we put $\mathcal{F}'_n = f_n(\mathcal{F}_n)$, $\mathcal{G}'_n = g_n(\mathcal{F}_n), r_n = \delta(\mathcal{F}'_n, \mathcal{G}'_n)$. Since $\lim_{n\to\infty} \|g_n - f_n\|_D = 0$, by Theorem 39.7 we first notice that $\lim_{n\to\infty} r_n = 0$. On the other hand, since $\max(\|f_n\|_D, \|g_n\|_D) \leq m \; \forall n \in \mathbb{N}$ and since $Mult(K[x])$ is locally compact, we can extract an increasing sequence of integers $(q(n)_{n\in N})$ such that the sequence $(\varphi_{\mathcal{F}'_{q(n)}})_{n\in N}$ converges in $MultK([x])$ to a limit $\varphi_{\mathcal{T}}$ with respect to the topology of simple convergence. But then, by Proposition 14.5 the sequence $(\varphi_{\mathcal{G}'_{q(n)}})_{n\in N}$ also converges to $\varphi_{\mathcal{T}}$, with respect to the topology of simple convergence, a contradiction to the hypothesis $\widehat{d}(\varphi_{\mathcal{F}'_n}, \varphi_{\mathcal{G}'_n}) \geq t$. □

Notation: We shall denote by $\mathcal{A}(K)$ the K-algebra of entire functions in K i.e. the power series with coefficients in K whose convergence radius is infinite. And we denote by $\mathcal{M}(K)$ the field of meromorphic functions in K, i.e. the field of fractions of $\mathcal{A}(K)$. Given $f \in \mathcal{M}(K)$, we will denote by $\pi(f)$ the set of poles of f.

Now, we will apply Theorem 40.1 to meromorphic functions. Lemma 40.4 is immediate:

Lemma 40.4: *Let $f \in \mathcal{M}(K)$ and let A be an affinoid such that $\pi(f) \cap A = \emptyset$. Then $f \in H(A)$.*

Proof: We can write f in the form $\frac{g}{h}$, where g, $h \in \mathcal{A}(K)$ and $h(x) \neq 0 \; \forall x \in A$, hence h is invertible in $H(A)$. □

Lemma 40.5: *Let A be an affinoid set, let $f \in H(A)$, let $B = f(A)$. Then $f_*(\Phi(A)) = \Phi(B)$.*

Proof: Let $\mathcal{G} \in \Phi(B)$. Let $(b_n)_{n\in\mathbb{N}}$ be a sequence of B thinner than $\mathcal{G}$ and let $(a_n)_{n\in\mathbb{N}}$ be a sequence of A such that $f(a_n) = b_n\ \forall n \in \mathbb{N}$. Since the sequence $(a_n)_{n\in\mathbb{N}}$ is bounded, by Corollary 10.8 we can extract a subsequence which is thinner than a circular filter $\mathcal{F}$. Then $\mathcal{F} \in \Phi(A)$. And then by Theorem 39.7 the sequence $(b_n)_{n\in\mathbb{N}}$ is thinner than $f_*(\mathcal{F})$, hence $f_*(\mathcal{F}) = \mathcal{G}$. □

Theorem 40.6: *Let $f \in \mathcal{M}(K)$ and let $\mathcal{F}$ be a circular filter on K which is not the filter of neighborhoods of a pole of f. Then $f(\mathcal{F})$ is a base of a circular filter $f_*(\mathcal{F})$ such that $\varphi_{f_*(\mathcal{F})}(P) = \varphi_{\mathcal{F}}(P\circ f)\ \forall P \in K[x]$. The function f^* from $Mult(K[x]) \setminus \pi(f)$ to $Mult(K[x])$ defined as $f^*(\varphi_{\mathcal{F}}) = \varphi_{f_*(\mathcal{F})}$ is a continuous function with respect to both topologies and is open with respect to the topology of simple convergence. Moreover, given $g \in \mathcal{M}(K)$, then $\delta(g_*(\mathcal{F}), f_*(\mathcal{F})) \leq \varphi_{\mathcal{F}}(f-g)$. And if $\varphi_{\mathcal{F}}(f-g) < diam(\sup(f_*(\mathcal{F}), g_*(\mathcal{F}))$, then $f_*(\mathcal{F}) = g_*(\mathcal{F})$.*

Proof: We can clearly reduce the proof to the case when $\mathcal{F}$ is a large circular filter, what we will assume. Let $\mathcal{F}'$ be the filter admitting for base $f(\mathcal{F})$. Since f has finitely many poles in any bounded set, we can find a base $\mathcal{C}$ of $\mathcal{F}$ consisting of infraconnected affinoid sets A such that $A \cap \pi(f) = \emptyset$, and then by Lemma 40.4 f belongs to $H(A)$. Let $B = f(A)$. Then B is an affinoid infraconnected set. Let $\epsilon = \delta(B, K \setminus B)$ and let $h \in R(A)$ satisfy $\|f - h\|_A < \epsilon$. Then it is easily seen that $h(A) = B$. Consequently, $\mathcal{F}'$ is the filter admitting for base $h(\mathcal{F})$, i.e. $h_*(\mathcal{F})$, whenever $h \in R(A)$ satisfying $\|f - h\|_A < \delta(f(A), K \setminus f(A))$, for any infraconnected affinoid set $A \in \mathcal{F}$ such that $f \in H(A)$. Thus, we can put $f_*(\mathcal{F}) = \mathcal{F}'$, and we check that given $P \in K[x]$ we have $\varphi_{f_*(\mathcal{F})}(P) = \varphi_{\mathcal{F}}(P \circ f)$. Now, consider $\varphi_{\mathcal{F}} \in Mult(K[x]) \setminus \pi(f)$, we can take a closed bounded infraconnected set D such that $\mathcal{F}$ is secant with D and such that $\pi(f) \cap D = \emptyset$. So, f belongs to $H(D)$ and we can apply Theorem 39.7 showing that the function f^* from $Mult(K[x]) \setminus \pi(f)$ to $Mult(K[x])$ defined as $f^*(\varphi_{\mathcal{H}}) = \varphi_{f_*(\mathcal{H})}$ is a continuous function at $\varphi_{\mathcal{F}}$ with respect to both topologies. In the same way, by Theorem 39.7 we can see that given $g \in \mathcal{M}(K)$, we have $\delta(g_*(\mathcal{F}), f_*(\mathcal{F})) \leq \varphi_{\mathcal{F}}(f - g)$ and if $\varphi_{\mathcal{F}}(f - g) < diam(\sup(f_*(\mathcal{F}), g_*(\mathcal{F}))$, then $f_*(\mathcal{F}) = g_*(\mathcal{F})$.

Now, we will show that f is open for the topology of simple convergence. Let $\varphi_{\mathcal{F}} \in Mult(K[x]) \setminus \pi(f)$. The family of sets $Mult(H(A), \|\ .\ \|_A)$, where A is a $\mathcal{F}$-affinoid, makes a base of neighborhoods of $\varphi_{\mathcal{F}}$. Therefore, in order to proving that f is open, it is sufficient to show that, given a $\mathcal{F}$-affinoid A, then $\{f^*(\varphi_{\mathcal{T}}) \mid \mathcal{T} \in \Phi(A)\}$ is a neighborhood of $f^*(\varphi_{\mathcal{F}})$. Thus, let A be a $\mathcal{F}$-affinoid containing no poles of f. Let $B = f(A)$ and let $\mathcal{G} = f_*(\mathcal{F})$. Then B lies in $\mathcal{G}$. Let us put $\mathcal{W} = Mult(H(A), \|\ .\ \|_A)$ and $\mathcal{Z} = Mult(H(B), \|\ .\ \|_B)$. Thus, $\mathcal{Z}$ is a neighborhood of $\varphi_{\mathcal{G}}$ in $Mult(K[x])$. Now, $\mathcal{Z} = \{\varphi_{\mathcal{H}} \mid \mathcal{H} \in \Phi(B)\}$, and by Lemma 40.5, $\{\varphi_{\mathcal{H}} \mid \mathcal{H} \in \Phi(B)\} = \{f^*(\varphi_{\mathcal{T}}) \mid \mathcal{T} \in \Phi(A)\} = f^*(\mathcal{W})$. Consequently, f^* maps the neighborhood $\mathcal{W}$ of $\varphi_{\mathcal{F}}$ onto the neighborhood $\mathcal{Z}$ of $f^*(\varphi_{\mathcal{F}})$, which proves that f^* is open with respect to the topology of simple convergence. □

Theorem 40.7: *Let $f \in \mathcal{M}(K)$. Then f^* is increasing from $Mult(K[x]) \setminus \pi(f)$ to $Mult(K[x])$ if and only if $f \in \mathcal{A}(K)$.*

Proof: Let $f \in \mathcal{A}(K)$. Let $\mathcal{F}$, $\mathcal{G} \in Mult(K[x])$ be such that $\mathcal{F} \prec \mathcal{G}$. Let $s = diam(\mathcal{G})$ and let $r = diam(\mathcal{F})$. We can find a center a of $\mathcal{G}$ such that $\delta(a, \mathcal{F}) < s$. Let $\rho = \delta(a, \mathcal{F})$. Without loss of generality we can assume $a = 0$. Let $f(x) = \sum_{n=0}^{\infty} a_n x^n$. Then we know that $\varphi_{\mathcal{F}}(f) \leq \sup_{n\in\mathbb{N}} |a_n|\rho^n \leq \sup_{n\in\mathbb{N}} |a_n|s^n = \varphi_{\mathcal{G}}(f)$, so f_* is increasing.

Now, consider $f \in \mathcal{M}(K) \setminus \mathcal{A}(K)$ and let a be a pole of f. Let $d(a, l)$ be a disk such that f admits neither zeros nor poles different from a inside $d(a, l)$. Without loss of generality we can assume $a = 0$. Let r, $s \in]0, l[$ be such that $r < s$, let $\mathcal{F}$ be the circular filter of center 0 and diameter r and let $\mathcal{G}$ be the circular filter of center 0 and diameter s. Let $g = \frac{1}{f}$. Then g belongs to $H(d(0, l))$ and is of the form $\sum_{n=0}^{\infty} a_n x^n$ inside $d(0, l)$. Thus we have $\varphi_{\mathcal{F}}(g) = \sup_{n\in\mathbb{N}} |a_n|r^n < \sup_{n\in\mathbb{N}} |a_n|s^n = \varphi_{\mathcal{G}}(g)$, hence $\varphi_{\mathcal{F}}(f) > \varphi_{\mathcal{G}}(f)$, which proves that f_* is not increasing. □

Remarks: Let us restrict to rational functions and to circular filters with center. As we saw in Chapter 13, a circular filter $\mathcal{F}$ of center a and diameter $r > 0$ defines a generic disk which is a transcendental element τ over K [15]. Let $h \in K(x)$ and let $\mathcal{G} = h_*(\mathcal{F})$. Let $|\ .\ |$ be the absolute value on $K(\tau)$ defined by $\mathcal{F}$ as $|f(\tau)| = \varphi_{\mathcal{F}}(f)\ \forall f \in K(x)$. Thus for every $f \in K(x)$ we have $|f(\tau)| = \varphi_{h_*(\mathcal{F})}(f) = \varphi_{\mathcal{F}}(f \circ h) = |f \circ h(\tau)|$, so in this way we can give another illustration of this mapping.

As noticed in Chapter 14, in [42], the author introduces a distance "*d*" on the set H_p of large circular filters which actually defines the same topology as δ on this set, and then given a rational function h, the author shows the continuity of h^* with respect to "*d*" on the space H_p.

Theorem 40.8 ([11]): *Let E be a closed bounded subset of K and $f \in K(x)$ and let $D = f^{-1}(E)$. Then $\Sigma(E) = f^*(\Sigma(D))$ and $\Sigma(D) = (f^*)^{-1}(\Sigma(E))$.*

Proof: If $\mathcal{F}$ is not secant with D, then $f_*(\mathcal{F})$ is not secant with E because $D = f^{-1}(E)$. Now, suppose that $\mathcal{F}$ is secant with D. Then $f_*(\mathcal{F})$ is obviously secant with E because so is the filter admitting for base $f(\mathcal{F})$. First we suppose $\mathcal{F}$ to be D-bordering. Suppose that $f_*(\mathcal{F})$ is not E-bordering. Then, $f_*(\mathcal{F})$ is not secant with $K \setminus E$ There exists $A \in f_*(\mathcal{F})$ such that $A \subset E$. But since $D = f^{-1}(E)$ of course we have $f^{-1}(A) \subset D$, but $f^{-1}(A) \in \mathcal{F}$, a contradiction to the hypothesis $\mathcal{F}$ *is* D*-bordering.*

Now, suppose that $\mathcal{F}$ is not D-bordering. Then it admits an element $A \subset D$. Hence $f(A) \subset E$ therefore $f_*(\mathcal{F})$ is not E-bordering. Consequently $\Sigma(E) = f^*(\Sigma(D))$ and $\Sigma(D) = (f^*)^{-1}(\Sigma(E))$. □

Corollary 40.9: *Let $\mathcal{G}$ be a circular filter of center a and diameter $r \in |K|$, let $h \in K(x)$ and let $D = h^{-1}(d(a,r))$. Then $(h^*)^{-1}(\{\mathcal{G}\}) = \Sigma(D)$.*

Remarks: In [41] E. Motzkin claims that the tree of a quasi-connected set is *conserved* through a conformal mapping. According to our definitions, this suggests that given a bounded closed infraconnected set D and an injective element of $H(D)$ might (under certain hypothesis) define a strictly monotonous continuous mapping from $\Phi(D)$ onto $\Phi(f(D))$. And under more restrictive hypothesis, f^{-1} should belong to $H(f(D))$.

Theorem 40.10 [11]: *Let $H(D)$ be a Krasner-Tate algebra of the form $K\{h\}[x]$, with $h \in K(x)$, $\deg(h) > 0$, and $D = h^{-1}(U)$. Let S be the set of D-bordering filters. Then S is finite and the restriction of Φ to S is a bijection from S onto the set of $\psi \in Mult(H(D), \|\ .\ \|_D)$ expanding the Gauss norm on $K\{h\}$ to $K\{h\}[x]$.*

Proof: Let $\mathcal{G}$ be the U-bordering circular filter. Then the Gauss norm on $K\{h\}$ is $\varphi_{\mathcal{G}}$. Let γ be the homomorphism from $H(U)$ into $H(D)$ defined as $\gamma(f) = f \circ h,\ f \in H(U)$. By Corollary 40.9 $\widehat{\gamma}^{-1}(\{\varphi_{\mathcal{G}}\})$ is the set of $\varphi_{\mathcal{F}}$ where $\mathcal{F}$ is a D-bordering filter. On the other hand, $h^{-1}(U)$ is an affinoid set, hence by Lemma 10.10 the set of D-bordering filters is finite. Hence, when considering $H(D)$ as a finite extension of $K\{h\}$, the extensions of the Gauss norm on $K\{h\}$ to $H(D)$ are the $\varphi_{\mathcal{F}}$ when $\mathcal{F}$ is a D-bordering filter. □

Remark: Let D be an infraconnected affinoid set and let $H(D) = K\{h\}[x]$, with $h \in K(x)$, $\deg_a(h) > 0$, and $D = h^{-1}(U)$. Since D has finitely many holes, so much the more it has finitely many envelopes of holes $d(a_j, r_j)$, $(1 \leq j \leq q)$. Let $\widetilde{D} = d(a_0, r_0)$. The D-bordering filters, and therefore the Shilov boundary of $K\{h\}[x]$, consist of the $q+1$ circular filters $\mathcal{F}_j$, $(j = 0, \ldots, q)$ of respective center a_j and diameter r_j, with $j = 0, \ldots, q$. Then in the Krasner-Tate algebra $H(D) = K\{h\}[x]$, the Gauss norm on $K\{h\}$ admits $q+1$ different extensions to $K\{h\}[x]$ which are the $\varphi_{\mathcal{F}_j}$. In particular the Gauss norm admits only one extension if and only if every holes of D have the same diameter as D itself.

On the other hand, it is well known that an absolute value defined on a complete field T admits a unique expansion to any algebraic extension. Here, we don't consider a complete field, but a complete ring $K\{h\}$, and a finite extension $K\{h\}[x]$. The Gauss norm on $K\{h\}$ can admit several distinct expansions to $K\{h\}[x]$ because the field of fractions F of $K\{h\}[x]$ is not complete, so an element of $K\{h\}[x]$ can admit a conjugate over F which actually lies in the completion of F.

Example: Let $h(x) = x + \frac{\alpha}{x}$, with $\alpha \in K$, such that $|\alpha| < 1$, let $r = |\alpha|$, and let $D = \{x \mid r \leq |x| \leq 1\}$. Then we check that $D = h^{-1}(U)$, and there are two D-bordering circular filters: the circular filter $\mathcal{F}'$ of center 0 and diameter r, and the circular filter $\mathcal{F}''$ of center 0 and diameter 1. Then, the Krasner-Tate algebra $H(D)$ is obviously equal to $K\{h\}[x]$, and the Gauss norm on $K\{h\}$ admits two extensions to $K\{h\}[x]$: $\varphi_{\mathcal{F}'}$ and $\varphi_{\mathcal{F}''}$.

Now, if we take $|\alpha| = 1$, then $\mathcal{F}' = \mathcal{F}''$: the Gauss norm on $K\{h\}$ has a unique extension to $H(D)$.

Chapter 41

Examples and Counterexamples

In this chapter, several typical examples let us answer natural questions. Consider a commutative uniform Banach K-algebra with unity A containing a dense K-algebra of finite type: A is not necessarily a K-affinoid algebra. It is also a natural question to ask whether a noetherian Banach K-algebra $H(D)$ such that $\|f\|_D \in |K| \ \forall f \in H(D)$, containing a dense Luroth K-subalgebra, is a K-affinoid algebra. The answer is no in the general case, again. Other examples will be useful to Chapter 43.

Theorem 41.1: *Let A be a noetherian commutative ultrametric Banach K-algebra with unity satisfying Property (p). Let $x \in A$ be such that $P(x) \neq 0 \ \forall P \in K[x]$. Then $sa(x)$ is infinite.*

Proof: Suppose $sa(x)$ is finite and equal to $\{a_1, \ldots, a_q\}$. Let $g = \prod_{j=1}^{q}(x - a_j)$. Thus we have $\|g\|_{sa} = 0$, hence by Property p), $\|g\|_{si} = 0$. Let I be the ideal of the $f \in A$ such that $fg^n = 0$ for some $n \in \mathbb{N}$, let $B = \frac{A}{I}$ and let θ be the canonical surjection from A onto B. Let $t = \theta(g)$. Clearly $t \neq 0$ because by hypothesis g is not nilpotent. We will show that t is regular in B. Let $z \in B$ be such that $tz = 0$ and let $w \in A$ be such that $\theta(w) = z$. Thus $wg \in I$, so there exists $n \in \mathbb{N}$ such that $(gw)g^n = 0$, thereby $w \in I$ and $z = 0$, so t is regular.

Since I is a closed ideal of A, B is a Banach K-algebra with respect to the quotient norm $\| \,.\, \|_q$. Now, since $\|g\|_{si} = 0$ and since $\|\theta(f)\|_q \leq \|f\| \ \forall f \in A$, in B we have $\lim_{n\to\infty} \sqrt[n]{\|t^n\|_q} = 0$. Consequently, there exists an increasing sequence of integers u_n such that $\lim_{n\to\infty} \frac{\|t^{u_n+1}\|_q}{\|t^{u_n}\|_q} = 0$ because otherwise there would exist $a > 0$ such that $\|t^{n+1}\|_q \geq a\|t^n\|_q$ $\forall n \in \mathbb{N}$ which would lead to $\lim_{n\to\infty} \sqrt[n]{\|t^n\|_q} \geq a$.

Since t is regular, the mapping ψ defined in B as $\psi(z) = tz$ is a continuous K-vector space isomorphism from B onto tB. But since a quotient of a noetherian ring B is noetherian, by Corollary 30.4 tB is closed in B with respect to $\| \,.\, \|_q$. Consequently, tB is complete for $\| \,.\, \|_q$, therefore by Banach's Theorem, ψ is bicontinuous.

Now, let $(l_n)_{n\in\mathbb{N}}$ be a sequence in K such that $1 \leq \|l_n t^{u_n}\|_q \leq 2$. Clearly, $\lim_{n\to\infty} \|t^{u_n+1}\|_q = 0$, so the sequence (y_n) defined as $y_n = l_n t^{u_n}$ satisfies $\|y_n\|_q \geq 1$, and $\lim_{n\to\infty} \|\psi(y_n)\|_q = 0$, a contradiction to the bicontinuity of ψ. □

We shall establish Proposition 41.2, already stated in [18] as proposition IV.8.

Proposition 41.2: *Let $(a_n)_{n\in\mathbb{N}}$ be a sequence in K such that $|a_n| < |a_{n+1}|$, $|a_n| \in |K|$, $\forall n \in \mathbb{N}$ and $\lim_{n\to\infty} |a_n| = r \in |K|$. For each $n \in \mathbb{N}$, let $d_n = |a_n|$, let $r_n \in |K|$ be such that $0 < r_n < d_n$, let $D_n = \{\xi \in K \mid |\xi - a_n| \leq r_n\}$ and let $D = C(0,r) \cup (\cup_{n=0}^{\infty} D_n)$. Then $H(D)$ is a non noetherian Banach K-algebra such that $\|f\|_D \in |K|$ $\forall f \in H(D)$, and its K-subalgebra $K[x, \frac{1}{x}]$ is dense in $H(D)$.*

Proof: Since D has infinitely many infraconnected components, by Corollary 16.13 we know that $H(D)$ is not noetherian. We will show that $\|f\|_D \in |K|$ $\forall f \in H(D)$. Let $f \in H(D)$ ($f \neq 0$) and let $h \in R(D)$ be such that $\|f - h\|_D < \|f\|_D$: of course $\|f\|_D = \|h\|_D$, so we just have to show that $\|h\|_D \in |K|$. We can find $s \in]0, r[$ such that h has neither any zeros, nor any poles inside $\Gamma(0, s, r)$.

Let $\mathcal{T} = \Gamma(0, s, r)$. Let $q \in \mathbb{N}$ be such that $r_q > s$ and $D' = \bigcup_{n=q+1}^{\infty} D_n$ and $D'' = D \setminus D'$. Suppose first that $\|h\|_D = \|h\|_{D''}$. Each set $C(0,r)$, D_n ($n \in \mathbb{N}$) is calibrated and strongly infraconnected. Hence, by Theorem 16.6 we check that there exists $a \in D''$ such that $\|h\|_{D''} = |h(a)|$, so $\|h\|_{D''} \in |K|$. We now suppose that $\|h\|_D > \|h\|_{D''}$, hence $\|h\|_D = \|h\|_{D'}$. Since h has neither any zeros nor any poles inside $\mathcal{T}$, by Corollary 12.4 we know that $|h(\xi)| = \varphi_{0,|\xi|}(h)$ $\forall \xi \in \mathcal{T}$. Next, we notice that $D' \subset \mathcal{T}$, and more precisely, $D' \subset \cup_{n=q+1}^{\infty} C(0, d_n)$. Consequently, $(\|h\|_{D'} = \sup_{\xi\in\mathcal{T}\cap D'} |h(\xi)| = \sup_{n>q} = \varphi_{0,|d_n|}(h)$. Further, by Lemma 12.7 the valuation function $v(h, \mu)$ inside the interval $] - \log r, -\log s[$ is of the form $t\mu - \log(\varphi_{0,r}(h))$, with $t \in \mathbb{Z}$. If $t \leq 0$, then $-\log(\|h\|_{D'}) = v(h, \log(r_{q+1}))$, hence by Theorem 12.9, $\|h\|_{D'} \in |K|$ because $r_{q+1} \in |K|$. Now, suppose $t > 0$. Then $-\log(\|h\|_{D'}) = \lim_{n\to\infty} v(h, -\log r_n) = v(h, -\log r)$. But of course, by Corollary 12.4 $v(h(\xi)) = v(h, -\log r)$ is true in all classes of $C(0,r)$, except at most in finitely many ones, which contradicts our hypothesis $\|h\|_D > \|h\|_{D''}$. This finishes showing that $\|f\|_D \in |K|\ \forall f \in H(D)$.

Now, we will show that the K-algebra $B = K[x, \frac{1}{x}]$ is dense in $H(D)$, which is not easy. Let A be the closure of B in $H(D)$. Actually it is sufficient to show that for every $a \in K \setminus D$, $\frac{1}{x-a}$ belongs to A. So, we fix $a \in K \setminus D$. By Lemma 16.4, we just have to show that there exists $h_a \in R(D) \cap A$ such that $|h_a(a)| > \|h_a\|_D$. When $|a| > r$, we just take $h_a = x$. And since $C(0,r) \subset D$, it only remains us to consider the case when $|a| < r$. Let q be the biggest of the integers n such that $|a| \geq d_n$. Then we can check that either $|a_n - a_l| \geq d_n\ \forall n \in \mathbb{N}$ or $|a_q - a_l| < d_q$.

We first suppose that $|a_n - a_l| \geq d_n\ \forall n \in \mathbb{N}$. Let $\lambda \in K$ be such that $|\lambda| = d_{q+1}$, and let $u = \frac{\lambda}{x}$. We will construct $f_0, \ldots, f_q \in R(D)$ satisfying

$$|f_j(\xi)| = 1\ \forall \xi \in K \text{ such that } |\xi - a_j| > d_j, \tag{41.1}$$

$$\left| u(\xi) \prod_{k=j}^{q} f_k(\xi) \right| \leq 1\ \forall \xi \in D_j. \tag{41.2}$$

In order to constructing such $f_0, \ldots, f_q$, we first define positive integers $t_0, \ldots t_q$ satisfying

$$\frac{|\lambda|}{d_j} \prod_{k=j+1}^{q} \left(\frac{d_k}{d_j}\right)^{t_k} \left(\frac{r_j}{d_j}\right)^{t_j} \leq 1\ \forall j = 0, \ldots, q-1 \tag{41.3}$$

$$\frac{|\lambda|}{d_q} \left(\frac{r_q}{d_q}\right)^{t_q} \leq 1. \tag{41.4}$$

Indeed, since $r_j < d_j\ \forall j$, we can easily define such integers t_j by a downing induction on j beginning at t_q.

Now we put $f_j = \left(\frac{x-a_j}{x}\right)^{t_j}$ $(0 \leq j \leq q)$. Clearly $f_0, \ldots f_q$ satisfy (41.1). We will check they satisfy (41.2). Indeed, let $\xi \in D_j$. We have $|f_k(\xi)| = \left(\frac{d_k}{d_j}\right)^{t_k}\ \forall k = j+1, \ldots q$, and $|f_j(\xi)| \leq \left(\frac{r_j}{d_j}\right)^{t_j}$, and $|u(\xi)| = \frac{|\lambda|}{d_j}$. Consequently we obtain

$$\left| u(\xi) \prod_{k=j}^{q} f_k(\xi) \right| \leq \frac{|\lambda|}{d_j} \prod_{k=j+1}^{q} \frac{d_k}{d_j} \left(\frac{r_j}{d_j}\right)^{t_j} \leq 1.$$

Thus we have constructed $f_0, \ldots, f_q$ satisfying (41.1) and (41.2).

Now, let $h_a = u\prod_{k=0}^{q} f_k$. We will check that h_a satisfies $\|h_a\|_D \leq 1 < |h_a(a)|$. Indeed, let $\xi \in C(0, r)$. We have $|u(\xi)| = \frac{|\lambda|}{r} < 1$ and $|f_j(\xi)| = 1\ \forall j = 0, \ldots, q$ thanks to (41.1). Consequently, when $\xi \in C(0, r)$ we have $|h_a(\xi)| < 1$. Now, if $\xi \in D_m$ with $m > q$, we have $|u(\xi)| = \frac{|\lambda|}{d_m} < 1$ and $|f_j(\xi)| = 1$. And finally, when $\xi \in D_m$ with $m \leq q$, then by (41.1) we have $|f_j(\xi)| = 1\ \forall j = 0, \ldots, q-1$, hence $|h_a(\xi)| = |u(\xi)\prod_{k=j}^{q} f_k(\xi)| \leq 1$, which finishes proving that $\|h_a\|_D \leq 1$. On the other hand, since by hypothesis $d_q \leq |a| < d_{q+1}$, we have $|u(\xi)| = \frac{|\lambda|}{|a|} > 1$, so by (41.1) we check that $|f_j(\xi)| = 1\ \forall j = 0, \ldots, q$, which shows that $|h_a(a)| > 1$.

We now suppose $|a_q - a| < d_q$. First, let us choose $w \in \mathbb{N}$ such that

$$\left(\frac{d_q}{d_{q+1}}\right)^w < \frac{r_q}{d_q}. \tag{41.5}$$

Let $\nu \in K$ be such that

$$|\nu| = (d_{q+1})^w \tag{41.6}$$

and let $y = \frac{\nu}{x^w}$. We will define $g_1, \ldots, g_q \in R(D)$ satisfying (41.1) and

$$\max(\|yg_q\|_{D_q}, 1) < |y(a)g_g(a)| \tag{41.7}$$

and

$$\left|y(\xi)\prod_{i=1}^{q} g_i(\xi)\right| \leq 1\ \forall \xi \in \bigcup_{i=0}^{q-1} D_i. \tag{41.8}$$

In order to constructing $g_0, \ldots, g_q$, we first choose positive integers $s_0, \ldots, s_q$ such that $s_q = 1$, and

$$\frac{|\nu|}{(d_n)^w}\prod_{i=n+1}^{q}\left(\frac{d_i}{d_n}\right)^{s_i}\left(\frac{r_n}{d_n}\right)^{s_n} \leq 1\ \forall n = 0, \ldots, q-1. \tag{41.9}$$

Then, for each $n \leq q$, we set $g_n = \left(\frac{x-a_n}{x}\right)^{s_n}$. It is easily seen that $g_0, \ldots, g_q$ satisfy (41.1). We will show that they satisfy (41.7) and (41.8). Let $n \leq$

$q-1$. Given $\xi \in D$, we have

$$|y(\xi)| = \frac{|\nu|}{(d_n)^w},\ |g_i(\xi)| = \left(\frac{d_i}{d_n}\right)^{s_i}\ \forall i = n+1, \ldots, q \text{ and } |g_n(\xi)| = \left(\frac{r_n}{d_n}\right)^{s_n}.$$

Thus, by (41.9) Relation (41.8) is satisfied. Finally by (41.5) we obtain

$$|y(a)g_q(a)| = \left(\frac{|\nu|}{(d_q)^w}\right)\left(\frac{|a-a_q|}{d_q}\right) > \left(\frac{r_q}{(d_q)^w}\right)\left(\frac{|a-a_q|}{d_q}\right) > 1.$$

On the other hand we have

$$\left(\frac{r_q}{(d_q)^w}\right)\left(\frac{|a-a_q|}{d_q}\right) > |y(\xi)g_q(\xi)|\ \forall \xi \in D_q,$$

which shows that g_q satisfies (41.8). Consequently, we have constructed $g_0, \ldots, g_q$ satisfying (41.7) and (41.8).

We can now conclude. Let $h_a(x) = y\prod_{n=0}^{q} g_n$. Clearly h_a satisfies $|h_a(a)| = |y(a)g_q(a)|$ and $|h_a(\xi)| = |y(\xi)g_q(\xi)|\ \forall \xi \in D_q$ hence by (41.7) we have $|h_a(a)| > \max(\|h_a\|_{D_q}, 1)$. On the other hand we have

$$|h_a(\xi)| = \left|y(\xi)\prod_{i=1}^{q} g_i(\xi)\right| \leq 1\ \forall \xi \in D_n\ \forall n \neq q.$$

Consequently,

$$|h_a(a)| > \|h_a\|_D.$$

□

Theorem 41.3: *Let A be a commutative ultrametric Banach K-algebra with unity, satisfying Property (q), containing a dense Luroth K-subalgebra. Then A satisfies Property (p).*

Proof: Let $B = K[h, x]$ be a dense Luroth K-subalgebra of A, with $h \in K(x)$. We will first show that $sa(x) = sp(x)$. let $D = sp(x)$, let $a \in D$, and let χ be defined on B by $\chi(f) = f(a)$. Suppose there exists $f \in B$ such that $\|f\| > |f(a)|$. Then by Theorem 6.13 $f(a) - f$ is invertible in A. Let $g = (f(a) - f)^{-1}$. Then, by Lemma 1.8 $x - a$ belongs to $K[g, x]$, hence to A, a contradiction. Consequently, we have $|\chi(f)| \leq \|f\|\ \forall f \in B$, which shows that χ is continuous, and therefore has continuation to A. So, $sa(x) = sp(x)$. This is immediately generalized to every $f \in B$ because given $f \in B$, by Theorem 23.4 we know that $sp(f) = f(sp(x))$ and $sa(f) = f(sa(x))$. Thus, we have $\|f\|_{sa} = \tau(f)\ \forall f \in B$. But since A satisfies

Property (q), actually we have $\|f\|_{sa} = \|f\|_{si}\ \forall f \in B$, and since B is dense in A, finally the equality holds in all A. □

Theorem 41.4: *Let $K[x]$ be provided with the ultrametric K-vector space norm*

$$\left\|\sum_{n=0}^{q} a_n x^n\right\| = \sup_{n\leq q} |a_n| \left(\frac{1}{n+1}\right)^n .$$

This norm is a K-algebra norm and the completion A of $K[x]$ for this norm is a Banach K-algebra satisfying Property (p), such that $P(x) \neq 0\ \forall P \in K[x]$ and such that $sp(x) = \{0\}$. Moreover, A is not noetherian.

Proof: By definition, the elements of A are power series $\sum_{n=0}^{\infty} \alpha_n x^n$ such that $\lim_{n\to\infty} |\alpha_n|(\frac{1}{n+1})^n = 0$. Consequently, given such a series $\sum_{n=0}^{\infty} \alpha_n x^n$, there exists $q \in \mathbb{N}$ such that

$$\left\|\sum_{n=0}^{\infty} \alpha_n x^n\right\| = |\alpha_q| \left(\frac{1}{q+1}\right)^q .$$

We first have to check that the norm is a K-algebra norm. Let $f = \sum_{n=0}^{\infty} a_n x^n$, $g = \sum_{n=0}^{\infty} b_n x^n$ and $fg = \sum_{n=0}^{\infty} c_n x^n$. Let $q \in \mathbb{N}$ be such that $\| \sum_{n=0}^{\infty} c_n x^n\| = |c_q|(\frac{1}{q+1})^q$. By noticing that the inequality

$$\left(\frac{1}{i+1}\right)^i \left(\frac{1}{q-i+1}\right)^{q-i} \leq \left(\frac{1}{q+1}\right)^q$$

is true for every $i \leq q$ we obtain

$$\begin{aligned} \|fg\| &\leq \sup_{0\leq i\leq q} \left(|a_i||b_{q-i}|\left(\frac{1}{i+1}\right)^i \left(\frac{1}{q-i+1}\right)^{q-i}\right) \\ &\leq \sup_{i\in\mathbb{N},\ j\in\mathbb{N}} \left(|a_i||b_{q-i}|\left(\frac{1}{i+1}\right)^i \left(\frac{1}{q-i+1}\right)^{q-i}\right) = \|f\|.\|g\|. \end{aligned}$$

Now, it easily seen that for all $b \in K$ we have $\lim_{n\to\infty} \|(bx)^n\| = 0$. In particular, for every $a \in K^*$ then, $\lim_{n\to\infty} \left\|(\frac{x}{a})^n\right\| = 0$. Consequently the series $\sum_{n=0}^{\infty}(\frac{x}{a})^n$ is the inverse of $1 - \frac{x}{a}$ which proves that $x - a$ is invertible in A for all $a \in K^*$. Therefore, $sp(x) = \{0\}$ and A is a local ring whose maximal ideal is xA and has codimension 1. More precisely, we can check that $\|g\|_{si} = 0\ \forall g \in xA$. Indeed let $f \in A$, let $\lambda \in K$ and let $\nu \in K$ be such that $\nu > \|f\|$. Then $\|(\lambda x f)^n\| \leq \|(\lambda\nu x)^n\|$, hence as we just saw, $\lim_{n\to\infty} \|(\lambda\nu x)^n\| = 0$. Since this is true for all $\lambda \in K$, by Theorem 6.9 we have $\|xf\|_{si} = 0$. Consequently, given any $f = \sum_{n=0}^{\infty} a_n x^n \in A$, we have

$\|\sum_{n=1}^{\infty} a_n x^n\|_{si} = 0$, therefore $\|\sum_{n=0}^{\infty} a_n x^n\|_{si} = |a_0|$. Now, consider the (unique) homomorphism χ from A onto K defined as $\chi(\sum_{n=0}^{\infty} a_n x^n) = a_0$: it shows that $\|f\|_{sa} = \|f\|_{si}$, so A satisfies property (p).

By Theorem 41.1, we notice that A is not noetherian. □

Notation: We will denote by (d) the following property, in a normed commutative K-algebra:
(d) *for all $f \in A$ such that $\|f\| \leq 1$, the sequence $\|f^n\|$ is bounded.*

According to Theorem VII.7 in [18] we have Theorem 41.5:

Theorem 41.5: *Let $\| \, . \, \|$ be the ultrametric K-vector space norm on $K[x]$ defined by $\|\sum_{n=0}^{q} a_n x^n\| = \sup_{0 \leq n \leq q}(n+1)|a_n|$ and let A be the completion of $K[x]$ for this norm. The norm is a K-algebra norm, and A is a Banach K-algebra isomorphic to a full K-subalgebra of $H(U)$, containing $R(U)$. Moreover, the spectral norm $\| \, . \, \|_{si}$ is induced by $\| \, . \, \|_U$. Further, A is principal, every ideal is generated by a polynomial whose zeros lie in U, and A satisfies Property (p) but not Property (d).*

Proof: First, we have to show that the K-vector space norm $\| \, . \, \|$ is a K-algebra norm. Let $f(x) = \sum_{j=0}^{m} a_j x^j$ and $g(x) = \sum_{j=0}^{n} b_j x^j$. Let $k, l \in \mathbb{N}$ be such that $\|f\| = |a_k|(k+1)$, $\|g\| = |b_l|(l+1)$. Let $f(x)g(x) = \sum_{j=0}^{m+n} c_j x^j$. Given $t \in \{0, \ldots, m+n\}$, and $i, j \in \mathbb{N}$ such that $i + j = t$, we have

$$|a_i b_j|(i+j+1) \leq |a_i|(i+1)|b_j|(j+1) \leq |a_k|(k+1)|b_l|(l+1) = \|f\|.\|g\|,$$

therefore $|c_t|(t+1) \leq \|f\|.\|g\|$ and finally $\|fg\| \leq \|f\|.\|g\|$.

The identical mapping from $(K[x], \| \, . \, \|)$ into $(K[x], \| \, . \, \|_U)$ being obviously continuous, the completion A of $(K[x], \| \, . \, \|)$ is a K-subalgebra of $H(U)$. For every $a \in K \setminus U$, it is seen that the series $\sum_{j=0}^{\infty}(\frac{x}{a})^n$ converges in A, so $x - a$ is invertible in A, and consequently $R(U)$ is included in A. Since $\|f\| \geq \| \, . \, \|_U$, of course we have $\|f\|_{si} \geq \| \, . \, \|_U$, (just because $\| \, . \, \|_U$ is semi-multiplicative). On the other hand, we have $\|x\|_{si} = 1$, hence

$$\left\| \sum_{j=0}^{\infty} a_j x^j \right\|_{si} \leq \sup_{0 \leq j \leq q} |a_j| = \left\| \sum_{j=0}^{\infty} a_j x^j \right\|_U,$$

and finally the equality holds for all $f \in A$. For every $a \in U$, we have a K-algebra homomorphism $\chi_a \in \mathcal{X}(A, K)$ defined as $\chi_a(f) = f(a)$, and this shows that $\|f\|_{sa} \geq \|f\|_U$, hence $\|f\|_{sa} \geq \|f\|_{si}$: thus A satisfies Property (p). Of course, A does not satisfy Property (d) because $\lim_{n\to\infty} \|x^n\| = +\infty$.

Let $f \in A$ be invertible in $H(U)$. By Theorem 17.6 f is of the form $\sum_{j=0}^{\infty} a_n x^n$, with $|a_0| > \sup_{n>0}(|a_n|)$. Since $\| \,.\, \|_{si}$ is induced by $\| \,.\, \|_U$, then we check that $|a_0| > \| \sum_{n=1}^{\infty} a_n x^n \|_{si}$, and therefore, by Theorem 6.13 f is also invertible in A. Consequently A is a full K-subalgebra of $H(U)$.

Now, we will show that each ideal of A is generated by a polynomial. First, we notice that for every $a \in U$, $g \in A$ we have $\|xg\| > \|g\|$, and therefore

$$\|(x-a)g\| > \|g\|. \tag{41.10}$$

Let $f \in A$ satisfy $f(a) = 0$ and let f_n be a sequence in $K[x]$ such that $\lim_{n\to\infty} \|f - f_n\| = 0$. Then we have

$$|f_n(a)| = |f(a) - f_n(a)| = |\chi_a(f - f_n)| \leq \|f - f_n\|_{si} \leq \|f - f_n\|.$$

Consequently, the sequence $(f_n - f(a))_{n\in\mathbb{N}}$ converges to f in A and obviously satisfies $f_n - f(a) = (x-a)g_n$, with $g_n \in K[x]$. Hence, by (41.10) we have $\|g_n\| \leq \|f_n - f_n(a)\|$, and therefore the sequence (g_n) converges in A. Let g be its limit. Thus, we have $f = (x-a)g$.

Consider now an element f of A. As an element of $H(U)$, it has a decomposition of the form $P(x)h(x)$, with P a polynomial whose zeros lie in U and h an invertible element of $H(U)$. By an immediate induction, we can see that h belongs to A, so this decomposition holds in A. And since A is a full K-subalgebra of $H(U)$, then h is invertible in A.

Now, according to a classical process, we can easily deduce that every ideal of A is generated by a polynomial whose zeros lie in U. Let I be an ideal of A, let P be the g.c.d. of polynomials which belong to I and let $f \in I$. As we have seen, f factorizes in the form $Q(x)h(x)$ with Q a polynomial whose zeros belong to U and h an invertible element of A. Thus Q also belongs to A, and therefore P divides Q. Hence P divides f in A, which proves that $I = PA$. □

Chapter 42

Associated Idempotents

Throughout the chapter, A is a commutative ultrametric Banach K-algebra with unity. We will show that given $t \in A$, if there exists a t-clear annulus, then A admits idempotents corresponding to the two subsets of $Mult(A, \| \, . \, \|)$ defined by the annulus.

Lemma 42.1: *Let $x \in A$, and let $r > \|x - a\|_{si}$ with $r \in |K|$, let $sp(x)$ admit an x-clear annulus $\mathcal{T} = \Gamma(a, r', r'')$. Let $s', s'' \in]r', r''[$ with $s' < s''$. The restriction of Θ_x to $R(d(a,r) \setminus \Gamma(a, s', s''), \| \, . \, \|)$ is continuous and expands to $H(d(a,r) \setminus \Gamma(a, s', s''))$.*

Proof: Let $\lambda \in C(0, \frac{1}{r})$, let $D = d(a,r) \setminus \Gamma(a, s', s'')$, let $b \in K$ such that $|b|^2 = s's''$ and let $c \in K$ such that $|c|^2 = s''r$. Let $t(x) = \lambda \frac{(x-a)(x-c)^2}{(x-b)^2}$. It is easily seen that $t(D) = U$, $D = t^{-1}(U)$. By Theorem 37.2 $H(D)$ is a Krasner-Tate algebra of the form $K\{t\}[x]$. Now, let $s = \|x\|_{si}$ and let $D' = d(a, r') \cup \Delta(a, r'', s)$. By hypothesis, for every $\psi \in Mult(A, \| \, . \, \|)$, $\mathcal{G}_{\psi_x}$ is clearly secant with D'. If $\mathcal{G}_{\psi_x}$ is secant with $d(a, r')$, we have $\psi(t) \leq \frac{r'}{s'}$, and if $\mathcal{G}_{\psi_x}$ is secant with $\Delta(a, r'', s)$, then $\psi(t) \leq \max(\frac{s''}{r''}, \frac{s}{r})$ hence anyway we have $\psi(t) \leq \max(\frac{r'}{s'}, \frac{r''}{s''}, \frac{s}{r}) < 1$, and therefore $\|t\|_{si} < 1$. Consequently, $\lim_{n\to\infty} \|t^n\| = 0$, and hence the restriction of Θ_x to $K[t, x]$, provided with the norm $\| \, . \, \|_D$ is continuous from $K[t, x]$ to A. Therefore Θ_x has continuation to a continuous homorphism from $K\{t\}[x]$ provided with the norm $\| \, . \, \|_D$ into A. But actually, $K\{t\}[x] = H(D)$. $\square$

Theorem 42.2: *Let $x \in A$ and suppose that there exists a x-clear annulus $\Gamma(a, r', r'')$. There exists an idempotent u such that:*

$\phi(u) = 1$, $\forall \phi \in Mult(A, \| \, . \, \|)$ such that $\phi(t) \leq r'$

and

$\phi(u) = 0$, $\forall \phi \in Mult(A, \| \cdot \|)$ *such that* $\phi(x) \le r''$.

Moreover, u is unique satisfying there properties.

Proof: Let $\mathcal{T} = \Gamma(a, r', r'')$. Let $r > \|x - a\|_{si}$, let $s', s'' \in]r', r''[\cap|K|$ with $s' < s''$ and let $D = d(a,r) \setminus \Gamma(a, s', s'')$. By Lemma 42.1 the restriction of Θ_x to $R(D)$ is continuous and expands to $H(D)$.

Now, by Theorem 17.15 $H(D)$ contains idempotents h and l such that $h(z) = 1\ \forall z \in d(a,s'), h(z) = 0\ \forall z \in d(a,r) \setminus d(a,s''^-)$, and $l(z) = 0\ \forall z \in d(a,s'), l(z) = 1\ \forall z \in d(a,r) \setminus d(a,s''^-)$. Therefore its image $A' = \Theta_x(H(D))$ contains an idempotent $u = \Theta_x(h)$. Let $\phi \in Mult(A, \| \cdot \|)$. Either $\mathcal{G}_{\phi_x}$ is secant with $d(a,s')$, and then $\phi(u) = \phi_x(h) = 1$, or $\mathcal{G}_{\phi_x}$ is secant with $d(a,r) \setminus d(a,s''^-)$, and then $\phi(u) = \phi_x(h) = 0$. Finally u is unique satisfying these relations by Theorem 6.17. □

Corollary 42.3: *Let* $t \in A$, *let* $a \in \widetilde{sp(t)}$, *let* $s = \|t - a\|_{si}$ *and let* $\mathcal{F}$ *be a circular filter of center a and diameter* $r < s$. *If there exists no* $\phi \in Mult(A, \| \cdot \|)$ *such that* $\varphi_{\mathcal{F}} = \phi_t$ *then there exists a unique idempotent* $u \in A$ *such that* $\psi(u) = 1\ \forall \psi \in Mult(A, \| \cdot \|)$ *such that* $\psi(t) < r$ *and* $\psi(u) = 0\ \forall \psi \in Mult(A, \| \cdot \|)$ *such that* $\psi(t) > r$.

Proof: Let $E = \{\phi \in Mult(A, \| \cdot \|) \mid \phi(t) < r\}$, and $F = \{\phi_t \in Mult(A, \| \cdot \|) \mid \phi_t(t) > r\}$. Then $\{E, F\}$ makes a partition of $\{\phi_t \mid \phi \in Mult(A, \| \cdot \|)\}$ by two open closed subsets, hence Theorem 42.2 shows the existence and the uniqueness of u. □

Corollary 42.4: *Suppose A has no non trivial idempotents, let* $t \in A$, *let* $a \in sp(t)$, *let* $s = \|t - a\|_{si}$, *let* $D = sp(t)$ *and let* $\mathcal{O}$ *be the strict t-spectral partition. For every* $h \in R(sp(t))$ *we have* $\|h(t)\|_{si} = \|h\|_{D,\mathcal{O}}$.

Proof: Since A has no non trivial idempotents, there exists no t-clear annulus, therefore by Theorem 23.10 we have $\|h(t)\|_{si} = \|h\|_{D,\mathcal{O}}\ \forall h \in R(sp(t))$. □

Proposition 42.5: *Let A have no non trivial idempotents. Let* $t \in A$, *let* $\phi \in Mult(A, \| \cdot \|)$, *let* $\phi_t = \varphi_{\mathcal{F}}$ *and assume that* $\mathcal{F}$ *is not secant with* $sp(t)$. *Let* $h \in K(t) \cap A$. *Then* $K[h,t]$ *is not dense in A.*

Proof: Let $D = sp(t)$. Suppose first that $\mathcal{F}$ is not secant with $\widetilde{D}$. Let $a \in D$. Then $\mathcal{F}$ is (a,s)-approaching, with $s > \tau(t-a)$. Let $r \in]\tau(t-a), s] \cap |K|$. Let $\mathcal{G}$ be the circular filter of center a and diameter r. Since A has no non trivial idempotents, by Theorem 42.2 there exists no t-clear annulus, hence there exists $\zeta \in Mult(A, \| \cdot \|)$ such that $\zeta_t = \varphi_{\mathcal{G}}$. Let $d(c, r^-)$ be a class of $C(a,r)$ which contains no poles of h and let $f \in K[h,t]$. Let $\mathcal{O}$ be the partition of $d(a,r)$ consisting of the family of classes of $d(a,r)$. The Mittag-Leffler decomposition of f in $H(\mathcal{O})$ is of the form $\sum_{j=0}^{q} f_j$, with $f_0 \in R(d(a,r))$, and $f_j \in R_0(K \setminus d(b_j, r^-))$ $(1 \leq j \leq q)$. Then by Theorem 21.1 we know that

$$\left\|\frac{1}{x-c} + f\right\|_{\mathcal{O}} \geq \left\|\frac{1}{x-c}\right\|_{\mathcal{O}}.$$

But on the other hand, for all $g(x) \in R(sp(t))$, we have $\|g\|_{\mathcal{O}} = \lim_{\mathcal{G}} |g(x)| = \phi_t(g) = \phi(g(t)) \leq \|g(t)\|_{si}$. Consequently, we obtain

$$\left\|\frac{1}{t-c} + f(t)\right\|_{si} \geq \left\|\frac{1}{x-c}\right\|_{\mathcal{O}},$$

which proves that $K[h,t]$ is not dense in A.

Suppose now that $\mathcal{F}$ is secant with $\widetilde{D}$. Let $l = diam(\mathcal{F})$. Then $\mathcal{F}$ must be secant with a hole $T = d(b, \rho^-)$ of D such that $l < \rho$. Hence $\mathcal{F}$ is (b,s)-approaching with $l \leq s < \rho$. Let $r \in]s, \rho[\cap|K|$ and let $\mathcal{G}$ be the circular filter of center b and diameter r. Since there exists no t-clear annulus, we can find $\zeta \in Mult(A, \| \, . \, \|)$ such that $\zeta_t = \varphi_{\mathcal{G}}$. We can now go on similarly to the previous case. Let $d(c, r^-)$ be a class of $C(b,r)$ which contains no poles of h and let $\mathcal{O}$ be the partition of $d(b,r)$ consisting of the family of classes of $d(b,r)$. Let $f \in K[h,t]$. By Theorem 21.1 we know that

$$\left\|\frac{1}{x-c} + f\right\|_{\mathcal{O}} \geq \left\|\frac{1}{x-c}\right\|_{\mathcal{O}}.$$

But on the other hand,

$$\left\|\frac{1}{t-c} + f(t)\right\|_{si} \geq \left\|\frac{1}{x-c}\right\|_{\mathcal{O}},$$

which proves that $K[h,t]$ is not dense in A again. □

Theorem 42.6: *Let A satisfy Property (q), let $x \in A$ and suppose that $sp(x)$ admits an empty annulus $\Gamma(a,r,s)$. There exists a unique idempotent u such that $\chi(u) = 1$ for all $\chi \in \mathcal{X}(A)$ such that $\chi(x-a) \leq r$, and $\chi(u) = 0$ for all $\chi \in \mathcal{X}(A)$ such that $\chi(x-a) \geq s$.*

Proof: Indeed, since A satisfies Property (q), by Proposition 28.1 we know that $Mult_m(A, \| \, . \, \|)$ is pseudo-dense in $Mult(A, \| \, . \, \|)$. Let $\phi \in Mult(A, \| \, . \, \|)$. Since $\Gamma(a,r,s) \cap sp(x) = \emptyset$, the circular filter $\mathcal{F} = \Psi^{-1}(\phi_x)$ of the restriction ϕ_x of ϕ to $K[x]$ must be secant either with $d(a,r)$, or with $K \setminus d(a, s^-)$. Consequently, either $\phi(x) \leq r$, or $\phi(x) \geq s$. Then$\Gamma(a,r,s)$, $\underline{sp(x)}$ admits a partition by two open closed subsets $\underline{sp(x) \cap \mathcal{I}(T)}$ and

$sp(x) \cap \mathcal{E}(T)$. Then the existence of u and the uniqueness come from Theorem 42.2. □

Corollary 42.7: *Let A satisfy Property (q) and have no idempotents differences from 0 and 1. Then for every $x \in A$ $sp(x)$ is infraconnected.*

By Theorem 8.13 we obtain this Corollary:

Corollary 42.8: *Let A satisfy Property (q), let $x \in A$ and suppose that $sp(x)$ has finitely many infraconnected components $D_1, \ldots, D_q$. For each $j = 1, \ldots, q$, there exists a unique idempotent u_j such that $\chi(u_j) = 1$ for all $\chi \in \mathcal{X}(A)$ such that $\chi(u) \in D_j$ and $\chi(u_j) = 0$ for all $\chi \in \mathcal{X}(A)$ such that $\chi(u) \notin D_j$.*

And by Theorem 25.2 we have these last Corollaries:

Corollary 42.9: *Let A be uniform, and have no maximal ideals of infinite codimension, let $x \in A$ and suppose that $sp(x)$ has finitely many infraconnected components $D_1, \ldots, D_q$. For each $j = 1, \ldots, q$, there exists a unique idempotent u_j such that $\chi(u_j) = 1$ for all $\chi \in \mathcal{X}(A)$ such that $\chi(u) \in D_j$ and $\chi(u_j) = 0$ for all $\chi \in \mathcal{X}(A)$ such that $\chi(u) \notin D_j$.*

Proof: Indeed, since A has no maximal ideals of infinite codimension, it obviously satisfy Property (o), and therefore, by Theorem 25.4 it satisfy Property (p). □

Corollary 42.10: *Let A be uniform and have no maximal ideals of infinite codimension, let $x \in A$ and suppose that $sp(x)$ has finitely many infraconnected components $D_1, \ldots, D_q$. For each $j = 1, \ldots, q$, there exists a unique idempotent u_j such that $\chi(u_j) = 1$ for all $\chi \in \mathcal{X}(A)$ such that $\chi(u) \in D_j$ and $\chi(u_j) = 0$ for all $\chi \in \mathcal{X}(A)$ such that $\chi(u) \notin D_j$.*

Corollary 42.11: *Let A be uniform and have no maximal ideals of infinite codimension and have no idempotents different from 0 and 1. Then $sp(x)$ is infraconnected.*

Theorem 42.12: *Let A have bounded normal ratio, have no maximal ideals of infinite codimension and have no idempotents different from 0 and 1. Then, for every $x \in A$, $sp(x)$ is infraconnected.*

Proof: Indeed, since A has bounded normal ratio, by Theorem 24.3 it satisfies Property (s). But then, since no maximal ideal has infinite codimension, then A satisfies Property (p), hence Property (q), and then by Corollary 42.7, since A has no non trivial idempotents, $sp(x)$ is infraconnected for all $x \in A$. □

Chapter 43

Krasner-Tate Algebras among Banach K-Algebras

Throughout the chapter, A is a commutative ultrametric Banach K-algebra with unity. Here we mean to characterize Krasner-Tate algebras among ultrametric K-Banach algebras through algebraic and topological properties denoted by $(a), (b), (c), (d), (e)$ [18]. In a final remark, we will show that these properties are logically independent, and therefore none of them is redundant.

Lemma 43.1: *Let A contain a dense Luroth K-subalgebra $K[t, h]$ with $h \in K(t) \cap A$ and be such that the strict t-spectral partition coincides with the canonical partition of $d(0, \|t\|_{si}) \setminus sp(t)$. Then A satisfies Property (q).*

Proof: Let $E = \{\chi \in \mathcal{X}(A) \mid \chi(t) \in K\}$. Let $\mathcal{O}$ be the strict t-spectral partition, and let $f \in K[h, t]$, with $h(x) \in K(x)$.

Since the strict t-spectral partition coincides with the canonical partition of $d(0, \|t\|_{si}) \setminus sp(t)$, by Chapter 18, in $R(sp(t))$, we have

$$\begin{aligned}\|f(x)\|_{sp(t),\mathcal{O}} = \|f(x)\|_{sp(t)} &= \sup\{|f(\lambda)| \mid \lambda \in sp(t)\}\\ &= \sup\{|\chi(f)| \mid \chi \in E\} = \tau(f) \leq \|f\|_{si}. \quad (43.1)\end{aligned}$$

On the other hand by Theorem 23.5 we have $\|f\|_{si} \leq \|f\|_{sp(t),\mathcal{O}}$, hence by (43.1) the equality $\|f\|_{si} = \tau(f)$ holds for all $f \in K[h, t]$. Then, since $K[h, t]$ is dense in A, we can generalize the equality to all A. Indeed, let $f \in A$, and let $(f_n)_{n\in\mathbb{N}}$ be a sequence in $K[h, t]$ such that $\lim_{n\to\infty} \|f_n - f\| = 0$. Let $\epsilon \in]0, \|f\|_{si}[$ and let $N \in \mathbb{N}$ be such that $\|f_n - f\| < \epsilon \ \forall n > N$. We notice that $\|f_n - f\|_{si} < \epsilon \ \forall n > N$, and therefore $\|f_n\|_{si} = \|f\|_{si} \ \forall n > N$. Since the equality holds in $K[h, t]$, we can find $\chi \in E$ such that $|\chi(f_n)| >$

$\|f_n\|_{si} - \epsilon$. Thus, the sequence $\|f_n\|_{si}$ is constant and equal to $\|f\|_{si}$ for all $n > N$ and we have $|\chi(f_n)| = |\chi(f)|$ on the other hand, the sequence $\chi(f_n)$ is a Cauchy sequence and therefore converges in K to $\chi(f)$. Hence $\chi(f)$ belongs to K and we have $|\chi(f)| > \|f\|_{si} - \epsilon$, which proves that $\|f\|_{si} = \tau(f)$ in the general case. □

Theorem 43.2: *If A contains a dense Luroth K-subalgebra and has no non trivial idempotents, then A satisfies Property (q).*

Proof: Suppose that A does not satisfy Property (q), and let $K[h,t]$ be a dense Luroth K-subalgebra, with $h \in K(t) \cap A$. By Lemma 43.1 the strict t-spectral partition does not coincide with the canonical partition of $sp(t)$. Let $T = d(b, \rho^-)$ be a hole of the strict t-spectral partition which is not a hole of $sp(t)$. There exists $\psi \in Mult(A, \| \,.\, \|)$ such that the circular filter $\mathcal{F} = \Phi^{-1}(\psi_t)$ is (b,r)-approaching, and therefore is not secant with $sp(t)$. Then by Proposition 42.5 we have a contradiction with the hypothesis $K[h,t]$ *is a dense Luroth K-subalgebra.* □

We are now able to characterize Krasner-Tate algebras among ultrametric Banach K-algebras through the following two theorems:

Theorem 43.3 ([18]): *A is a Krasner-Tate algebra if and only if it satisfies the following five properties:*

(a) A is noetherian,
(b) A is reduced,
(c) $\|f\|_{si} \in |K|\ \forall f \in A$,
(d) for every $f \in A$ such that $\|f\|_{si} = 1$, the sequence $\|f^n\|$ is bounded,
(e) A contains a dense Luroth K-subalgebra.

Theorem 43.4: *Let A have no non trivial idempotents and admit infinitely many maximal ideals of codimension 1. Then A is a Krasner-Tate algebra if and only if it satisfies the following four properties:*

(b) A is reduced,
(c) $\|f\|_{si} \in |K|\ \forall f \in A$,
(d) For every $f \in A$ such that $\|f\|_{si} = 1$, the sequence $\|f^n\|$ is bounded,
(e) A contains a dense Luroth K-subalgebra.

Proof of Theorems 43.3 and 43.4: A Krasner-Tate algebra obviously satisfies Properties $(a), (b), (c), (d), (e)$. So, we just have to consider a K-Banach algebra satisfying these five properties, and prove that it is a

Krasner-Tate algebra. Let $B = K[h, x]$ be a dense Luroth K-subalgebra, with $h \in K(x)$ and let $D = sp(x)$.

First, we assume that A has no non trivial idempotents. By Theorem 43.2 A satisfies Property (q). Hence, by Theorem 41.3 A satisfies Property (p) in both Theorems 43.3 and 43.4. Consequently, by Theorem 41.1 D is infinite in the hypotheses of Theorems 43.3. Now, assuming the hypotheses of Theorem 43.4, suppose that D is finite: let $D = \{a_1, \dots, a_s\}$. For each $j = 1, \dots, s$, there exists $\chi_j \in \mathcal{X}(A)$ such that $\chi_j(x) = a_j$ and then $\chi_j(h) = h(a_j)$. Given any $\xi \in \mathcal{X}(A)$, its restriction to B is equal to the restriction of one of the χ_j say χ_k. Then, since B is dense in A, we have $\xi = \chi_k$. Consequently, A only has s maximal ideals of codimension 1, a contradiction. Thus we have proven that D is infinite in both Theorems 43.3 and 43.4.

Next, B is isomorphic to a K-subalgebra of $R(D)$, so we will consider that $B \subset R(D) \subset A$. We can check that for every $f \in R(D)$, we have $sp_A(f) = f(D)$. Indeed, since $R(D) \subset A$, we have $sp_A(f) \subset f(D)$ because $f(D) = sp_{R(D)}(f)$. But given $\lambda \in f(D)$, and $a \in D$ such that $f(a) = \lambda$, there exists $\chi \in \mathcal{X}(A)$ such that $\chi(x) = a$, hence $\chi(f) = f(a) = \lambda$. Therefore $\lambda \in sp_A(f)$ and finally $sp_A(f) = f(D)$. Consequently, we have $\tau_A(f) = \|f\|_D$, and therefore by property q) we have $\|f\|_D = \|f\|_{si}$ $\forall f \in R(D)$. Consequently, A and $H(D)$ are the completions of $R(D)$ for the norms $\| \, . \, \|$ and $\| \, . \, \|_D$, respectively.

On the other hand, since A has no non trivial idempotents, by Theorem 43.2 it satisfies Property (q). Then by Corollary 42.7 $sp_A(f)$, hence D, is infraconnected. Let $d(a, r) = \widetilde{D}$, with $r \in |K|$. Since B is dense in $H(D)$, by Lemma 16.15, D has finitely many holes: $d(a_j, r_j^-)$ $(1 \le j \le q)$. Since $\|x - a\|_{si} = \|x - a\|_D = r$, by Property c) we notice that r lies in $|K|$. Similarly, since

$$\left\|\frac{1}{x - a_j}\right\|_{si} = \left\|\frac{1}{x - a_j}\right\|_D = \frac{1}{r_j} \ (1 \le j \le q),$$

we have $r_j \in |K|$. Hence by Theorem 37.2 $H(D)$ is a Krasner-Tate algebra $K\{h\}[x]$ where $h \in R(D)$ satisfies $\deg(h) > 0$, $U = h(D)$, $D = h^{-1}(U)$. Then, by Property (d) the canonical mapping ψ from $K[h, x]$ into A defined as $\psi(F(h, x)) = F((h(x), x)$ is continuous. Consequently, it has continuation to a continuous homomorphism $\widetilde{\psi}$ from $H(D)$ to A, and this homomorphism is obviously bicontinuous because $\|f\|_D = \|\psi(f)\|_{si} \le \|\psi(f)\|$. Thus, $H(D)$ is a K-subalgebra of A, containing B, hence is dense in A

and such that both norms $\| \,.\, \|_D$ and $\| \,.\, \|$ are equivalent on $H(D)$. Consequently, $A = H(D)$. This finishes proving that A is the Krasner-Tate algebra $H(D)$ when A has no non trivial idempotents. In particular, this finishes the proof of Theorem 43.4.

We now consider the general case in the proofs of Theorems 43.3. Clearly A has finitely many idempotents because A is noetherian. So, we can find a family of idempotents $u_1, \dots, u_k$ satisfying $u_i u_j = 0 \ \forall i \neq j$, and $\sum_{i=1}^{k} u_i = 1$. For each $i = 1, \dots, k$, we put $A_i = u_i A$ and $D_i = sp(u_i x)$. Then A is isomorphic to the direct product $A_1 \times \dots \times A_k$. Then we can check that each K-algebra A_i $(1 \leq i \leq k)$ also satisfies Properties $(b), (c), (d), (e)$. Let us fix $k \leq q$, and suppose first that D_k is infinite. Then $u_i B$ is a Luroth K-algebra again because $u_k x$ is transcendental over K. Consequently, $u_k A$ satisfies Property (e) and we can make the same reasoning as we did when D is infraconnected and we conclude that $H(D_k)$ is a Krasner-Tate algebra. Suppose now that D_k is finite. Since it is infraconnected, it is a singleton $\{a\}$. Consequently, $u_k A$ is of the form $\frac{K[Y]}{(Y-a)^s K[Y]}$. But since A is reduced, we have $s = 1$, hence $u_k B$ is isomorphic to K and so is $u_k A$. Finally, if we denote by $D_1, \dots, D_m$ the infinite infraconnected components of D and by $D_{m+1}, \dots, D_q$ those which are singletons, we can see that A is isomorphic to $A_1 \times \dots \times A_m \times K^{q-m}$ which itself is a Krasner-Tate algebra isomorphic to $H(\cup_{j=1}^{q} D_j)$. □

By Theorem 43.3 we obtain Corollary 43.5:

Corollary 43.5: *If A is uniform, it is a Krasner-Tate algebra if and only if it satisfies the following four properties:*

(a) *A is noetherian,*
(b) *A is reduced,*
(c) $\|f\|_{si} \in |K| \ \forall f \in A$,
(e) *A contains a dense Luroth K-subalgebra.*

Corollary 43.6: *A reduced K-affinoid algebra is a Krasner-Tate algebra if and only if it contains a dense Luroth K-algebra.*

Theorem 43.7: *Let A be a K-affinoid algebra without non zero divisors of zero, containing a K-subalgebra of finite type B, dense in A, whose field of fractions is a pure degree one transcendence extension of K. Then A is a Krasner-Tate algebra if and only if A is integrally closed.*

Proof: We just have to assume that A is integrally closed and show that A is a Krasner-Tate algebra. Let $K(x)$ be the field of fractions of B and let

$B = K[y_1, \ldots, y_q]$. For each $j = 1, \ldots, q$, we put $y_j = f_j(x)$ ($f_j \in K(x)$). There obviously exists a linear fractional function h such that $\deg(f_1 \circ h) > 0$. So, we put $z = h^{-1}(x)$ and $g_j = f_j \circ h$ $\forall j = 1, \ldots, q$. So, z is obviously integral over B, and therefore belongs to A. Let $B' = B[z]$. Then B' is a Luroth K-algebra hence by Corollary 43.5 A is a Krasner-Tate algebra. □

Remark: By Proposition 41.2 there exist Banach K-algebras $H(D)$ satisfying Properties $(b), (c), (d), (e)$ but not satisfying Property (a).

By Proposition 37.9, there exist K-affinoid algebras satisfying $(a), (c), (d), (e)$ but not satisfying Property (b).

If $|K| \neq \mathbb{R}$, any Banach K-algebras $H(d(0, r))$, with $r \notin |K|$ satisfies Properties $(a), (b), (d), (e)$ but does not satisfy Property (c).

Theorem 41.5 shows the existence of commutative ultrametric Banach K-algebras with unity satisfying Properties $(a), (b), (c), (e)$, but not satisfying Property (d).

There obviously exist commutative ultrametric Banach K-algebras with unity satisfying Properties $(a), (b), (c), (d)$ but not satisfying Property (e): we can just consider T_n for $n > 1$.

Thus, none of the five Properties $(a), (b), (c), (d), (e)$ is a consequence of the four others in the category of commutative ultrametric Banach K-algebras with unity. However, if K is such that $|K| = [0, +\infty[$, Condition (c) is obviously trivial and must be cut.

Remark: We don't know examples of noetherian uniform Banach K-algebras without non zero divisors of zero, satisfying Property (c) and containing a dense K-subalgebra of finite type, which are not K-affinoid algebras.

References

1. Amice, Y. *Les nombres p-adiques*, P.U.F. (1975).
2. Berkovich, V. *Spectral Theory and Analytic Geometry over Non-archimedean Fields.* AMS Surveys and Monographs 33, (1990).
3. Bosch, S., Guntzer, U., and Remmert, R *Non-Archimedean analysis,* Grundlerhen Wissenschaften, Bd. 261, Springer, Berlin-Heidelberg-New York, (1984).
4. Bourbaki, N. *Topologie générale, Ch.3* Actualités scientifiques et industrielles, Hermann, Paris.
5. Bourbaki, N. *Theorie spectrale Ch. 1 et 2.*, Actualités scientifiques et industrielles, Hermann, Paris.
6. Bourbaki, N. *Algebre commutative, Ch. 2.*, Actualités scientifiques et industrielles, Hermann, Paris.
7. Boussaf, K. and Escassut, A. *Absolute values on algebras of analytic elements,* Annales Mathématiques Blaise Pascal 2, n2 (1995).
8. Boussaf, K. *Shilov Boundary of a Krasner Banach algebra $H(D)$* Italian Journal of pure and applied Mathematics, N.8, pp. 75–82 (2000).
9. Boussaf, K., Hemdahoui, M., and Mainetti, N. *Tree structure on the set of multiplicative semi-norms of Krasner algebras $H(D)$.* Revista Matematica Complutense, vol XIII, N. 1, pp. 85–109 (2000).
10. Boussaf *Image of circular filters*, International Journal of Mathematics, Game Theory and Algebra, Volume 10, N. 5, pp. 365–372.
11. Boussaf, K., Escassut, A., and Mainetti, N. *Mappings in the tree $Mult(K[x])$.* To appear in Bulletin of the Belgian Mathematical Society Simon Stevin.
12. Cherry, W. *Non-Archimedean analytic curves in Abelian varieties,* Mathematishe Annalen 300, pp. 396–404 (1994).
13. Decomps-Guilloux, A. and Motzkin, E. *Fonctions analytiques univalentes dans un corps ultramétrique complet algébriquement clos.* C.R.A.S. Paris t. 268, pp. 1531–1533 (1969).
14. Diarra, B. *Ultraproduits ultramétriques de corps valués*, Annales Scientifiques de l'Université de Clermont II, Série Math., Fasc. 22, pp. 1–37, (1984).

15. Dwork, B. *Lectures on p-adic differential equations.* Springer-Verlag, (1982).
16. Escassut, A. *Algèbres d'éléments analytiques au sens de Krasner dans un corps valué non archimédien complet algébriquement clos*, C.R.A.S. Paris, A 270, pp. 758–761 (1970).
17. Escassut, A. *Compléments sur le prolongement analytique dans un corps valué non archimédien, complet algébriquement clos*, C.R.A.S.Paris, A 271, pp. 718–721 (1970).
18. Escassut, A. *Algèbres de Banach ultramétriques et algèbres de Krasner-Tate*, Asterisque n. 10, pp. 1–107, (1973).
19. Escassut, A. *Algèbres d'éléments analytiques en analyse non archi- médienne*, Indag. math.,t.36, pp. 339–351 (1974).
20. Escassut, A. *T-filtres, ensembles analytiques et transformation de Fourier p-adique*, Ann. Inst. Fourier 25, n 2, pp. 45–80, (1975).
21. Escassut, A. *Spectre maximal d'une algèbre de Krasner*, Colloquium Mathematicum, XXXVIII, fasc. 2, pp. 339–357, (1978).
22. Escassut, A. *Eléments spectralement injectifs et générateurs universels dans une algèbre de Tate*, Collectanae Mathematica (Barcelona), Vol XXVIII, Fasc.2, pp. 131–148 (1977).
23. Escassut, A. *The ultrametric spectral theory*, Periodica Mathematica Hungarica, Vol. 11, (1), pp. 7–60, (1980).
24. Escassut, A. *Maximum principle for Analytic Elements and Lubin-Hensel Theorem for series with Analytic Coefficients*, Ann. Mat. Pura Appl. vol. CXXXV, pp. 265–278 (1983).
25. Escassut, A. *Integrally closed algebras of analytic elements*, Communication in Algebra, 19, pp. 1565–1584 (1991).
26. Escassut, A. and Diarra, B. *Non integrally closed algebras $H(D)$*, p-adic functional analysis, Lecture Notes in Pure and Applied Mathematics n. 137, pp. 63–74 (1992).
27. Escassut, A. and Sarmant, M.-C. *Sufficient Conditions for injectivity of Analytic Elements*, Bull. Sci. Math. vol. 118, n. 1, pp. 29–46, (1994).
28. Escassut, A. and Sarmant, M.-C. *Injectivity, Mittag-Leffler Series and Motzkin Products*, Annales des Sciences Mathématiques du Quebec, 16, (2), pp. 155–173 (1992).
29. Escassut, A. and Mainetti, N *Spectral semi-norm of a p-adic Banach algebra,* Bulletin of the Belgian Mathematical Society, Simon Stevin, vol. 8, pp. 79–61, (1998).
30. Escassut, A. *Analytic Elements in p-adic Analysis*, World Scientific Publishing Inc., Singapore (1995).
31. Escassut, A. *Shilov boundary for ultrametric algebras*, preprint.
32. Garandel, G. *Les semi-normes multiplicatives sur les algèbres d'éléments analytiques au sens de Krasner*, Indag. Math., 37, n. 4, pp. 327–341, (1975).
33. Guennebaud, B. *Algèbres localement convexes sur les corps valués*, Bull. Sci. Math. 91, pp. 75–96, (1967).
34. Guennebaud, B. *Sur une notion de spectre pour les algèbres normées ultramétriques*, thèse Université de Poitiers, (1973).

35. Krasner, M. *Prolongement analytique uniforme et multiforme dans les corps valués complets: préservation de l'analycité par la convergence uniforme, Théorème de Mittag-Leffler généralisé pour les éléments analytiques,* C.R.A.S. Paris, A 244, pp. 2570–2573, (1957).
36. Krasner, M. *Prolongement analytique uniforme et multiforme dans les corps valués complets. Les tendances géométriques en algèbre et théorie des nombres,* Clermont-Ferrand, pp. 94–141 (1964). Centre National de la Recherche Scientifique (1966), (Colloques internationaux de C.N.R.S. Paris, 143).
37. Mainetti, N. *Spectral properties of p-adic Banach algebras.* Lecture Notes in Pure and Applied Mathematics n. 207, pp. 189–210, (1999).
38. Mainetti, N. *Sequential Compactness of some Analytic Spaces,* Journal of Analysis, 8, pp. 39–54 (2000).
39. Mainetti, N. *Metrizability of some analytic affine spaces* Lecture Notes in Pure and Applied Mathematics, Marcel Dekker, (2001).
40. Morita, Y. *On the induced h-structure on an open subset of the rigid analytic space* $P^1(k)$. Mathematische Annalen 242, pp. 47–58, (1979).
41. Motzkin, E. *Un invariant conforme p-adique,* Groupe d'Etude d'Analyse Ultramétrique de l'IHP, 1981–82,
42. Rivera-Letellier, J. *Espace hyperbolique p-adique et dynamique des fonctions rationnelles,* preprint.
43. Robert, Alain *Advanced calculus for users* North-Holland, Amsterdam, New York, Oxford, Tokyo (1989).
44. Van Rooij, A.C.M. *Non-Archimedean Functional Analysis,* Marcel Decker, Inc. (1978).
45. Salmon, P. *Sur les séries formelles restreintes,* Bulletin de la Société Mathématique de France, 92, pp. 385–410 (1964).
46. Sarmant, M.-C. et Escassut A. *T-suites idempotentes,* Bull. Sci. Math. 106, pp. 289–303, (1982).
47. Sarmant, M.-C. *Prolongement analytique à travers un T-filtre,* Studia Scientiarum Mathematicarum Hungarica 22, pp. 407–444, (1987).
48. Schikhof, W.H. *Ultrametric calculus. An introduction to p-adic analysis,* Cambridge University Press (1984).
49. Tate, J. *Rigid analytic spaces,* Inventiones Mathematicae 12, pp. 257–289 (1971).

Definitions Index

Notation Index